ANNALS *of* THE NEW YORK ACADEMY OF SCIENCES

T0188428

EDITOR-IN-CHIEF
Douglas Braaten

ASSOCIATE EDITOR
Rebecca E. Cooney

PROJECT MANAGER
Steven E. Bohall

EDITORIAL ADMINISTRATOR
Daniel J. Becker

Artwork and design by Ash Ayman Shairzay

The New York Academy of Sciences
7 World Trade Center
250 Greenwich Street, 40th Floor
New York, NY 10007-2157

annals@nyas.org
www.nyas.org/annals

The New York Academy of Sciences

Published by Blackwell Publishing
On behalf of the New York Academy of Sciences

Boston, Massachusetts
2012

ANNALS *of* THE NEW YORK ACADEMY OF SCIENCES

VOLUME
1258

ISSUE

Barriers and Channels Formed by Tight Junction Proteins II

ISSUE EDITORS

Michael Fromm and Jörg-Dieter Schulzke

Charité, Campus Benjamin Franklin, Berlin, Germany

TABLE OF CONTENTS

Inflammation and gastrointestinal function

1   TNF-α–induced intestinal epithelial cell shedding: implications for intestinal barrier function
Alastair J.M. Watson and Kevin R. Hughes

9   Tight junctions on the move: molecular mechanisms for epithelial barrier regulation
Le Shen

19   HIV infection and the intestinal mucosal barrier
Hans-Jörg Epple and Martin Zeitz

25   Zonulin, regulation of tight junctions, and autoimmune diseases
Alessio Fasano

34   Myosin light chain kinase: pulling the strings of epithelial tight junction function
Kevin E. Cunningham and Jerrold R. Turner

43   Defective tight junctions in refractory celiac disease
Michael Schumann, Sarah Kamel, Marie-Luise Pahlitzsch, Lydia Lebenheim, Claudia May, Michael Krauss, Michael Hummel, Severin Daum, Michael Fromm, and Jörg-Dieter Schulzke

52   Microbial butyrate and its role for barrier function in the gastrointestinal tract
Svenja Plöger, Friederike Stumpff, Gregory B. Penner, Jörg-Dieter Schulzke, Gotthold Gäbel, Holger Martens, Zanming Shen, Dorothee Günzel, and Joerg R. Aschenbach

Epithelial transport, barrier modulation, and food components

60 SUMOylation of claudin-2
Christina M. Van Itallie, Laura L. Mitic, and James M. Anderson

65 Proof of concept for claudin-targeted drug development
Hidehiko Suzuki, Masuo Kondoh, Azusa Takahashi, and Kiyohito Yagi

71 Loss of enteral nutrition in a mouse model results in intestinal epithelial barrier dysfunction
Yongjia Feng, Matthew W. Ralls, Weidong Xiao, Eiichi Miyasaka, Richard S. Herman, and Daniel H. Teitelbaum

78 The protein C pathway in intestinal barrier function: challenging the hemostasis paradigm
Silvia D'Alessio, Marco Genua, and Stefania Vetrano

86 Analysis of absorption enhancers in epithelial cell models
Rita Rosenthal, Miriam S. Heydt, Maren Amasheh, Christoph Stein, Michael Fromm, and Salah Amasheh

93 Calcium regulation of tight junction permeability
Markus Bleich, Qixian Shan, and Nina Himmerkus

100 Effects of quercetin studied in colonic HT-29/B6 cells and rat intestine *in vitro*
Maren Amasheh, Julia Luettig, Salah Amasheh, Martin Zeitz, Michael Fromm, and Jörg-Dieter Schulzke

108 Claudin-based paracellular proton barrier in the stomach
Atsushi Tamura, Yuji Yamazaki, Daisuke Hayashi, Koya Suzuki, Kazuhiro Sentani, Wataru Yasui, and Sachiko Tsukita

115 Regulation of epithelial proliferation by tight junction proteins
Attila E. Farkas, Christopher T. Capaldo, and Asma Nusrat

Tight junctions in intestinal and renal epithelia

125 Barrier dysfunction and bacterial uptake in the follicle-associated epithelium of ileal Crohn's disease
Åsa V. Keita and Johan D. Söderholm

135 The protozoan pathogen *Toxoplasma gondii* targets the paracellular pathway to invade the intestinal epithelium
Caroline M. Weight and Simon R. Carding

143 Ion transport and barrier function are disturbed in microscopic colitis
Christian Barmeyer, Irene Erko, Anja Fromm, Christian Bojarski, Kristina Allers, Verena Moos, Martin Zeitz, Michael Fromm, and Jörg-Dieter Schulzke

149 Enteropathogenic *E. coli* effectors EspG1/G2 disrupt tight junctions: new roles and mechanisms
Lila G. Glotfelty and Gail A. Hecht

159 Abnormal intestinal permeability in Crohn's disease pathogenesis
Christopher W. Teshima, Levinus A. Dieleman, and Jon B. Meddings

166 A dynamic paracellular pathway serves diuresis in mosquito Malpighian tubules
Klaus W. Beyenbach

177   Impaired paracellular ion transport in the loop of Henle causes familial
            hypomagnesemia with hypercalciuria and nephrocalcinosis
            Lea Haisch and Martin Konrad

185   The yin and yang of claudin-14 function in human diseases
            Jianghui Hou

---

191   Erratum for Ann. N. Y. Acad. Sci. 1253: 206–221

Ann. N.Y. Acad. Sci. ISSN 0077-8923

ANNALS OF THE NEW YORK ACADEMY OF SCIENCES
Issue: *Barriers and Channels Formed by Tight Junction Proteins*

TNF-α–induced intestinal epithelial cell shedding: implications for intestinal barrier function

Alastair J.M. Watson and Kevin R. Hughes

Department of Medicine, Norwich Medical School, University of East Anglia, Norwich Research Park, Norwich, England

Address for correspondence: Alastair J.M. Watson, M.D., F.R.C.P. (Lond), DipABRSM, Department of Medicine, Norwich Medical School, Rm 2.14 Elizabeth Fry Building, University of East Anglia, Norwich Research Park, Norwich NR4 7TJ. alastair.watson@uea.ac.uk

Although epithelial cells are continuously shed from the surface of the intestine, the intestinal epithelium maintains the integrity of the epithelial barrier. This is achieved through a highly dynamic process involving reorganization of tight junction and adherens junction proteins. This process both ejects the cell from the epithelial monolayer and plugs the gap left after the cell is shed. Inflammatory insults can trigger a disturbance of these barrier functions by increasing rates of cell shedding. Epithelial cell shedding and loss of barrier can be visualized by confocal laser endomicroscopy in humans. A simple grading system of confocal laser endomicroscopic images can stratify inflammatory bowel disease (IBD) patients in remission into those who will relapse over the subsequent six months and those who will not. Here, we review the mechanisms governing maintenance of these barrier functions and the implications of inflammation-induced barrier dysfunction in IBD.

Keywords: intestine; epithelium; tight junction; confocal laser endomicroscopy; barrier function; inflammatory bowel disease

Introduction

The intestinal epithelium forms a barrier against the intestinal contents and the wider environment,[1] allowing entry of selected molecules for nutrition and programming of the mucosal immune system, but excluding toxins and most microorganisms. The epithelium is constantly renewed by new cells derived from stem cells at the crypt base, which migrate upwards to either the villus of the small intestine or crypt mouth of the colon, from where they are shed. The entire intestinal epithelium is renewed every five to seven days, a process which may have evolved to enable epithelial cells to continually function at maximum metabolic efficiency.[2–5] However, the shedding of the epithelium presents a major challenge to the intestinal barrier, potentially causing a breach at the sites of cell shedding.[6] As this barrier remains intact in health, mechanisms must have evolved to maintain barrier function at these sites.[7] Loss of barrier function has significant mechanistic implications for disease pathogenesis.[8,9]

Cell shedding in health

Tight junctions between epithelial cells are a major component of the epithelial barrier and are known to be highly dynamic, opening and closing in response to a number of signaling pathways to control the passive absorption of water, ions, and other solutes.[10] For an epithelial cell to be shed from the intestinal epithelial monolayer, reorganization of the tight junction proteins is required. Early electron microscopy studies suggested that tight junctions might rearrange beneath the shedding cell.[11] Subsequently, it has proven possible to study cell shedding and subcellular redistribution of the tight junction protein zonula occludens protein-1 (ZO-1) *in vivo*. Using transgenic mice expressing a fusion protein of ZO-1 and monomeric red fluorescent protein (mRFP1), coupled with nuclear staining using Hoechst 33342 or 33258 and Lucifer yellow as an intestinal permeability probe, the dynamics of cell shedding in three dimensions can be visualized.[12] The first identifiable event with this imaging system was condensation of ZO-1 at the tight junction,

doi: 10.1111/j.1749-6632.2012.06523.x

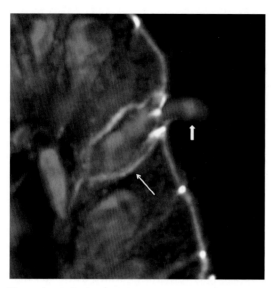

Figure 1. Phalloidin staining of actin redistribution around a shedding cell (thin arrow). The nucleus is the process of being extruded is shown (thick arrow. ZO-1). Adapted from Ref. 7.

followed by redistribution toward the basal pole of the cell. ZO-1 forms a "funnel" around and beneath the shedding cell which when viewed *en face* appears as a "purse string" structure. This ZO-1 redistribution process is completed in approximately 25 minutes. Actin is also redistributed along with ZO-1, and this can be easily visualized with phalloidin staining (Fig. 1). Movement of the cell, as defined by movement of the cell nucleus, does not start until 15 minutes after increased condensation of ZO-1 can be identified at the junction, with loss from the monolayer into the lumen taking approximately 15 minutes. Once shed, a "gap" is left in the monolayer, filled with material including ZO-1, which subsequently shrinks and redistributes back to the tight junction over the following 20 minutes. Thus, while completion of cell shedding (as characterized by monitoring loss of nuclear staining from the monolayer) takes approximately 15 minutes— as defined by redistribution of ZO-1—the process takes 45 minutes.

Although Lucifer yellow sometimes enters the apical region of the intercellular space adjoining the shedding cell, it is always prevented from reaching the basal pole by the ZO-1–containing material, demonstrating maintenance of barrier integrity during cell shedding.[6]

Initial studies identified small numbers of gaps or discontinuities in the epithelium as defined by the lack of intravital staining of the cell nucleus; the majority of resultant gaps could be clearly associated with shedding cells.[6] Subsequent experiments showed that gaps are more often seen in the *en face* view than the lateral view.[12] It also became clear that the dyes Hoechst 333258 and to a lesser extent Hoechst 33342 do not always stain the nuclei of goblet cells, potentially leading to the misidentification of some gaps. These difficulties were overcome in later studies using additional probes for tight junction components, including ZO-1, and confining analysis to events in which cell shedding can be unequivocally identified.[7]

In summary, these studies confirmed the Madara "zipper" model of maintenance of barrier function during cell shedding. In this model, which is based on data from electron microscopy studies, it was observed that the cells neighboring the shedding cell extend processes beneath the cell about to be shed. These processes included tight junction elements. As the cell is shed, the processes from the neighboring cells come together in a process that was likened by Madara to a "zipper" being drawn up, thereby maintaining the epithelial barrier at the shedding site.[11]

TNF-α–induced cell shedding

Tumor necrosis factor-α (TNF-α), an important inflammatory cytokine, increases cell shedding in mice,[13] although the degree of stimulation of cell shedding is highly variable from experiment to experiment. This may partially be explained by the highly variable rates of endogenous cell shedding in the murine intestine. In contrast to the healthy gut, after administration of TNF-α, barrier function is disturbed by an increase in cell shedding and by the simultaneous shedding of two or more cells at certain sites, producing microerosions that cannot be sealed by tight junction protein redistribution.[14]

The redistribution of tight junction proteins has been studied in detail after TNF-α administration. Although low doses of TNF-α do induce some structural and functional changes in the tight junctions, high doses (7.5 μg) are required to reliably induce cell shedding. Studies were undertaken in mRFP-1 ZO-1 mice described earlier, and also in a further transgenic mouse strain in which the tight junction protein, occludin, is conjugated to enhanced green fluorescent protein.[7] The redistribution of ZO-1 was very similar to that seen in the healthy mouse intestine. Similar results were obtained with occludin, though the results were less clear, as expression of occludin is not restricted to the tight junction.[7] Immunohistochemical studies showed that the tight junction proteins claudin-7 and claudin-15 form a funnel around the shedding cell, suggesting that all tight junction proteins may be redistributed during cell shedding.[7,15] Furthermore, E-cadherin, a component of the adherens junction, is also redistributed.[16]

The very rapid redistribution of transmembrane proteins occludin, claudins, and E-cadherin to cover the entire basolateral membranes is remarkable, and suggests that membrane traffic might be involved. This was confirmed with dynasore, an inhibitor of dynamin, the guanosine triphosphate hydrolase responsible for endocytosis, which trapped cells in a partially extruded position unable to complete shedding.[7,17]

Antibody studies for activated caspase-3 show that during cell shedding, the cell undergoes apoptosis.[3,4] It has long been debated whether apoptosis is a cause or consequence of cell shedding.[18,19] Infusion of the intestine with the pan caspase inhibitor N-(2-quinolyl)valyl-aspartyl-(2,6-difluorophenoxy)methyl ketone inhibits TNF-α–induced cell shedding. Thus, in the case of TNF-α–induced cell shedding, it is clear that TNF-α first induces apoptosis, which subsequently induces cell shedding as a secondary event. Unfortunately, low rates of endogenous cell shedding make such studies in healthy intestine a considerable technical challenge and have therefore not been described. Interestingly, studies in human small intestine have shown that acyl-CoA synthetase is differentially expressed along the crypt/villus axis, and that such differential expression may sensitize enterocytes at the villus tip to apoptotic cell death, via the death receptor TRAIL R1.[20] This provides one possible mechanistic explanation for loss of cells at the villus tip.

Actomyosin contraction and redistribution of actin is also important in initiating cell shedding. cytochalasin D, an inhibitor of actin polymerization, reduces cell shedding by 60%.[21] Myosin IIC, but not IIA or IIB, is redistributed around the shedding cell and the myosin II motor inhibitor blebbistatin also strongly inhibits shedding.[22] Involvement of actomyosin contraction suggests that the actin regulatory proteins Rho-associated kinase (ROCK)

and myosin light chain kinase (MLCK) also participate.[4,23–25] Consistent with this idea, increased phosphorylation of myosin light chains was observed on the lateral walls adjacent to the shedding cell while the Rho-kinase inhibitor Y27632 and MLCK[−/−] mice significantly reduced cell shedding.[26] Microtubule polymerization is also important in the early stages of cell shedding as the microtubule depolymerizing agent colcemid inhibits the early stages of this process.[5,7]

Cell shedding after apoptotic stimuli

Studies using *in vitro* monolayer culture models of cell extrusion provide further insight into the absolute requirement for tight regulation of these processes *in vivo*. Using MDCK II and 16 HBE-14o cell lines, it was shown that epithelia can trigger extrusion after intrinsic or extrinsic apoptotic stimuli, and that these extrusion pathways are partially dependent upon mitochondrial outer membrane permeabilization and caspase activation. Importantly, knockdown of caspases results in cells dying by necrosis and subsequently being removed by passive cell movement.[19] Thus, caspase activation seems to be a conserved step in the apoptotic pathway and not simply a result of TNF-α–induced epithelial shedding. Caspase activation might in turn control tight junction and adherens junction reorganization to facilitate maintenance of barrier function, although *in vivo* studies suggest that ZO-1 redistribution provides the earliest indicator of cell shedding.[12] Furthermore, synthesis of sphingosine-1-phosphate is important for cell shedding, as inhibition of sphingosine kinase inhibits cell shedding in zebrafish.[27] Future studies should determine whether the redistribution of ZO-1 precedes, or occurs coincident with, mitochondrial dysfunction, and/or caspase activation.

Further recent studies using zebrafish have also provided insight into the role of the tumor suppressor gene APC, a member of Wnt family of signaling molecules. APC localizes near sites of actomyosin contraction during apical and basal extrusion, and in association with the microtubules, is critical for determining the direction of cell extrusion from the monolayer.[28] In these studies, the stimulus for cell shedding is not known but is part of zebrafish development. The authors argue that inward shedding into the intestinal wall may be a mechanism through which APC mutation encourages cancer growth and metastasis in colon cancer.

Together, these studies suggest a tentative ordering of events in which cell shedding is initiated by apoptosis with contributions from MLCK, myosin ATPase, sphingosine kinase activity, and microtubule events. This may be coincident with, or followed by, redistribution of microtubules, tight junction, and adherences junction proteins along the lateral membranes into a funnel-like structure around the shedding cell. Completion and resolution of the shedding requires ROCK, MLCK, and dynamin. An important caveat is that none of the inhibitors completely prevented cell shedding, suggesting the existence of alternative escape pathways.[7]

Cell shedding in humans

The development of an endoscopic device incorporating a confocal microscope has enabled studies of cell shedding in humans and produced images similar to those from paraffin sections of fixed intestinal biopsies stained with hematoxylin and eosin.[3] Initial studies were undertaken using acriflavine as the fluorescent probe.[29] Acriflavine promiscuously labels all cellular elements, but fluorescence from the nucleus is more efficient than from the rest of the cell, making it an excellent tool for cell shedding studies. Indeed, epithelial gaps, probably resulting from cell shedding, could be easily identified. As discussed, epithelial gaps can be difficult to definitively distinguish from goblet cells. However, at confocal endomicroscopy, it was possible to show that goblet cells have a specific appearance when the focal plane is positioned at the apical surface of the cell. Such analysis generates a composite image showing the mouth of the goblet cell, the surrounding cytoplasm and the endogenous mucin, thus distinguishing goblet cells from epithelial gaps. A further distinguishing feature is that the acriflavine efficiently labels the nuclei of goblet cells and does not have the poor labeling efficiency of the Hoechst dyes. To gain further evidence that epithelial gaps were not misidentified goblet cells, we imaged in the intestine of mice with an intestine specific deletion of the transcription factor Math1 (Math1$^{\Delta intestine}$).[30] These mice lack goblet cells in the distal small intestine. Using a handheld confocal endomicroscope (Optiscan, Notting Hill, Victoria, Australia) and electron microscopy, presence of epithelial gaps was

shown in these mice along with examples of shedding cells creating epithelial gaps.

Recently, studies have been undertaken with fluorescein as a fluorescence probe.[14] Unlike acriflavine, which is sprayed onto the epithelial surface via the endoscope, fluorescein is administered intravenously. Fluorescein does not provide the subcellular resolution of acriflavine, but is able to image epithelial cells during the process of cell shedding and can be used as a probe for barrier defects. Although delivered intravenously, a little dye leaks out of submucosal capillaries in the intestine and will label epithelial cells. Some dye will also track between the epithelial cells in the lateral intercellular spaces. Fluorescein cannot traverse the tight junction at the apical border of the epithelial cell and only leaks into the lumen at sites of barrier loss. This use of fluorescein with confocal endomicroscopy is the only method to visualize sites of epithelial barrier loss in a clinical setting. Alternative methods of assessing intestinal barrier function include absorption of small molecular weight saccharides or ^{51}Cr-labelled ethylenediaminetetraacetic acid and measurement of their appearance in the urine.[31] These methods have proved useful in providing an integrated measurement of barrier function along a long segment of gut, but provide no information of the precise site of barrier loss.[32] Electrophysiological methods can give very precise functional information, but again do not provide visual information.[33–35]

In the healthy gut, the fluorescein signal from the lumen should be absent or of lower intensity than the adjacent epithelium. When the barrier is impaired, such as sites of microerosions, fluorescein will rapidly enter the lumen, making it intensely fluorescent. Sometimes, "plumes" of fluorescein can be seen flooding out of the epithelium, suggesting there is pressure within the villus forcing the fluorescein out (Fig. 2).

There is a potential difficulty with the fluorescein method for detecting barrier defects, because it can only visualize efflux from the villus interior, across the epithelium, and into the lumen.[36–38] Most disease pathogenesis scenarios involve influx of material from the intestinal lumen into the lamina propria; in other words in the opposite direction to fluorescein efflux. This issue was addressed in mouse experiments in which Alexa-fluor 647 conjugated dextran was administered intravenously to

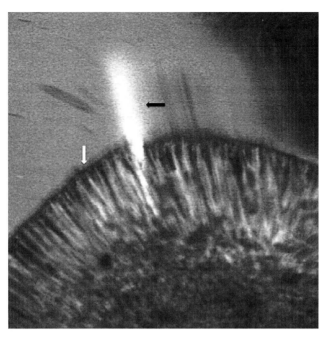

Figure 2. Efflux of fluorescein (black arrow) at a site of local barrier loss. Image from confocal endomicroscopy of the terminal ileum of a patient with Crohn's disease. Efflux of fluorescein only occurs at a single cell position, probably at a site of cell shedding, where sealing of the resulting gap has failed. The tight junction prevents efflux of fluorescein where the epithelium is intact (white arrow). Adapted from Ref. 14.

visualize the villus vasculature and potential efflux from the intestine into the lumen. Simultaneously, FITC-conjugated dextrans were applied to the luminal surface in a solution of 300 mOsm/L, to assess influx across the intestinal barrier. When the luminal osmolarity was isotonic relative to plasma, the direction of flow was outward into the lumen in the majority of cases (76%) of cell shedding, with a significant minority showing inward flow (18%), and some (6%) showing bidirectional flow (interpreted as free diffusion in both directions). Reducing luminal osmolarity to 246 mOsm/L increased both inward and bidirectional flow at these leaking sites, showing that direction of flow is critically dependant upon luminal osmolarity. Thus, the outward flow seen with fluorescein at confocal endomicroscopy of humans seems a reasonable surrogate for areas of inward flow. Although classically the intestinal lumen is considered to isotonic, there are likely to be areas of incomplete mixing, resulting in a wide range of osmolarities within the intestinal lumen. Thus, in individuals with barrier loss, the degree of influx of undesirable luminal molecules will vary both with position in the gut and also with time.

Barrier loss and cell shedding in inflammatory bowel disease

It remains unknown whether barrier dysfunction is a cause or effect of inflammatory bowel disease (IBD). Using a mouse model that enables targeted disruption of intestinal epithelial tight junctions via constitutive activation of MLCK, increases in paracellular permeability of up to 65% were shown to predispose and contribute to the development of subclinical immune activation and hallmarks of IBD. These included a 40% increase in lamina propria CD4[+] lymphocytes and redistribution of CD11c[+] cells (possibly dendritic cells) along with mild increases in IFN-γ and TNF-α.[39] Importantly, mice showed a worse clinical outcome in an adoptive transfer model of colitis. Further studies demonstrated that constitutive activation of MLCK produced increases in mucosal IL-13, resulting in increases in levels of claudin-2 and consequently, an increase in the number of small pores.[40] Thus, the ability of the tight junction to maintain epithelial barrier integrity is vital to health.

Defects in intestinal barrier function have been reported previously in IBD,[41] using absorption of polysaccharides to measure barrier function. It was therefore of interest to determine if defects in intestinal barrier could be detected in patients with Crohn's disease or ulcerative colitis using the confocal endomicroscopy/fluourescein technique described earlier. A retrospective study of IBD patients was undertaken, which demonstrated substantial increases in epithelial gaps, microerosions, and fluorescein leakage in these patients compared to age matched controls. Results were used to create a three-tier grading system, the Watson score, which reflected severity of epithelial shedding and barrier loss. A Watson grade of I is normal physiological cell shedding, defined as single cells being shed. In preliminary versions of the grade, we found that the number of cells being shed did not predict subsequent relapse, providing there was no fluorescein leakage or microerosions created by multiple cells being shed from a single site. Thus, the number of shedding cells being lost is not part of Watson grade I. Evidence of loss of barrier function when fluorescein leaks into the lumen is designated Watson grade II (functional defect). Microerosions caused by multiple epithelial cells being shed from a single site revealing the lamina propria to the lumen is designated Watson grade III (structural defect).[14]

The Watson grading system was used to evaluate 58 patients with IBD using images from the terminal ileum (47 with ulcerative colitis, 11 with Crohn's disease). At the time of endomicroscopy, these patients were in clinical remission and had normal macroscopic appearances of the terminal ileum. Twenty-four patients suffered a flare of their disease within 12 months after the confocal laser endomicroscopic (CLE) examination. Kaplan–Meier analysis was undertaken of relapse for the 12-month period after CLE, after stratification by their Watson grade. Patients who subsequently had a flare of their disease had significantly more fluorescein leakage and microerosions, thus, a higher Watson grade than patients who did not suffer a flare $P < 0.001$. The sensitivity, specificity and accuracy for the Watson scores II/III to predict flares were 62.5% (CI 95%: 40.8–80.4), 91.2% (CI 95%: 75.2–97.7), and 79% (CI 95%: 57.7–95.5), respectively. Additional studies demonstrating that patients with IBD show a greater gap density distribution than control patients (median gap density, 61 per 1,000 cells vs. 18 per 1,000 cells; $P < 0.001$) support this work.[42]

Overall, it seems that CLE will prove useful in predicting the likelihood of inflammatory relapse and may therefore be a useful new tool for management of targeting biological and immunosuppressive therapy to IBD patients at risk of relapse.

Concluding remarks and future directions

Although recent advances have provided valuable insights into many aspects of cell shedding and barrier function in inflammation and infection, significant questions remain. It is interesting that rates of cell shedding increase in response to inflammatory stimuli, but this raises the critical question of how such increases in rates of cell loss are balanced by proliferation in cells within the transit amplifying region and/or stem cell compartment, and how signals might be propagated from the sites of shedding to the crypt. It seems likely that epithelial renewal is highly dynamic and is regulated according to microbial load, inflammatory stimuli, or other external factors. Recent studies in *Drosophila* suggest a molecular link between the disturbance of gut microbiota, the immune response, and proliferation of intestinal stem cells.[43] Delineating the mechanisms governing such intracellular messaging in mammals will accelerate our understanding of many chronic inflammatory diseases such as IBD.

Observations in mRFP-1 ZO-1 transgenic mice that 15% of cells neighboring the shedding cell also begin the extrusion process some 5–10 minutes later raise further questions about the potential methods of cell-to-cell communication between shedding cells and their local environment.[12] It is interesting that although this latter study suggests that neighboring cells do not produce any local force to facilitate extrusion of shedding cells, conflicting *in vitro* studies suggest a significant involvement of neighbors in this process.[44] It nevertheless seems likely that some cell-to-cell communication between extruding cells and their neighbors does take place. Understanding these methods of communication (via gap junctions, cytokines, or otherwise) should provide new perspectives on the process of cell shedding in health and disease.

Other interesting observations in zebrafish that possess APC, a gene commonly associated with colon cancer and which is implicated in a cell's decision of whether to extrude basally or apically, raise the possibility that mutations of the APC gene, as frequently observed in colon cancer, may lead to increased rates of basal extrusion, leading to metastasis.[44] Curiously, studies in jejuna of mice have only described apical shedding with no description of such events occurring basally into the lamina propria of the villus. Ischemic reperfusion models in human small intestine have demonstrated a surprising ability for jejunal tissue to rapidly clear apoptotic cells from the epithelium to avoid an inflammatory response caused by unresolved necrotic cells.[45] These studies indicate the importance of rapid apical extrusion during periods of insult/injury. Defective apical extrusion may thus lead to basally oriented extrusion and/or phagocytosis of potentially necrotic cells. It is tempting to speculate that higher incidences of colon versus small intestinal cancer may be partially regulated by mutations in the APC gene and the subsequent failure of colonic apical extrusion.

In conclusion, maintaining the integrity of the epithelial barrier is clearly crucial in controlling episodes of intestinal disease. Failure of barrier function leads to entry of noxious agents and aberrant stimulation of the intestinal immune system. An improved understanding of the mechanisms which regulate cell shedding, cell replacement from the crypts, and barrier function in normal and inflamed or infected tissue is required to drive advances in our understanding of the pathogenesis of diseases such as IBD. Future directions should also address the role of various drugs/probiotic therapies in these pathways and how such substances may be used to improve clinical outcomes of disease, either via enhancement of mucosal repair processes or by modulating reductions in cell shedding.

Conflicts of interest

The authors declare no conflicts of interest.

References

1. Montrose, M.H. 2003. The future of GI and liver research: editorial pefrspectives: I. Visions of epithelial research. *Am. J. Physiol. Gastrointest. Liver Physiol.* **284:** G547–550.
2. Potten, C.S. & M. Loeffler. 1990. Stem cells: attributes, cycles, spirals, pitfalls and uncertainties. Lessons for and from the crypt. *Development* **110:** 1001–1020.
3. Bullen, T.F. *et al.* 2006. Characterization of epithelial cell shedding from human small intestine. *Lab. Invest.* **86:** 1052–1063.
4. Rosenblatt, J., M.C. Raff & L.P. Cramer. 2001. An epithelial cell destined for apoptosis signals its neighbors to extrude it by an actin- and myosin-dependent mechanism. *Curr. Biol.* **11:** 1847–1857.
5. Slattum, G., K.M. McGee & J. Rosenblatt. 2009. P115 RhoGEF and microtubules decide the direction apoptotic

6. Watson, A.J. *et al.* 2005. Epithelial barrier function in vivo is sustained despite gaps in epithelial layers. *Gastroenterology* **129:** 902–912.
7. Marchiando, A.M. *et al.* 2011. The epithelial barrier is maintained by in vivo tight junction expansion during pathologic intestinal epithelial shedding. *Gastroenterology* **140:** 1208–1218.
8. Meddings, J. 2008. The significance of the gut barrier in disease. *Gut* **57:** 438–440.
9. Marchiando, A.M., W.V. Graham & J.R. Turner. 2010. Epithelial barriers in homeostasis and disease. *Annu. Rev. Pathol.* **5:** 119–144.
10. Shen, L. *et al.* 2011. Tight junction pore and leak pathways: a dynamic duo. *Annu. Rev. Physiol.* **73:** 283–309.
11. Madara, J.L. 1990. Maintenance of the macromolecular barrier at cell extrusion sites in intestinal epithelium: physiological rearrangement of tight junctions. *J. Membr. Biol.* **116:** 177–184.
12. Guan, Y. *et al.* 2011. Redistribution of the tight junction protein ZO-1 during physiological shedding of mouse intestinal epithelial cells. *Am. J. Physiol. Cell Physiol.* **300:** C1404–1414.
13. Kiesslich, R. *et al.* 2007. Identification of epithelial gaps in human small and large intestine by confocal endomicroscopy. *Gastroenterology* **133:** 1769–1778.
14. Kiesslich, R. *et al.* 2012. Local barrier dysfunction identified by confocal laser endomicroscopy predicts relapse in inflammatory bowel disease. *Gut* In press.
15. Furuse, M. *et al.* 2002. Claudin-based tight junctions are crucial for the mammalian epidermal barrier: a lesson from claudin-1-deficient mice. *J. Cell. Biol.* **156:** 1099–1111.
16. Hermiston, M.L. & J.I. Gordon. 1995. Inflammatory bowel disease and adenomas in mice expressing a dominant negative N-cadherin. *Science* **270:** 1203–1207.
17. Marchiando, A.M. *et al.* 2010. Caveolin-1-dependent occludin endocytosis is required for TNF-induced tight junction regulation in vivo. *J. Cell. Biol.* **189:** 111–126.
18. Abreu, M.T. *et al.* 2000. Modulation of barrier function during Fas-mediated apoptosis in human intestinal epithelial cells. *Gastroenterology* **119:** 1524–1536.
19. Andrade, D. & J. Rosenblatt. 2011. Apoptotic regulation of epithelial cellular extrusion. *Apoptosis* **16:** 491–501.
20. Gassler, N. *et al.* 2007. Regulation of enterocyte apoptosis by acyl-CoA synthetase 5 splicing. *Gastroenterology* **133:** 587–598.
21. Hartwig, J.H. & T.P. Stossel. 1979. Cytochalasin B and the structure of actin gels. *J. Mol. Biol.* **134:** 539–553.
22. Straight, A.F. *et al.* 2003. Dissecting temporal and spatial control of cytokinesis with a myosin II Inhibitor. *Science* **299:** 1743–1747.
23. Clayburgh, D.R. *et al.* 2004. A differentiation-dependent splice variant of myosin light chain kinase, MLCK1, regulates epithelial tight junction permeability. *J. Biol. Chem.* **279:** 55506–55513.
24. Clayburgh, D.R. *et al.* 2005. Epithelial myosin light chain kinase-dependent barrier dysfunction mediates T cell activation-induced diarrhea in vivo. *J. Clin. Invest.* **115:** 2702–2715.

cells extrude from an epithelium. *J. Cell. Biol.* **186:** 693–702.

25. Tamada, M. *et al.* 2007. Two distinct modes of myosin assembly and dynamics during epithelial wound closure. *J. Cell. Biol.* **176:** 27–33.

26. Uehata, M. *et al.* 1997. Calcium sensitization of smooth muscle mediated by a Rho-associated protein kinase in hypertension. *Nature* **389:** 990–994.

27. Gu, Y. *et al.* 2011. Epithelial cell extrusion requires the sphingosine-1-phosphate receptor 2 pathway. *J. Cell. Biol.* **193:** 667–676.

28. Eisenhoffer, G.T. & J. Rosenblatt. 2011. Live imaging of cell extrusion from the epidermis of developing zebrafish. *J. Vis. Exp.* **52. pii:** 2689.

29. Deinert, K. *et al.* 2007. In-vivo microvascular imaging of early squamous-cell cancer of the esophagus by confocal laser endomicroscopy. *Endoscopy* **39:** 366–368.

30. Shroyer, N.F. *et al.* 2005. Gfi1 functions downstream of Math1 to control intestinal secretory cell subtype allocation and differentiation. *Genes. Dev.* **19:** 2412–2417.

31. Meddings, J.B. *et al.* 1993. Sucrose: a novel permeability marker for gastroduodenal disease. *Gastroenterology* **104:** 1619–1626.

32. Arrieta, M.C., L. Bistritz & J.B. Meddings. 2006. Alterations in intestinal permeability. *Gut* **55:** 1512–1520.

33. Schulzke, J.D. *et al.* 1987. Adaptation of the jejunal mucosa in the experimental blind loop syndrome: changes in paracellular conductance and tight junction structure. *Gut* **28:** 159–164.

34. Munkholm, P. *et al.* 1994. Intestinal permeability in patients with Crohn's disease and ulcerative colitis and their first degree relatives. *Gut* **35:** 68–72.

35. Hilsden, R.J., J.B. Meddings & L.R. Sutherland. 1996. Intestinal permeability changes in response to acetylsalicylic acid in relatives of patients with Crohn's disease. *Gastroenterology* **110:** 1395–1403.

36. Xavier, R.J. & D.K. Podolsky. 2007. Unravelling the pathogenesis of inflammatory bowel disease. *Nature* **448:** 427–434.

37. Colombel, J.F., A.J. Watson & M.F. Neurath. 2008. The 10 remaining mysteries of inflammatory bowel disease. *Gut* **57:** 429–433.

38. Khor, B., A. Gardet & R.J. Xavier. 2011. Genetics and pathogenesis of inflammatory bowel disease. *Nature* **474:** 307–317.

39. Su, L. *et al.* 2009. Targeted epithelial tight junction dysfunction causes immune activation and contributes to development of experimental colitis. *Gastroenterology* **136:** 551–563.

40. Weber, C.R. *et al.* 2010 Epithelial myosin light chain kinase activation induces mucosal interleukin-13 expression to alter tight junction ion selectivity. *J. Biol. Chem.* **285:** 12037–46.

41. Wyatt, J. *et al.* 1993. Intestinal permeability and the prediction of relapse in Crohn's disease. *Lancet* **341:** 1437–1439.

42. Liu, J.J. *et al.* 2011. Increased epithelial gaps in the small intestines of patients with inflammatory bowel disease: density matters. *Gastrointest. Endosc.* **73:** 1174–1180.

43. Buchon, N. *et al.* 2009. Invasive and indigenous microbiota impact intestinal stem cell activity through multiple pathways in Drosophila. *Genes. Dev.* **23:** 2333–2344.

44. Marshall, T.W. *et al.* 2011. The tumor suppressor adenomatous polyposis coli controls the direction in which a cell extrudes from an epithelium. *Mol. Biol. Cell* **22:** 3962–3970.

45. Matthijsen, R.A. *et al.* 2009. Enterocyte shedding and epithelial lining repair following ischemia of the human small intestine attenuate inflammation. *PLoS One* **4:** e7045.

Ann. N.Y. Acad. Sci. ISSN 0077-8923

ANNALS OF THE NEW YORK ACADEMY OF SCIENCES
Issue: *Barriers and Channels Formed by Tight Junction Proteins*

Tight junctions on the move: molecular mechanisms for epithelial barrier regulation

Le Shen

Department of Surgery, The University of Chicago, Chicago, Illinois

Address for correspondence: Le Shen, M.D., Ph.D., Department of Surgery, The University of Chicago, 5841 S. Maryland Ave., MC1089, Chicago, IL 60637. leshen@uchicago.edu

Increasing evidence suggests that the tight junction is a dynamically regulated structure. Cytoskeletal reorganization, particularly myosin light chain phosphorylation–induced actomyosin contraction, has increasingly been recognized as a mediator of physiological and pathophysiological tight junction regulation. However, our understanding of molecular mechanisms of tight junction modulation remains limited. Recent studies using live cell and live animal imaging techniques allowed us to peek into the molecular details of tight junction regulation. At resting conditions, the tight junction is maintained by dynamic protein–protein interactions, which may provide a platform for rapid tight junction regulation. Following stimulation, distinct forms of tight junction protein reorganization were observed. Tumor necrosis factor (TNF-α) causes a myosin light chain kinase (MLCK)–mediated barrier regulation by inducing occludin removal from the tight junction through caveolar endocytosis. In contrast, MLCK- and CK2-inhibition–caused tight junction regulation is mediated by altered zonula occludens (ZO)-1 protein dynamics and requires ZO-1–mediated protein–protein interaction, potentially through regulating claudin function. Although some of the molecular details are missing, studies summarized above point to modulating protein localization and dynamics that are common mechanisms for tight junction regulation.

Keywords: tight junction; protein dynamics; fluorescent recovery after photobleaching; endocytosis; epithelial barrier function

Introduction

Epithelia form boundaries that separate liquid-filled tissue compartments throughout the body. The cell layer is not only critical to forming physically separate spaces, but also is crucial to defining the fluid composition of these spaces. For example, the renal tubular epithelium forms a barrier to maintain the distinct luminal ion and macromolecular content relative to the interstitium, which is critical to supporting transport across the epithelium, and is essential to defining urine composition and normal waste removal and water balance. In the gut, similar requirements exist to allow efficient nutrient absorption and water handling. However, because many microorganisms normally reside in the gut lumen, the intestinal epithelium must also effectively form a barrier to prevent bacteria from entering the lamina propria.

The intestinal epithelial cells are critical to forming such a barrier. Without selective channels and transporters, the plasma membrane would be essentially impermeable to water and solutes. Thus, a continuous epithelial cell layer is critical to maintaining epithelial barrier. In addition, the space between individual epithelial cells also needs to be sealed. This function is carried out by the apical junctional complex, containing the tight junction, adherens junction, and desmosome. While the adherens junction and desmosome do not directly seal the space between epithelial cells, they are critical to providing the adhesive force to ensure the integrity of the cell layer.[1,2] Once cell–cell adhesion is established, the tight junction seals the space through its protein components, including the transmembrane proteins claudins and occludin, and the cytoplasmic adaptor proteins such as ZO-1 (Fig. 1A).[3–5] Because ZO-1 and related proteins bind to a large number

doi: 10.1111/j.1749-6632.2012.06613.x

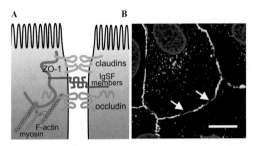

Figure 1. Components and dynamics of the tight junction. (A) The tight junction is composed of multiple interacting transmembrane and cytoplasmic proteins that are linked to the actin cytoskeleton. (B) In epithelial monolayers expressing both red fluorescent protein-tagged occludin (red) and EGFP-tagged clauidn-1 (green). Following photobleaching of both fluorescent proteins (shown with arrows), significantly more fluorescent recovery occurred for occludin than claudin-1, making the bleached region appear red. Bar: 5 μm.

of transmembrane tight junction proteins and the actin filaments through distinct protein domains, it was thought that these proteins may form stable protein complexes.[3,5–7] However, such sealing is not absolute. For example, Na^+ must be able to pass the tight junction to enter the intestinal lumen to establish a Na^+ gradient to drive nutrient absorption, and following digestion, glucose, and other nutrients cross the tight junction from the luminal side to maximize absorption.[8–11] Thus, the intestinal tight junction must balance the need to form a barrier and the need to support paracellular transport to ensure the proper function of the gastrointestinal tract.

Intestinal tight junction regulation occurs in health and disease

The tight junction must adapt to the different needs of providing a seal and allowing paracellular transport under distinct physiological and pathophysiolgical conditions.[12–14] An increasing number of stimuli are found to regulate tight junction function and there is a growing understanding about how defective paracellular sealing or transport may contribute to inherited and acquired human diseases.[9,15–19] For example, activation of Na^+-nutrient cotransport, such as Na^+-glucose cotransport mediated by sodium-glucose cotransporter (SGLT)-1, increases paracellular permeability and allows paracellular water and nutrient absorption.[9–11,20–23] Such enhanced paracellular absorption is thought

to supplement the saturable transcellular pathways to allow efficient nutrient absorption.

Increased paracellular permeability is well-recognized in patients with inflammatory bowel disease, particularly Crohn's disease.[24,25] A fraction of patients with quiescent Crohn's disease has increased intestinal permeability. When stratified by this criteria, patients with increased permeability are more likely to relapse compared to patients with normal intestinal permeability, indicating that intestinal permeability may participate in disease progression.[26] Furthermore, such intestinal permeability increases are also present in a subset of healthy first-degree relatives of Crohn's disease patients, leading to the hypothesis that reduced intestinal barrier function contributes to Crohn's disease pathogenesis.[27,28] This hypothesis is supported by a recent study showing that mice with intrinsic increases in tight junction permeability have subclinical inflammation and accelerated and exacerbated naive CD4 cell adoptive transfer–induced intestinal inflammation.[29] Such increased tight junction permeability is at least partially mediated by TNF, as neutralizing antibody for TNF normalizes intestinal permeability defects in Crohn's disease patients and TNF can directly induce increases in tight junction permeability in both cultured intestinal epithelial cells and mouse epithelium.[30–32]

MLCK mediates acute physiological and pathophysiolgical tight junction regulation

The mechanisms for tight junction regulation are under close investigation. Studies of humans and mice have demonstrated that tight junction protein expression and targeting are critical to determining tight junction function. For example, IL-13–mediated expression of tight junction protein claudin-2 is associated with increased intestinal permeability in ulcerative colitis patients, but the effects of IL-13 on barrier function do not occur rapidly.[33,34]

In contrast, physiological regulation of the tight junction by SGLT-1 activation and pathophysiolgical tight junction regulation by proinflammatory cytokines, such as TNF, LIGHT (lymphotoxin-like inducible protein that competes with glycoprotein D for herpes virus entry on T cells), and IL-1β, can happen in a matter of hours without changing tight junction protein expression.[35–39] Early morphological studies showed that Na^+-glucose cotransport activation induces perijunctional actomyosin condensation and is associated with actomyosin contraction, indicated by phosphorylation of myosin light chain (MLC), the regulatory component of actomyosin function.[11,38] Further studies demonstrated that MLCK is responsible for MLC phosphorylation and SGLT-1–induced barrier regulation.[38] Similarly, TNF also activates MLCK to increase tight junction permeability and remarkably, specific MLCK inhibition either pharmacologically or genetically blocks TNF-induced tight junction regulation in cell culture and animal models.[35,40,41] Furthermore, intestinal cell lines or animals with intestinal epithelial specific constitutively active MLCK expression have increased paracellular permeability, confirming MLCK activity alone is sufficient to regulate tight junction function.[29,42] Furthermore, increased MLCK activity has been associated with other forms of tight junction regulation, including enteropathogenic *E. coli* infection.[43] Thus, MLCK-mediated actin cytoskeletal reorganization has both physiological and pathophysiological relevance to tight junction barrier regulation.

Caveolar endocytosis of occludin is critical for cytokine-induced, MLCK-dependent TJ reorganization

Despite our understanding of the role of MLCK in regulating the tight junction function, the molecular mechanisms for tight junction regulation downstream of actomyosin contraction remain poorly understood. Although activation of actomyosin contraction by MLCK alters tight junction protein distribution in detergent insoluble glycoprotein rich microdomains, little is known about how such change occurs and the functional significance for these changes.[35,42] To understand the molecular basis for cytokine-induced tight junction regulation, we examined the distribution of individual tight junction proteins before and after exposure to T cell–derived cytokines.[41] Most strikingly, occludin disappeared from the tight junction in small intestinal epithelial cells and appeared in intracellular vesicles. This occludin removal from the tight junction is MLCK dependent, as both pharmacological and genetic inhibition of MLCK blocked this change.[41] Remarkably, this tight junction reorganization closely resembles the acute occludin endocytosis from the tight junction induced by the actin depolymerizing drug latrunculain A in cul-

tured epithelial monolayers.[44] In this form of tight junction reorganization, latrunculin A–induced occludin removal is through dynamin II–dependent caveolar endocytosis, and inhibiting such endocytosis blocks latrunculin A–induced tight junction disruption.[44]

We then thought to test if and how the caveolar-dependent pathway contributes to cytokine-induced tight junction regulation. We first demonstrated that LIGHT, a cytokine closely related to TNF, increases tight junction permeability in an MLCK-dependent mechanism in cultured epithelial monolayers.[36] This cytokine also induced occludin redistribution from the tight junction to caveolin-1–containing vesicles, which was blocked by drugs inhibiting caveolar function, but not drugs inhibiting clathrin-mediated endocytosis and macropinocytosis.[36] Such studies demonstrated a role for caveolar-mediated processes in cytokine-mediated tight junction regulation, but the direct evidence for occludin endocytosis and if occludin removal itself is critical for cytokine-mediated tight junction regulation remained uncertain.

To address this question, we took advantage of the *in vivo* model for cytokine-induced barrier regulation. When live animal imaging of the small intestinal epithelium was performed on mice with intestinal epithelial-specific transgenic expression of enhanced green fluorescent protein (EGFP)–tagged occludin, we found TNF-induced focal accumulation of occludin and evidence of endocytosis, indicated by vesicle budding from sites of accumulation.[45] Using inhibitors for endocytosis, we demonstrated that this internalization is through the caveolar pathway.[45] Furthermore, pharmacological inhibitors for caveolar endocytosis and caveolin-1 deficiency both block TNF-induced occludin endocytosis and loss of tight junction function.[45] Studies using the occludin transgenic animal showed EGFP-occludin overexpression can maintain large amounts of occludin at the tight junction, which inhibited TNF-induced increase in tight junction permeability, showing occludin itself is a critical component for TNF-induced tight junction regulation (Fig. 2).[45] This finding is consistent with the report that occludin participates in TNF-induced tight junction regulation in canine kidney epithelial cells, although the transepithelial resistance (TER) is increased rather than decreased in this cell type following TNF treatment.[46]

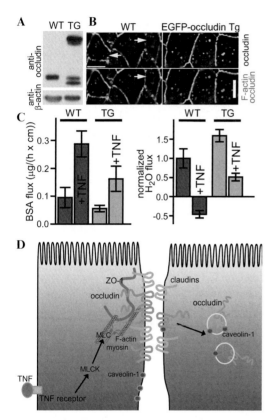

Figure 2. Occludin overexpression limits TNF-induced tight junction regulation. (A) Overexpression of EGFP-occludin in intestinal epithelium determined by SDS-PAGE immunoblot. Jejunal epithelial cells were isolated from wild-type and EGFP-occludin transgenic mice and were subjected to immunoblot. (B) EGFP-occludin expression preserves occludin localization at the tight junction following TNF treatment. Jejunal tissues from wild-type and EGFP-occluidn transgenic mice treated with 5 μg TNF for 120 min were sectioned and stained for occludin (top, and green in merge), and F-actin (red in merge). Regions of the tight junction lacking occludin developed in wild-type, but not EGFP-occludin transgenic mice. Bar: 10 μm. (C) Intestinal epithelial EGFP-occludin overexpression limits TNF-induced barrier loss and water secretion in small intestine. *In vivo* perfusion assays show increased paracellular BSA flux (left) and water secretion (right) in wild-type mice. TNF-induced paracellular BSA flux (left) was attenuated, while TNF-induced water secretion did not occur in EGFP-occludin transgenic mice. (D) Schematic presentation of the pathways for TNF to regulate tight junction in intestinal epithelial cells. TNF activates MLCK in intestinal epithelial cells to phosphorylate MLC and cause actomyosin contraction (left side). These events lead to tight junction reorganization to cause occludin to internalize in a caveolin-1–dependent fashion (right side). The original data are adapted from Ref. 45.

MLCK induces TJ regulation through changing ZO-1 protein dynamics

Although occludin endocytosis occurs prominently in cytokine-induced tight junction regulation, it is not seen in all forms of MLCK-mediated tight junction regulation. For example, no occludin internalization can be observed in cells with activated SGLT-1, and direct inhibition of MLCK failed to demonstrate changes in occludin internalization, although both stimuli affect tight junction function. These findings suggest the tight junction organizational changes, if any, are subtle following these stimuli. Understanding the underlying mechanisms has been hampered by the lack of experimental tools to study protein distribution and protein–protein interactions *in situ*.

The first breakthrough came when it was discovered that distinct tight junction proteins have unique dynamic behaviors when assessed by fluorescent recovery after photobleaching (FRAP) experiments. By using epithelial cell lines expressing well-validated tight junction proteins with N-terminally fused EGFP, we demonstrated that transmembrane proteins claudin-1 and occludin have very different FRAP behaviors at the tight junction: while ~60% of claudin-1 molecules are immobile at the tight junction, only about ~30% of occludin molecules are immobile (Fig. 1B).[47] Furthermore, a large fraction of the cytoplasmic protein ZO-1 undergoes constant exchange between the cytoplasmic and tight junction pools.[47] Thus, despite the ability of ZO-1 to directly bind to claudin-1, occludin, other tight junction proteins, and the actin cytoskeleton, these proteins do not always form a stable protein complex. Furthermore, this study suggested that tight junction protein dynamics can be altered by a range of stimuli. For example, cholesterol depletion blocked occludin, but not ZO-1, fluorescent recovery while ATP depletion selectively inhibited ZO-1 fluorescent recovery.[47] These findings show that tight junction protein dynamic behavior, which reflects protein–protein interactions and protein–lipid interactions, can be modulated. Such modulations may be critical for functional regulation of the tight junction, as both cholesterol depletion and ATP depletion also decrease paracellular barrier function. Thus, studying the biochemical–functional relationships of the tight junction by combining measurements of

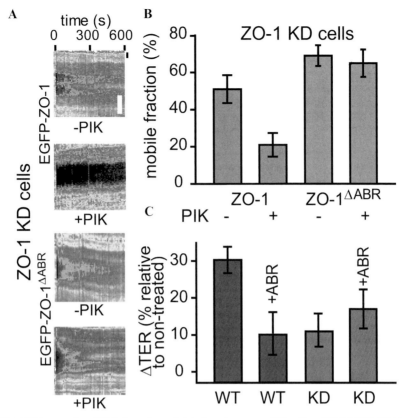

Figure 3. MLCK inhibition regulates ZO-1 dynamics and tight junction permeability. (A and B) MLCK inhibition by peptide inhibitor PIK decreases wild-type ZO-1 protein, but not ZO-1 mutant lacking the ABR exchange at the tight junction. ZO-1 knockdown of Caco-2 monolayers expressing either the EGFP-ZO-1 or EGFP-ZO-1 mutant lacking the ABR were treated with 300 μM MLCK inhibitor PIK and then subjected to FRAP experiments. Representative kymographs of ZO-1 protein fluorescent recovery in FRAP experiments are shown in A. Bar: 5 μm. Mobile fractions of the wild-type ZO-1 and ZO-1 mutants lacking the ABR with or without MLCK inhibition were determined on the basis of these experiments (B). (C) The ZO-1 ABR domain overexpression affects the MLCK inhibition–induced TER increase in ZO-1–sufficient, but not ZO-1 knockdown, cells. Treating wild-type Caco-2 monolayers with 300 μM PIK increased TER, which was blocked by ZO-1 ABR expression. In contrast, ZO-1 knockdown cells had minimal TER increases following PIK treatment, which could not be blocked by ZO-1 ABR expression. The original data are adapted from Ref. 48.

protein dynamics and paracellular barrier function may provide us mechanistic insights toward tight junction regulation.

We then applied such an approach to understand how MLCK may affect tight junction protein dynamics and function.[48] When the dynamic behaviors for representative tight junction proteins were measured in cells with active or inhibited MLCK by FRAP experiments, altered exchange of ZO-1 to and from the tight junction was observed, but the dynamic behaviors of occludin, and claudin-1 were unchanged.[48] Following MLCK inhibition, ZO-1 had decreased fluorescent recovery, indicating it had increased anchoring at the tight junction

(Fig. 3A and B).[48] The same results were seen *in vivo*, after MLCK inhibition, in mice with transgenic expression of fluorescently tagged ZO-1.[48] Furthermore, in knockout animals lacking the epithelial-specific isoform of MLCK, ZO-1 anchoring are insensitive to MLCK inhibition.[48] In addition, in mice with constitutively active MLCK expression in intestinal epithelial cells, ZO-1 recovery is faster.[48] These results demonstrate a critical role for MLCK to regulate ZO-1 dynamics. Using the model epithelium, we further showed that ZO-1 anchoring depends on its ability to directly bind to the actin cytoskeleton, as the ZO-1 mutant lacking its actin-binding region (ABR) failed to alter its dynamic behavior following MLCK inhibition (Fig. 3A and B).[48,49] In addition, ZO-1 ABR expression blocks MLCK inhibition caused ZO-1 anchoring at the tight junction.[48] With the results described above, we were able to test the contribution of ZO-1 and its ABR to tight junction permeability regulation. MLCK inhibition caused a rapid decrease in tight junction permeability in ZO-1–sufficient cells, however, in cells lacking ZO-1, MLCK inhibition failed to regulate tight junction function, indicating a critical role for ZO-1 in MLCK-dependent tight junction regulation (Fig. 3C).[48] While expression of the ZO-1 ABR domain was able to block the MLCK inhibition-induced tight junction regulation in ZO-1–sufficient cells, it failed to affect tight junction function in ZO-1-knockdown cells following MLCK inhibition (Fig. 3C),[48] demonstrating that the ability of ZO-1 ABR to affect tight junction function depends on endogenous ZO-1 expression. Taken together, the FRAP and barrier function studies demonstrate that ABR-mediated ZO-1 anchoring is responsible for decreased tight junction permeability following MLCK inhibition.

CK2 regulates tight junction through affecting tight junction protein complex formation

The study discussed above demonstrated that ZO-1 anchoring is critical for MLCK inhibition-caused decrease in tight junction permeability. However, it did not identify how ZO-1, a cytoplasmic protein, may affect permeability across the tight junction. One can hypothesize that ZO-1 could do so by regulating the function of transmembrane proteins at the tight junction. Such evidence comes from studies of how CK2 inhibition regulates tight junction function. CK2 is a kinase that directly phosphorylates occludin,[50-52] and CK2 inhibition results in decreased tight junction permeability to small, but not large ions.[53] Knocking down occludin blocked CK2 inhibition–induced barrier regulation, indicating a critical role for occludin in CK2-mediated tight junction regulation.[53] In occludin knockdown cells, exogenous expression of wild-type occludin, but not occludin with nonphosphorylable or phosphomimetic mutations at S408, a major phosphorylation site for CK2-mediated occludin phosphorylation, were able to reverse the phenotype, indicating a

critical role of occludin S408 phosphorylation in CK2-mediated tight junction regulation.[53] When the dynamic behavior of tight junction proteins were measured, occludin, ZO-1, claudin-1, and claudin-2 FRAP behavior were altered following CK2 inhibition.[53] Although these molecules have dramatically different FRAP behaviors at steady state, CK2 inhibition caused decreased occludin dynamics with increased claudin-1, claudin-2, and ZO-1 dynamics, making the FRAP behavior for these molecules to converge.[53] Such results can be explained by formation of a large protein complex containing occludin, ZO-1, claudin-1, and claudin-2 following CK2 inhibition.[53] As occludin does not directly bind to claudins, we tested the idea that ZO-1, which binds to claudins through its first PDZ domain and occludin through its U5-GuK region,[54-58] is responsible for complex formation.[53] Indeed, ZO-1 knockdown blocked CK2 inhibition-caused alterations in claudin-2 dynamics, and expression of wild-type ZO-1, but not ZO-1 mutants that cannot bind to occludin or claudins, were able to reverse the phenotype.[53] Such findings suggest that ZO-1 may serve as an adaptor molecule to transduce the CK2-mediated occludin phosphorylation signal to claudins.

Guided by these FRAP findings, functional studies were performed. Knocking down occludin or ZO-1 individually blocked CK2 inhibition–induced tight junction regulation.[53] While separate knockdown of claudin-1 or -2 attenuated CK2-mediated tight junction regulation, double knockdown of claudin-1 and -2 abolished this effect, indicating a cooperative role for these claudins.[53] Furthermore, occludin C-terminal tail with S408A mutation, a protein mimics the dephosphorylated occludin generated by CK2 inhibition, has increased association with ZO-1, claudin-1, and claudin-2 than wild-type occludin tail and occludin tail with S408D mutation, suggesting a multiple protein complex formation among occludin, ZO-1, claudin-1, and claudin-2 following CK2 inhibition (Fig. 4A).[53] Such a complex formation depends on ZO-1, as claudin-2 cannot be recruited to occludin in the absence of ZO-1 when the occludin C-terminal tail S408A mutant in protein lysates without ZO-1 expression (Fig. 4B).[53] Together with FRAP studies, these findings suggest tight junction protein complex formation is responsible for CK2 inhibition–induced tight junction regulation.

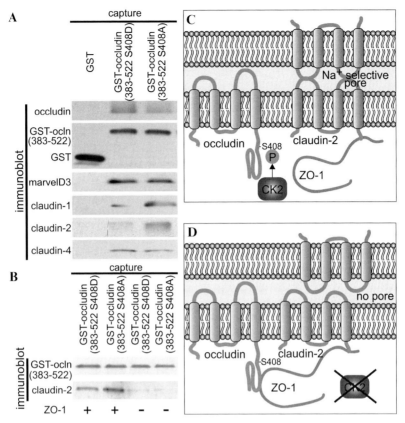

Figure 4. CK2 inhibition alters tight junction protein interaction and barrier function. (A) GST-occludin C-terminal tails (383–522) immobilized on glutathione-agarose beads were used to capture proteins from Caco-2 lysates. Recovered proteins were assessed by SDS-PAGE immunoblot. (B) GST-occludin C-terminal tails (383–522) were used to capture control and ZO-1 knockdown Caco-2 lysates. Recovered proteins were assessed as in A. (C) In the presence of CK2 activity, claudin-2 forms Na^+ permeable channels and does not form a complex with occludin. (D) When CK2 activity is blocked, occludin is not phosphorylated at S408, which allows it to bind to ZO-1. Occludin, ZO-1, and claudin-2 form a multimolecular complex to disrupt claudin-2 function, which leads to decreased paracelllar ion permeability. The original data are adapted from Ref. 53.

Modulating tight junction protein interactions as a general mechanism for tight junction regulation

MLCK inhibition and CK2 inhibition are very different stimuli, and they regulate tight junction protein dynamics in distinct manners. MLCK inhibition selectively anchors ZO-1 at the tight junction, without changing the FRAP behavior of claudin-1 and occludin.[48] However, how ZO-1 is anchored at the tight junction and the transmembrane component for this form of tight junction regulation still needs to be defined. In contrast, CK2 inhibition modulates occludin, ZO-1, claudin-1, and claudin-2 dynamics to allow formation of a large protein complex.[53] In this form of barrier regulation, the

direct contribution of occludin in forming the barrier remains to be addressed. As claudin-1 and -2 double knockdown made CK2 inhibition unable to regulate barrier function even in cells expressing occludin and ZO-1, it is likely occludin is primarily a signaling intermediate rather than the molecule that directly seals the paracellular space for CK2 to regulate paracellular permeability.

Despite these differences, these two studies presented a unified theme in the mechanisms for tight junction regulation. Both MLCK inhibition and CK2 inhibition (1) selectively decrease paracellular permeability of small, but not large ions, (2) affect ZO-1 dynamic behavior, and (3) require ZO-1 to regulate barrier. Furthermore, in contrast to wild-type ZO-1, ZO-1 mutants lacking the protein domains that are responsible for interacting with its upstream binding partners, that is, actin cytoskeleton or occludin, do not alter their FRAP behaviors following stimulation, and these protein domains are responsible for tight junction barrier regulation. Although some of the molecular details are still missing, such studies demonstrate a common modality for barrier regulation: ZO-1 functions as an adaptor to receive signals from the actin cytoskeleton or other tight junction proteins, which in turn regulates the ability of its transmembrane binding partners, such as claudins to regulate epithelial barrier.

These studies demonstrate that we are on the verge of understanding the molecular mechanisms for acute tight junction regulation. Using similar approaches, the molecular mechanisms for tight junctions to be affected by a variety of stimuli can be dissected. Detailed understanding of tight junction protein–protein interactions at the resting state and their regulation following stimulation at amino acid residues or even at the atomic level will not only provide us with scientific insight, but also will provide targets for specific modulation of tight junction functions for disease treatment and prevention.

Acknowledgments

The author is grateful for continued support of his mentor, Jerrold R. Turner and acknowledges Christopher R. Weber, David Raleigh, Amanda Marchiando, and Dan Yu for contributing graphs and helpful discussions. Work from the group was supported by National Institute of Diabetes and Digestive and Kidney Diseases Grants R01DK61931, R01DK68271, and P01DK67887 (JRT). The author is supported by a Crohn's Colitis Foundation of America (CCFA) Career Development Award and was supported by a CCFA Research Fellowship Award.

Conflicts of interest

The author declares no conflicts of interest.

References

1. Hermiston, M.L. & J.I. Gordon. 1995. In vivo analysis of cadherin function in the mouse intestinal epithelium: essential roles in adhesion, maintenance of differentiation, and regulation of programmed cell death. *J. Cell Biol.* **129:** 489–506.
2. Hermiston, M.L. & J.I. Gordon. 1995. Inflammatory bowel disease and adenomas in mice expressing a dominant negative N-cadherin. *Science* **270:** 1203–1207.
3. Tsukita, S., M. Furuse & M. Itoh. 2001. Multifunctional strands in tight junctions. *Nat. Rev. Mol. Cell Biol.* **2:** 285–293.
4. Mitic, L.L. & J.M. Anderson. 1998. Molecular architecture of tight junctions. *Annu. Rev. Physiol.* **60:** 121–142.
5. Gonzalez-Mariscal, L., A. Betanzos, P. Nava & B.E. Jaramillo. 2003. Tight junction proteins. *Prog. Biophys. Mol. Biol.* **81:** 1–44.
6. Schneeberger, E.E. & R.D. Lynch. 2004. The tight junction: a multifunctional complex. *Am. J. Physiol.—Cell Physiol.* **286:** C1213–C1228.
7. Guillemot, L., S. Paschoud, P. Pulimeno, *et al.* 2008. The cytoplasmic plaque of tight junctions: a scaffolding and signalling center. *Biochim. Biophys. Acta* **1778:** 601–613.
8. Tamura, A., H. Hayashi, M. Imasato, *et al.* 2011. Loss of claudin-15, but not claudin-2, causes Na+ deficiency and glucose malabsorption in mouse small intestine. *Gastroenterology* **140:** 913–923.
9. Pappenheimer, J.R. 1987. Physiological regulation of transepithelial impedance in the intestinal mucosa of rats and hamsters. *J. Membr. Biol.* **100:** 137–148.
10. Pappenheimer, J.R. & K.Z. Reiss. 1987. Contribution of solvent drag through intercellular junctions to absorption of nutrients by the small intestine of the rat. *J. Membr. Biol.* **100:** 123–136.
11. Madara, J.L. & J.R. Pappenheimer. 1987. Structural basis for physiological regulation of paracellular pathways in intestinal epithelia. *J. Membr. Biol.* **100:** 149–164.
12. Turner, J.R. 2006. Molecular basis of epithelial barrier regulation: from basic mechanisms to clinical application. *Am. J. Pathol.* **169:** 1901–1909.
13. Yu, D. & J.R. Turner. 2007. Stimulus-induced reorganization of tight junction structure: the role of membrane traffic. *Biochim. Biophys. Acta* **1778:** 709–716.
14. Turner, J.R. 2009. Intestinal mucosal barrier function in health and disease. *Nat. Rev. Immunol.* **9:** 799–809.
15. Shen, L. & J.R. Turner. 2006. Role of epithelial cells in initiation and propagation of intestinal inflammation. Eliminating the static: tight junction dynamics exposed. *Am. J. Physiol.—Gastrointest. Liver Physiol.* **290:** G577–G582.
16. Nusrat, A., J.R. Turner & J.L. Madara. 2000. Molecular physiology and pathophysiology of tight junctions: IV. Regulation of tight junctions by extracellular stimuli: nutrients, cytokines, and immune cells. *Am. J. Physiol.—Gastrointest. Liver Physiol.* **279:** G851–G857.
17. Simon, D.B., Y. Lu, K.A. Choate, *et al.* 1999. Paracellin-1, a renal tight junction protein required for paracellular Mg2+resorption. *Science* **285:** 103–106.
18. Carlton, V.E., B.Z. Harris, E.G. Puffenberger, *et al.* 2003. Complex inheritance of familial hypercholanemia with associated mutations in TJP2 and BAAT. *Nat. Genet.* **34:** 91–96.
19. Pappenheimer, J.R. & K. Volpp. 1992. Transmucosal impedance of small intestine: correlation with transport of sugars and amino acids. *Am. J. Physiol.* **263:** C480–C493.
20. Pappenheimer, J.R. 1993. On the coupling of membrane digestion with intestinal absorption of sugars and amino acids. *Am. J. Physiol.* **265:** G409–G417.

21. Turner, J.R. & J.L. Madara. 1995. Physiological regulation of intestinal epithelial tight junctions as a consequence of Na$^+$-coupled nutrient transport. *Gastroenterology* **109:** 1391–1396.

22. Atisook, K., S. Carlson & J.L. Madara. 1990. Effects of phlorizin and sodium on glucose-elicited alterations of cell junctions in intestinal epithelia. *Am. J. Physiol.* **258:** C77–C85.

23. Atisook, K. & J.L. Madara. 1991. An oligopeptide permeates intestinal tight junctions at glucose-elicited dilatations. Implications for oligopeptide absorption. *Gastroenterology* **100:** 719–724.

24. Hollander, D., C.M. Vadheim, E. Brettholz, *et al.* 1986. Increased intestinal permeability in patients with Crohn's disease and their relatives. A possible etiologic factor. *Ann. Intern. Med.* **105:** 883–885.

25. May, G.R., L.R. Sutherland & J.B. Meddings. 1993. Is small intestinal permeability really increased in relatives of patients with Crohn's disease? *Gastroenterology* **104:** 1627–1632.

26. Wyatt, J., H. Vogelsang, W. Hubl, *et al.* 1993. Intestinal permeability and the prediction of relapse in Crohn's disease. *Lancet* **341:** 1437–1439.

27. Katz, K.D., D. Hollander, C.M. Vadheim, *et al.* 1989. Intestinal permeability in patients with Crohn's disease and their healthy relatives. *Gastroenterology* **97:** 927–931.

28. Buhner, S., C. Buning, J. Genschel, *et al.* 2006. Genetic basis for increased intestinal permeability in families with Crohn's disease: role of CARD15 3020insC mutation? *Gut* **55:** 342–347.

29. Su, L., L. Shen, D.R. Clayburgh, *et al.* 2009. Targeted epithelial tight junction dysfunction causes immune activation and contributes to development of experimental colitis. *Gastroenterology* **136:** 551–563.

30. Marini, M., G. Bamias, J. Rivera-Nieves, *et al.* 2003. TNF-alpha neutralization ameliorates the severity of murine Crohn's-like ileitis by abrogation of intestinal epithelial cell apoptosis. *Proc. Natl. Acad. Sci. USA* **100:** 8366–8371.

31. Suenaert, P., V. Bulteel, L. Lemmens, *et al.* 2002. Anti-tumor necrosis factor treatment restores the gut barrier in Crohn's disease. *Am. J. Gastroenterol.* **97:** 2000–2004.

32. Zeissig, S., C. Bojarski, N. Buergel, *et al.* 2004. Downregulation of epithelial apoptosis and barrier repair in active Crohn's disease by tumour necrosis factor alpha antibody treatment. *Gut* **53:** 1295–1302.

33. Weber, C.R., D.R. Raleigh, L. Su, *et al.* 2010. Epithelial myosin light chain kinase activation induces mucosal interleukin-13 expression to alter tight junction ion selectivity. *J. Biol. Chem.* **285:** 12037–12046.

34. Heller, F., P. Florian, C. Bojarski, *et al.* 2005. Interleukin-13 is the key effector Th2 cytokine in ulcerative colitis that affects epithelial tight junctions, apoptosis, and cell restitution. *Gastroenterology* **129:** 550–564.

35. Wang, F., W.V. Graham, Y. Wang, *et al.* 2005. Interferon-gamma and tumor necrosis factor-alpha synergize to induce intestinal epithelial barrier dysfunction by up-regulating myosin light chain kinase expression. *Am. J. Pathol.* **166:** 409–419.

36. Schwarz, B.T., F. Wang, L. Shen, *et al.* 2007. LIGHT signals directly to intestinal epithelia to cause barrier dysfunction via cytoskeletal and endocytic mechanisms. *Gastroenterology* **132:** 2383–2394.

37. Al-Sadi, R.M. & T.Y. Ma. 2007. IL-1beta causes an increase in intestinal epithelial tight junction permeability. *J. Immunol.* **178:** 4641–4649.

38. Turner, J.R., B.K. Rill, S.L. Carlson, *et al.* 1997. Physiological regulation of epithelial tight junctions is associated with myosin light-chain phosphorylation. *Am. J. Physiol.* **273:** C1378–C1385.

39. Mullin, J.M., K.V. Laughlin, C.W. Marano, *et al.* 1992. Modulation of tumor necrosis factor-induced increase in renal (LLC-PK1) transepithelial permeability. *Am. J. Physiol.* **263:** F915–F924.

40. Zolotarevsky, Y., G. Hecht, A. Koutsouris, *et al.* 2002. A membrane-permeant peptide that inhibits MLC kinase restores barrier function in in vitro models of intestinal disease. *Gastroenterology* **123:** 163–172.

41. Clayburgh, D.R., T.A. Barrett, Y. Tang, *et al.* 2005. Epithelial myosin light chain kinase-dependent barrier dysfunction mediates T cell activation-induced diarrhea in vivo. *J. Clin. Invest.* **115:** 2702–2715.

42. Shen, L., E.D. Black, E.D. Witkowski, *et al.* 2006. Myosin light chain phosphorylation regulates barrier function by remodeling tight junction structure. *J. Cell Sci.* **119:** 2095–2106.

43. Yuhan, R., A. Koutsouris, S.D. Savkovic & G. Hecht. 1997. Enteropathogenic Escherichia coli-induced myosin light chain phosphorylation alters intestinal epithelial permeability. *Gastroenterol.* **113:** 1873–1882.

44. Shen, L. & J.R. Turner. 2005. Actin depolymerization disrupts tight junctions via caveolae-mediated endocytosis. *Mol. Biol. Cell* **16:** 3919–3936.

45. Marchiando, A.M., L. Shen, W.V. Graham, *et al.* 2010. Caveolin-1-dependent occludin endocytosis is required for TNF-induced tight junction regulation in vivo. *J. Cell Biol.* **189:** 111–126.

46. Van Itallie, C.M., A.S. Fanning, J. Holmes & J.M. Anderson. 2010. Occludin is required for cytokine-induced regulation of tight junction barriers. *J. Cell Sci.* **123:** 2844–2852.

47. Shen, L., C.R. Weber & J.R. Turner. 2008. The tight junction protein complex undergoes rapid and continuous molecular remodeling at steady state. *J. Cell Biol.* **181:** 683–695.

48. Yu, D., A.M. Marchiando, C.R. Weber, *et al.* 2010. MLCK-dependent exchange and actin binding region-dependent anchoring of ZO-1 regulate tight junction barrier function. *Proc. Natl. Acad. Sci. USA* **107:** 8237–8241.

49. Fanning, A.S., T.Y. Ma & J.M. Anderson. 2002. Isolation and functional characterization of the actin binding region in the tight junction protein ZO-1. *FASEB J.* **16:** 1835–1837.

50. Smales, C., M. Ellis, R. Baumber, *et al.* 2003. Occludin phosphorylation: identification of an occludin kinase in brain and cell extracts as CK2. *FEBS Lett.* **545:** 161–166.

51. Cordenonsi, M., F. Turco, F. D'Atri, *et al.* 1999. Xenopus laevis occludin. Identification of in vitro phosphorylation sites by protein kinase CK2 and association with cingulin. *Eur. J. Biochem.* **264:** 374–384.

52. Cordenonsi, M., E. Mazzon, L. De Rigo, *et al.* 1997. Occludin dephosphorylation in early development of Xenopus laevis. *J. Cell Sci.* **110:** 3131–3139.

53. Raleigh, D.R., D.M. Boe, D. Yu, *et al.* 2011. Occludin S408 phosphorylation regulates tight junction protein interactions and barrier function. *J. Cell Biol.* **193:** 565–582.

54. Schmidt, A., D.I. Utepbergenov, G. Krause & I.E. Blasig. 2001. Use of surface plasmon resonance for real-time analysis of the interaction of ZO-1 and occludin. *Biochem. Biophys. Res. Commun.* **288:** 1194–1199.

55. Muller, S.L., M. Portwich, A. Schmidt, *et al.* 2005. The tight junction protein occludin and the adherens junction protein alpha-catenin share a common interaction mechanism with ZO-1. *J. Biol. Chem.* **280:** 3747–3756.

56. Fanning, A.S., B.P. Little, C. Rahner, *et al.* 2007. The Unique-5 and -6 Motifs of ZO-1 Regulate Tight Junction Strand Localization and Scaffolding Properties. *Mol. Biol. Cell* **18:** 721–731.

57. Fanning, A.S., B.J. Jameson, L.A. Jesaitis & J.M. Anderson. 1998. The tight junction protein ZO-1 establishes a link between the transmembrane protein occludin and the actin cytoskeleton. *J. Biol. Chem.* **273:** 29745–29753.

58. Itoh, M., M. Furuse, K. Morita, *et al.* 1999. Direct binding of three tight junction-associated MAGUKs, ZO-1, ZO-2, and ZO-3, with the COOH termini of claudins. *J. Cell Biol.* **147:** 1351–1363.

Ann. N.Y. Acad. Sci. ISSN 0077-8923

ANNALS OF THE NEW YORK ACADEMY OF SCIENCES
Issue: *Barriers and Channels Formed by Tight Junction Proteins*

HIV infection and the intestinal mucosal barrier

Hans-Jörg Epple and Martin Zeitz

Department of Gastroenterology, Infectious Diseases and Rheumatology, Charité – Universitätsmedizin Berlin, Campus Benjamin Franklin, Berlin, Germany

Address for correspondence: Hans-Jörg Epple, M.D., Medical Clinic, Gastroenterology, Infectious Diseases and Rheumatology, Charité – Universitätsmedizin Berlin, Campus Benjamin Franklin, Hindenburgdamm 30, 12200 Berlin, Germany. hans-joerg.epple@charite.de

HIV infection induces a barrier defect of the intestinal mucosa, which is closely linked to immune activation and CD4 T cell depletion. The HIV-induced barrier defect is initiated in early acute and maintained through chronic infection. In acute infection, increased epithelial permeability is associated with increased epithelial apoptosis possibly caused by perforin-expressing cytotoxic T cells. In chronic infection, mucosal production of inflammatory cytokines is associated with increased epithelial permeability, epithelial apoptosis, and alterations of epithelial tight junctions. In addition to HIV-induced immune-mediated effects, viral proteins have the potential to directly affect epithelial barrier function. After prolonged viral suppression by antiretroviral therapy, there is, at least partial, restoration of the HIV-associated intestinal mucosal barrier defect despite persisting alterations of the mucosal immune system.

Keywords: HIV infection; mucosal barrier; microbial translocation; transepithelial resistance; tight junctions

Introduction

Containing the majority of its cellular targets, the intestinal mucosa is a main target organ for HIV infection.[1–8] Independent of the route of transmission, HIV infection causes severe depletion of mucosal CD4$^+$ T cells and significant alterations within the mucosal immune system. At least partly owing to its effects on the local immune system, HIV also induces changes in the structure of the intestinal mucosa, resulting in a defect of the mucosal barrier function. In a seminal paper published in 2006, Brenchley *et al.* described the finding of raised serum levels of lipopolysaccharides (LPS) in untreated HIV-infected patients associated with markers of systemic immune activation.[9] As chronic immune activation is considered the driving force of progressive CD4$^+$ T cell depletion in chronic HIV infection, the LPS and immune activation data were linked to a unifying model,[10] which subsequently has been widely adopted as a key mechanism of HIV immunopathogenesis. According to this model, HIV infection causes a barrier defect of the intestinal mucosa by mucosal CD4$^+$ T cell deple-

tion. The HIV-induced barrier defect in turn allows for increased translocation of microbial antigens, which—together with other mechanisms—results in a perpetual state of heightened immune activation and hence progressive CD4$^+$ T cell depletion.[10,11]

When this model was first proposed, there was little direct data on mucosal barrier function in HIV infection. Most studies were performed in the pre–highly active antiretroviral therapy (HAART) era between the late 1980s and mid-1990s in patients with advanced infection. They addressed intestinal mucosal barrier function to find a key to the problem of pathogen-negative diarrhea.[12–15] Intestinal permeability was assessed by quantification of urine recovery of orally ingested nonabsorbable carbohydrates. Although oral test meal studies serve well in giving an approximate estimate of the overall small intestinal permeability, this technique does not allow for a quantitative and mechanistic analysis of the mucosal barrier function. Therefore, Brenchley *et al.*,[9] and several subsequent studies on the topic,[16–20] used increased serum levels of LPS or microbial DNA as a surrogate for an assumed

doi: 10.1111/j.1749-6632.2012.06512.x

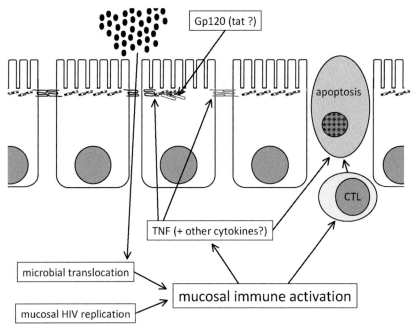

Figure 1. Mucosal immune activation is the key mechanism of HIV-induced intestinal mucosal barrier impairment. In acute infection, mucosal perforin-expressing cytotoxic T cells are associated with increased epithelial apoptosis. In chronic infection, mucosal production of TNF-α and other cytokines leads to altered membrane composition of tight junction proteins. In chronic infection, mucosal cytokines also induce epithelial apoptosis. In addition to immune-mediated effects on epithelial barrier function, viral proteins, such as gp120, exert effects on epithelial tight junctions via receptor-mediated tubulin disruption.

intestinal barrier defect, allowing for increased microbial translocation. Thus, the hypothesis that HIV infection causes disruption of the intestinal epithelial barrier was based primarily on indirect evidence. Apart from the aforementioned oral test meal studies, there was only one paper presenting *in vitro* data on intestinal barrier function in HIV-infected patients.[21] In this study, barrier function of small intestinal epithelium was assessed by alternating current impedance spectroscopy and mucosal-to-serosal lactulose fluxes. Using this quantitative approach, decreased epithelial resistance and increased paracellular permeability were found in the duodenal epithelium of HIV-infected patients with diarrhea and low CD4 T cell counts.[21]

Therefore in 2006, when the mucosal barrier became a major focus of HIV research, it was unclear at what stage of the infection the mucosal barrier defect develops, which mechanisms contribute to epithelial barrier dysfunction, and whether or not the HIV-induced barrier defect could be reversed

by effective antiretroviral therapy. In this review, we would like to summarize data published on these questions.

Onset of mucosal barrier dysfunction in HIV infection

As mentioned earlier, several oral test meal studies investigated small bowel permeability to nonabsorbable mono- and disaccharides in symptomatic and asymptomatic patients with chronic HIV infection in the pre-HAART era;[12–15] consistently, an increase of intestinal permeability was found. Furthermore, in 1998, an electrophysiological study described an increased paracellular permeability in the duodenum of untreated patients with advanced HIV infection.[21] Thus, there is evidence for a small intestinal barrier defect in (advanced) chronic infection. However, as none of the studies systematically investigated barrier function in acute or early HIV infection, they were not designed to identify at what stage of the infection the mucosal barrier defect is initiated.

Recently, plasma LPS levels determined in patients with acute or chronic infection seemed to indicate, that the integrity of the intestinal epithelium is well preserved during the acute phase and lost only during the chronic phase of the infection.[9] On the other hand, a gene expression study reported evidence, albeit indirect, for impairment of the intestinal mucosal barrier already detectable during acute infection.[22] Using immunohistochemistry and microarray analysis, the authors observed high local viral replication associated with increased expression of genes related to immune activation and decreased expression of genes related to mucosal repair in duodenal biopsies obtained from four patients with acute HIV infection.[22]

To the best of our knowledge, there is only one study to date directly assessing intestinal mucosal barrier function in patients with acute HIV infection.[23] In this study, duodenal mucosa obtained from eight acutely infected patients before or during seroconversion was analyzed in vitro. Impedance spectroscopy and mannitol flux measurement indicated that at this early stage of the infection a mucosal barrier defect is already present.

Mechanisms of HIV-induced mucosal barrier function

Early in the epidemic, alterations in mucosal structure and function had been recognized in patients with chronic HIV infection. These changes were later termed HIV-enteropathy,[24,25] although the term was also used to denote the clinical condition of pathogen-negative diarrhea. Histologically, HIV-enteropathy is characterized by partial villous atrophy with increased numbers of apoptotic epithelial cells clustering at the villus tips.[26] In pigtail macaques, which are a model for pathogenic lentiviral infection, there seem to be breaches in the intestinal epithelial lining, allowing for increased microbial translocation.[18] However, no such changes have been observed in the human intestinal mucosa of HIV-infected patients. On the other hand, increased epithelial apoptosis is already present in acute infection as shown in a simian immunodeficiency virus (SIV) model[27] and in patients with primary HIV infection (Fig. 1).[23] In accord with these data, increased expression of epithelial activated caspase-3 protein, together with increased expression of genes related to immune activation and downregulated expression of genes related to

epithelial repair have been found in patients with primary HIV infection.[22]

The sealing properties of the intestinal epithelium are achieved by a monolayer of columnar epithelial cells, which are closely connected by intercellular tight junctions to form a moderately tight monolayer. Therefore, the normal function of the intestinal mucosal barrier not only depends on intact epithelial cells but also on the composition and function of the tight junctions. The first evidence that HIV perturbs the tight junctions of intestinal epithelia came from a study using cultured T84 colon cells. In this model, supernatants from HIV-infected macrophages induced epithelial apoptosis and, as shown by electrophysiological methods, alterations of the tight junctions.[28] Using the same model, increased monolayer permeability and disruption of tight junction proteins claudin-1, claudin-2, claudin-4, occludin, and ZO-1 were found after exposition of the T84 cells against laboratory HIV-1 strains.[29] An immunohistochemical study on small intestinal and rectal biopsies reported reduced epithelial tubulin staining in HIV-infected patients. The finding was interpreted as a sign for microtubule depolymerization in the context of the cytoskeletal changes associated with perijunctional actin–myosin ring contraction, which represents an established mechanism for pathological increase of transepithelial permeability.[30] A more recent study investigated barrier function and tight junction protein expression in duodenal mucosa obtained from patients with chronic HIV infection. In this study, increased expression of the pore-forming claudin-2 and reduced expression of the sealing claudin-1 indicated an altered tight junction protein composition as a contributing factor to increased epithelial permeability.[26] Thus, both cellular (increased epithelial apoptosis) and paracellular (altered tight junction composition) changes contribute to the barrier defect of the intestinal epithelium associated with HIV infection.

Presently, our understanding of the complicated and intricate interplay between the epithelial barrier function and the various cells of the local immune system is only beginning to evolve. Still, there is little doubt that the defect of mucosal barrier function in HIV infection is at least partly caused by the strong effect of the infection on the mucosal immune system. As mentioned previously, it has been hypothesized that the intestinal barrier defect is a consequence of the massive depletion of HIV target cells, such as activated effector CD4$^+$ T cells, from the mucosa.[10] However, a recent study performed in patients with acute HIV infection, observed increased rather than decreased mucosal CD4$^+$ T cells despite evidence for significant barrier impairment of the small intestinal mucosa in these patients.[23] Therefore, mucosal CD4$^+$ T cell depletion likely does not play a causative role in the initiation of the HIV-induced mucosal barrier defect during acute infection.

HIV infection causes not only CD4$^+$ T cell depletion, but also numerous other immunological changes within the intestinal mucosa. As shown in acute and chronic SIV and HIV infection, there is marked mucosal immune activation and inflammation in the infected host.[31–36] Notably, in patients with acute—but not in those with chronic—infection, a significant perforin expression of mucosal CD8$^+$ T cells was found.[23] As perforin expression was associated with increased numbers of apoptotic epithelial cells, the authors hypothesized that the mucosal barrier defect initiated during early acute HIV infection could be secondary to epithelial apoptosis induced by perforin-expressing cytotoxic T cells.[23] Accordingly, increased gut epithelial cell apoptosis was also observed in acute SIV infection together with increased expression of Fas on enterocytes and Fas-ligand on lamina propria lymphocytes.[27]

Whereas an increased mucosal density of perforin-expressing cytotoxic T cell has been observed in early acute infection,[23,37] there is a loss of perforin expression in later stages of the disease.[37,38] Therefore, other mechanisms besides CTL-induced epithelial apoptosis maintain the mucosal barrier defect after termination of the acute phase of the infection. In untreated patients with chronic infection, mucosal production/expression of TNF-α, IL-2, IL-4, and IL-13 was associated with epithelial apoptosis, altered tight junction composition, and a barrier defect, and the same cytokines induced increased epithelial permeability when applied to rat jejunal mucosa.[26] This finding is in line with previous studies demonstrating induction of mucosal barrier impairment by inflammatory cytokines such as TNF-α in intestinal mucosal samples.[39–42] Taken together, there is strong evidence that mucosal immune activation triggers an intestinal mucosal barrier defect via the action of cytotoxic T cells in acute and via inflammatory mucosal cytokines in chronic infection.

In addition to effects elicited via local immune cells, HIV and viral proteins also exert direct effects on epithelial integrity. First hints came from the observation that incubation with HIV-1 or its glycoprotein (gp) 120 decreased the transepithelial electrical resistance of cultured colonic cells (HT-29D4).[43] This result was reproduced and further analyzed in a study employing T84 cells.[44] The decrease in TER correlated with disruption of tight junction proteins (claudin-1, claudin-2, claudin-4, occludin, and ZO-1), and both effects could be inhibited by neutralization of gp120. Intriguingly, incubation with HIV-1 also evoked increased production of inflammatory cytokines including TNF-α from the epithelial cells [44] indicating a cross-link between epithelial effects directly triggered by viral proteins and virus-induced alterations of the mucosal immune system as discussed earlier. The epithelial effect of gp120 has been attributed to be mediated by the orphan G protein–coupled receptor GPR15/Bob, which was characterized as a noninfection-inducing HIV coreceptor. It is expressed on enterocytes and, after binding of gp120, induces microtubule loss via calcium signaling.[45] Controversy exists regarding the effect of the HIV transactivator factor, Tat, on barrier function of intestinal epithelia. As opposed to inactivated HIV-1 and gp120, Tat did not decrease the transepithelial resistance in T84 and CaCo-2 cells.[29,46] On the other hand, Tat affected the microtubule and actin cytoskeleton and induced apoptosis in cultured intestinal cells.[47,48]

Effect of HAART on HIV-induced barrier dysfunction

There are conflicting data as to whether or not gut mucosal CD4$^+$ T cells can be restored by effective antiretroviral therapy.[26,49–54] However, there is no doubt that even after prolonged viral suppression, significant changes of the intestinal mucosal immune system persist. HIV induces irreversible fibrosis with extensive collagen deposition in ileal Peyer's patches[51] and—although to a far lesser degree than in untreated infection—there is ongoing mucosal immune activation in treated patients as reflected by increased numbers of activated mucosal CD8$^+$ T cells.[26] Still, surrogate marker studies using LPS or other bacterial markers indicated that effective combination antiretroviral therapy leads to reduction of microbial translocation.[10,17,55] A more in-depth analysis evaluated barrier function in duodenal mucosal samples of patients with chronic HIV infection by electrophysiological, histological, and molecular methods. Whereas untreated patients showed significant barrier impairment and villous atrophy, both parameters were not different from HIV-negative controls in patients on suppressive antiretroviral therapy.[26] Thus, available data indicate that restoration of the intestinal mucosal barrier function is possible in patients on suppressive antiretroviral therapy despite persistence of structural and functional abnormalities of the mucosal immune system.

Summary

Altered protein composition of the tight junctions and increased epithelial apoptosis have been identified as structural correlates of the HIV-induced mucosal barrier impairment. Its central trigger mechanism is represented by mucosal immune activation, as reflected by increased mucosal density of cytotoxic T cells in acute and increased mucosal cytokine production in chronic infection. Once the barrier defect is initiated, microbial translocation and mucosal immune activation build up a mutually enhancing vicious cycle. However, the strong mucosal effects of suppressive antiviral therapy illustrate the major importance of HIV replication for the perpetuation of both mucosal barrier defect and mucosal immune activation.

References

1. Guadalupe, M. *et al*. 2003. Severe CD4$^+$ T-cell depletion in gut lymphoid tissue during primary human immunodeficiency virus type 1 infection and substantial delay in restoration following highly active antiretroviral therapy. *J. Virol.* **77:** 11708–11717.
2. Li, Q. *et al*. 2005. Peak SIV replication in resting memory CD4$^+$ T cells depletes gut lamina propria CD4$^+$ T cells. *Nature* **434:** 1148–1152.
3. Mattapallil, J.J. *et al*. 2005. Massive infection and loss of memory CD4$^+$ T cells in multiple tissues during acute SIV infection. *Nature* **434:** 1093–1097.
4. Mehandru, S. *et al*. 2004. Primary HIV-1 infection is associated with preferential depletion of CD4+ T lymphocytes from effector sites in the gastrointestinal tract. *J. Exp. Med.* **200:** 761–770.
5. Veazey, R.S. *et al*. 1998. Gastrointestinal tract as a major site of CD4$^+$ T cell depletion and viral replication in SIV infection. *Science* **280:** 427–431.
6. Schneider, T. *et al*. 1995. Loss of CD4$^+$ T lymphocytes in patients infected with human immunodeficiency virus type

1 is more pronounced in the duodenal mucosa than in the peripheral blood. Berlin Diarrhea/Wasting Syndrome Study Group. *Gut* **37:** 524–529.
7. Zeitz, M. *et al*. 1998. HIV/SIV enteropathy. *Ann. N. Y. Acad. Sci.* **859:** 139–148.
8. Kewenig, S. *et al*. 1999. Rapid mucosal CD4(+) T-cell depletion and enteropathy in simian immunodeficiency virus-infected rhesus macaques. *Gastroenterology* **116:** 1115–1123.
9. Brenchley, J.M. *et al*. 2006. Microbial translocation is a cause of systemic immune activation in chronic HIV infection. *Nat. Med.* **12:** 1365–1371.
10. Brenchley, J.M., D.A. Price & D. C. Douek. 2006. HIV disease: fallout from a mucosal catastrophe? *Nat. Immunol.* **7:** 235–239.
11. Sodora, D.L. & G. Silvestri. 2010. HIV, mucosal tissues, and T helper 17 cells: where we come from, where we are, and where we go from here. *Curr. Opin. HIV AIDS* **5:** 111–113.
12. Keating, J. *et al*. 1995. Intestinal absorptive capacity, intestinal permeability and jejunal histology in HIV and their relation to diarrhoea. *Gut* **37:** 623–629.
13. Lim, S.G., I.S. Menzies, C.A. Lee, *et al*. 1993. Intestinal permeability and function in patients infected with human immunodeficiency virus. A comparison with coeliac disease. *Scand. J. Gastroenterol.* **28:** 573–580.
14. Obinna, F.C. *et al*. 1995. Comparative assessment of small intestinal and colonic permeability in HIV-infected homosexual men. *AIDS* **9:** 1009–1016.
15. Ott, M., B. Lembcke, S. Staszewski, *et al*. 1991. Intestinal permeability in patients with acquired immunodeficiency syndrome (AIDS). *Klin. Wochenschr.* **69:** 715–721.
16. Ancuta, P. *et al*. 2008. Microbial translocation is associated with increased monocyte activation and dementia in AIDS patients. *PLoS One* **3:** e2516.
17. Jiang, W. *et al*. 2009. Plasma levels of bacterial DNA correlate with immune activation and the magnitude of immune restoration in persons with antiretroviral-treated HIV infection. *J. Infect. Dis.* **199:** 1177–1185.
18. Klatt, N. R. *et al*. 2010. Compromised gastrointestinal integrity in pigtail macaques is associated with increased microbial translocation, immune activation, and IL-17 production in the absence of SIV infection. *Mucosal Immunol.* **3:** 387–398.
19. Marchetti, G. *et al*. 2008. Microbial translocation is associated with sustained failure in CD4+ T-cell reconstitution in HIV-infected patients on long-term highly active antiretroviral therapy. *AIDS* **22:** 2035–2038.
20. Baroncelli, S. *et al*. 2009. Microbial translocation is associated with residual viral replication in HAART-treated HIV$^+$ subjects with <50copies/ml HIV-1 RNA. *J. Clin. Virol.* **46:** 367–370.
21. Stockmann, M. *et al*. 1998. Duodenal biopsies of HIV-infected patients with diarrhoea exhibit epithelial barrier defects but no active secretion. *AIDS* **12:** 43–51.
22. Sankaran, S. *et al*. 2008. Rapid onset of intestinal epithelial barrier dysfunction in primary human immunodeficiency virus infection is driven by an imbalance between immune response and mucosal repair and regeneration. *J. Virol.* **82:** 538–545.

23. Epple, H. J. *et al.* 2010. Acute HIV infection induces mucosal infiltration with CD4+ and CD8+ T cells, epithelial apoptosis, and a mucosal barrier defect. *Gastroenterology* **139:** 1289–1300.

24. Kotler, D.P., H.P. Gaetz, M. Lange, *et al.* 1984. Enteropathy associated with the acquired immunodeficiency syndrome. *Ann. Intern. Med.* **101:** 421–428.

25. Ullrich, R. *et al.* 1989. Small intestinal structure and function in patients infected with human immunodeficiency virus (HIV): evidence for HIV-induced enteropathy. *Ann. Intern. Med.* **111:** 15–21.

26. Epple, H.J. *et al.* 2009. Impairment of the intestinal barrier is evident in untreated but absent in suppressively treated HIV-infected patients. *Gut* **58:** 220–227.

27. Li, Q. *et al.* 2008. Simian immunodeficiency virus-induced intestinal cell apoptosis is the underlying mechanism of the regenerative enteropathy of early infection. *J. Infect. Dis.* **197:** 420–429.

28. Schmitz, H. *et al.* 2002. Supernatants of HIV-infected immune cells affect the barrier function of human HT-29/B6 intestinal epithelial cells. *AIDS* **16:** 983–991.

29. Nazli, A. *et al.* Exposure to HIV-1 directly impairs mucosal epithelial barrier integrity allowing microbial translocation. *PLoS Pathog* **6:** e1000852.

30. Clayton, F., S. Kapetanovic & D.P. Kotler. 2001. Enteric microtubule depolymerization in HIV infection: a possible cause of HIV-associated enteropathy. *AIDS* **15:** 123–124.

31. Kotler, D.P., S. Reka & F. Clayton. 1993. Intestinal mucosal inflammation associated with human immunodeficiency virus infection. *Dig. Dis. Sci.* **38:** 1119–1127.

32. Reka, S., M.L. Garro & D.P. Kotler. 1994. Variation in the expression of human immunodeficiency virus RNA and cytokine mRNA in rectal mucosa during the progression of infection. *Lymphokine Cytokine Res.* **13:** 391–398.

33. Olsson, J. *et al.* 2000. Human immunodeficiency virus type 1 infection is associated with significant mucosal inflammation characterized by increased expression of CCR5, CXCR4, and beta-chemokines. *J. Infect. Dis.* **182:** 1625–1635.

34. McGowan, I. *et al.* 2004. Increased HIV-1 mucosal replication is associated with generalized mucosal cytokine activation. *J. Acquir. Immune Defic. Syndr.* **37:** 1228–1236.

35. Abel, K., D.M. Rocke, B. Chohan, *et al.* 2005. Temporal and anatomic relationship between virus replication and cytokine gene expression after vaginal simian immunodeficiency virus infection. *J. Virol.* **79:** 12164–12172.

36. Li, Q. *et al.* 2009. Glycerol monolaurate prevents mucosal SIV transmission. *Nature* **458:** 1034–1038.

37. Quigley, M.F. *et al.* 2006. Perforin expression in the gastrointestinal mucosa is limited to acute simian immunodeficiency virus infection. *J. Virol.* **80:** 3083–3087.

38. Shacklett, B.L. *et al.* 2004. Abundant expression of granzyme A, but not perforin, in granules of CD8+ T cells in GALT: implications for immune control of HIV-1 infection. *J. Immunol.* **173:** 641–648.

39. Zeissig, S. *et al.* 2004. Downregulation of epithelial apoptosis and barrier repair in active Crohn's disease by tumour necrosis factor alpha antibody treatment. *Gut* **53:** 1295–1302.

40. Schmitz, H. *et al.* 1996. Tumor necrosis factor-alpha induces Cl⁻ and K+ secretion in human distal colon driven by prostaglandin E2. *Am. J. Physiol.* **271:** G669–G674.

41. Heller, F. *et al.* 2005. Interleukin-13 is the key effector Th2 cytokine in ulcerative colitis that affects epithelial tight junctions, apoptosis, and cell restitution. *Gastroenterology* **129:** 550–564.

42. Grotjohann, I., H. Schmitz, M. Fromm & J.D. Schulzke. 2000. Effect of TNF alpha and IFN gamma on epithelial barrier function in rat rectum *in vitro*. *Ann. N. Y. Acad. Sci.* **915:** 282–286.

43. Delezay, O. *et al.* 1997. Direct effect of type 1 human immunodeficiency virus (HIV-1) on intestinal epithelial cell differentiation: relationship to HIV-1 enteropathy. *Virology* **238:** 231–242.

44. Nazli, A. *et al.* 2011. Exposure to HIV-1 directly impairs mucosal epithelial barrier integrity allowing microbial translocation. *PLoS Pathog* **6:** e1000852.

45. Clayton, F. *et al.* 2001. Gp120-induced Bob/GPR15 activation: a possible cause of human immunodeficiency virus enteropathy. *Am. J. Pathol.* **159:** 1933–1939.

46. Canani, R.B. *et al.* 2003. Effects of HIV-1 Tat protein on ion secretion and on cell proliferation in human intestinal epithelial cells. *Gastroenterology* **124:** 368–376.

47. Canani, R.B. *et al.* 2006. Inhibitory effect of HIV-1 Tat protein on the sodium-D-glucose symporter of human intestinal epithelial cells. *AIDS* **20:** 5–10.

48. Buccigrossi, V. *et al.* 2011. The HIV-1 transactivator factor (tat) induces enterocyte apoptosis through a redox-mediated mechanism. *PLoS One* **6:** e29436.

49. Guadalupe, M. *et al.* 2006. Viral suppression and immune restoration in the gastrointestinal mucosa of human immunodeficiency virus type 1-infected patients initiating therapy during primary or chronic infection. *J. Virol.* **80:** 8236–8247.

50. Mehandru, S. *et al.* 2006. Lack of mucosal immune reconstitution during prolonged treatment of acute and early HIV-1 infection. *PLoS Med.* **3:** e484.

51. Estes, J. *et al.* 2008. Collagen deposition limits immune reconstitution in the gut. *J. Infect. Dis.* **198:** 456–464.

52. Sheth, P.M. *et al.* 2008. Immune reconstitution in the sigmoid colon after long-term HIV therapy. *Mucosal Immunol.* **1:** 382–388.

53. Macal, M. *et al.* 2008. Effective CD4+ T-cell restoration in gut-associated lymphoid tissue of HIV-infected patients is associated with enhanced Th17 cells and polyfunctional HIV-specific T-cell responses. *Mucosal Immunol.* **1:** 475–488.

54. Schulbin, H. *et al.* 2008. Cytokine expression in the colonic mucosa of HIV-infected individuals before and during nine months of antiretroviral therapy. *Antimicrob. Agents Chemother.* **52:** 3377–3384.

55. Marchetti, G. *et al.* 2011. Microbial translocation predicts disease progression of HIV-infected antiretroviral-naive patients with high CD4+ cell count. *AIDS* **25:** 1385–1394.

Ann. N.Y. Acad. Sci. ISSN 0077-8923

ANNALS OF THE NEW YORK ACADEMY OF SCIENCES

Issue: *Barriers and Channels Formed by Tight Junction Proteins*

Zonulin, regulation of tight junctions, and autoimmune diseases

Alessio Fasano

Mucosal Biology Research Center and Center for Celiac Research, University of Maryland School of Medicine, Baltimore, Maryland

Address for correspondence: Alessio Fasano, M.D., Mucosal Biology Research Center, University of Maryland School of Medicine, 20 Penn Street HSF II Building, Room S345, Baltimore, MD 21201. afasano@mbrc.umaryland.edu

Recent studies indicate that besides digestion and absorption of nutrients and water and electrolytes homeostasis, another key function of the intestine is to regulate the trafficking of environmental antigens across the host mucosal barrier. Intestinal tight junctions (TJs) create gradients for the optimal absorption and transport of nutrients and control the balance between tolerance and immunity to nonself antigens. To meet diverse physiological challenges, intestinal epithelial TJs must be modified rapidly and in a coordinated fashion by regulatory systems that orchestrate the state of assembly of the TJ multiprotein network. While considerable knowledge exists about TJ ultrastructure, relatively little is known about their physiological and pathophysiological regulation. Our discovery of zonulin, the only known physiologic modulator of intercellular TJs described so far, has increased our understanding of the intricate mechanisms that regulate the intestinal epithelial paracellular pathway and has led us to appreciate that its upregulation in genetically susceptible individuals leads to autoimmune diseases.

Keywords: tight junctions; intestine; autoimmunity; zonulin

Introduction

Improved hygiene leading to reduced exposure to microorganisms has been implicated as one possible cause for the "epidemic" of immune-mediated diseases, particularly autoimmune diseases, in industrialized countries during the past 3–4 decades.[1] Collectively, autoimmune diseases are highly prevalent in the United States, affecting between 14.7 and 23.5 million people—up to 8% of the population.[2] The social and financial burdens imposed by these chronic, debilitating diseases include poor quality of life, high health care costs, and substantial loss of productivity. For example, in 2002, the average annual medical costs for treating type 1 diabetes (T1D) in the United States were an estimated $6.7 billion.[3] In less than a decade, these costs jumped to $14.4 billion.[4] Apart from genetic makeup and exposure to environmental triggers, a third key element, i.e., increased intestinal permeability, which may be influenced by the composition of the gut microbiota, has been proposed in the pathogenesis of these dis-

eases.[5–8] Intestinal permeability, together with antigen (Ag) sampling by enterocytes and lamina propria dendritic cells, regulates molecular trafficking between the intestinal lumen and the submucosa, leading to either tolerance or immunity to nonself Ag.[9–11] However, the dimensions of the paracellular space (10–15 Å) suggest that solutes with a molecular radius exceeding 15 Å (~3.5 kDa) (including proteins) are normally excluded and taken up by the transcellular pathway. Intercellular tight junctions (TJs) tightly regulate paracellular Ag trafficking. TJs are extremely dynamic structures that operate in several key functions of the intestinal epithelium under both physiological and pathological circumstances.[6] However, despite major progress in our knowledge on the composition and function of the intercellular TJs, the mechanism(s) by which they are regulated is(are) still incompletely understood.

In the past decade, we have focused our research effort on the discovery and characterization of zonulin as the only human protein discovered to date that is known to reversibly regulate intestinal

doi: 10.1111/j.1749-6632.2012.06538.x

permeability by modulating intercellular TJs.[12–14] Zonulin expression is augmented in autoimmune conditions associated with TJ dysfunction, including celiac disease (CD) and T1D.[13,15] Both animal studies[16] and human trials[17] using the zonulin synthetic peptide inhibitor AT1001 (now named Larazotide acetate) have established that zonulin is integrally involved in the pathogenesis of autoimmune diseases. Zonulin can be used as a biomarker of impaired gut barrier function for several autoimmune, neurodegenerative, and tumoral diseases,[18] and can be a potential therapeutic target for the treatment of these devastating conditions.

The zonulin system

Identification of zonulin as prehaptoglobin 2
Through proteomic analysis of human sera, we have recently identified zonulin as prehaptoglobin (HP)2,[19] a molecule that, to date, has only been regarded as the inactive precursor for HP2, one of the two genetic variants (together with HP1) of human HPs. Mature human HPs are heterodimeric plasma glycoproteins composed of α and β polypeptide chains that are covalently associated by disulfide bonds and in which only the β chain is glycosylated[20]

(Fig. 1). While the β chain (36 kDa) is constant, the α chain exists in two forms, i.e., α1 (∼9 kDa) and α2 (∼18 kDa). The presence of one or both of the α chains results in the three human HP phenotypes, i.e., HP1-1 homozygote, HP2-1 heterozygote, and HP2-2 homozygote.

Despite this multidomain structure, the hemoglobin only function assigned to HPs, to date, is to bind Hb to form stable HP–Hb complexes thereby preventing Hb-induced oxidative tissue damage.[21] In contrast, no function has ever been described for their precursor forms. The primary translation product of mammalian Hp mRNA is a polypeptide that dimerizes contranslationally and is proteolitically cleaved while still in the endoplasmic reticulum.[22] Conversely, zonulin is detectable in an uncleaved form in human serum, adding an extremely intriguing aspect of the multifunctional characteristics of HPs. HPs are unusual secretory proteins in that they are proteolytically processed in the endoplasmic reticulum, the subcellular fraction of which we detected the highest zonulin concentration.[23] Wicher and Fries[22] found that Cr1LP mediates this cleavage in a specific manner, since the enzyme did not cleave the proform of

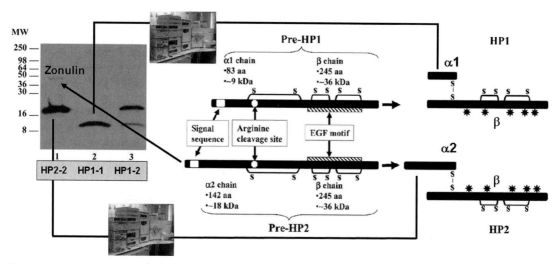

Figure 1. Western blotting using zonulin cross-reacting anti-Zot polyclonal antibodies on CD patient sera showed three main patterns: sera showing an 18 kDa immunoreactive band and a fainter 45 kDa band (lane 1), sera showing only a 9 kDa band (lane 2), and sera showing both the 18 and 9 kDa bands (lane 3). The cartoon shows the structure of both pre-HP1 and pre-HP2 and their mature proteins. HPs evolved from a complement-associated protein (mannose-binding lectin-associated serine protease, MASP), with their α-chain containing a complement control protein (CCP), while the β-chain is related to chymotrypsin-like serine proteases (SP domain) containing an epidermal growth factor-like motif. The gene encoding the α2-chain of pre-HP2 originated in India almost 2 million years ago through a chromosomal aberration (unequal crossing over) of HP1. Pre-HPs are translated as single-chain precursor proteins. Pre-HPs may be proteolytically cleaved intracellularly into α- and β-chains that remain disulfide linked, referred to as cleaved, two-chain, mature HPs.

complement C1s, a protein similar to pre-HP2, particularly around the cleavage site. Therefore, it is conceivable to hypothesize that the activity of Cr1LP modulates the zonulin pool.

Zonulin functional characterization

Since we have reported previously that the key biological effect of zonulin is to affect the integrity of intercellular TJs, we specifically focused our efforts on demonstrating that the recombinant pre-HP2 alters intestinal permeability. Indeed, *ex vivo* experiments showed that recombinant pre-HP2 induced a time- and dose-dependent reduction in TEER when added to murine small intestinal mucosa.[19] These results were validated independently in an *in vivo* intestinal permeability assay in which zonulin, but not its cleaved form, induced a significant and reversible increase in both gastroduodenal and small intestinal permeability.[19] The evidence that zonulin cleaved in its α2 and β subunits lost the permeability activity further supports the notion that pre-HP2 (alias, zonulin) and mature HP2 exert two different biological functions most likely related to the different folding of the protein in its cleaved or uncleaved form.

Structural analysis of zonulin revealed similarities with several growth factors that, like zonulin, affect intercellular TJ integrity.[24,25] Our data showing that zonulin but not its cleaved subunits activate EGFR[19] and that its effect on TEER was prevented by the EGFR tyrosine kinase inhibitor AG-1478[19] suggest that zonulin is properly folded to activate EGFR and, therefore, to cause TJ disassembly only in its uncleaved form. Several G protein–coupled receptors (GPCR), including Proteinase Activated Receptor 2 (PAR$_2$), transactivate EGFR.[26] Zonulin prokaryotic counterpart Zot active peptide FCIGRL (AT1002) has structural similarities with PAR$_2$-activating peptide (AP), SLIGRL, and causes PAR$_2$-dependent changes in TEER, a finding that we have demonstrated in WT, but not PAR2$^{-/-}$ mice.[27] Therefore, it was not totally unexpected that experiments in Caco2 cell, in which PAR$_2$ was silenced, showed decreased EGFR Y1068 phosphorylation in response to recombinant zonulin compatible with PAR$_2$-dependent transactivation of EGFR.[19] To further establish a role for PAR$_2$ in EGFR activation in response to zonulin, we conducted small intestinal barrier function studies using segments isolated from either C57BL/6 WT or PAR$_2$$^{-/-}$ mice.

As expected, zonulin decreased TEER in intestinal segments from C57BL/6 WT mice, while it failed to reduce TEER in small intestinal segments from PAR$_2$$^{-/-}$ mice,[19] so linking zonulin-induced PAR$_2$-dependent transactivation of EGFR with barrier function modulation.

To summarize, we have reported for the first time the novel characterization of zonulin as pre-HP2, a multifunctional protein that, in its intact single chain form, regulates intestinal permeability through PAR$_2$-mediated EGFR transactivation, while in its cleaved two-chain form acts as a Hb scavenger (Fig. 2).

Stimuli that cause zonulin release in the gut

Among the several potential intestinal luminal stimuli that can trigger zonulin release, we identified small intestinal exposure to bacteria or gluten as the two more powerful triggers.[18,23,28] Enteric infections have been implicated in the pathogenesis of several pathological conditions, including allergic, autoimmune, and inflammatory diseases, by causing impairment of the intestinal barrier. We have generated evidence that small intestines exposed to enteric bacteria secrete zonulin.[28] This secretion was independent of either the animal species from which the small intestines were isolated or the virulence of the microorganisms tested, occurred only on the luminal aspect of the bacteria-exposed small intestinal mucosa, and was followed by an increase in intestinal permeability coincident with the disengagement of the protein ZO-1 from the tight junctional complex.[28] This zonulin-driven opening of the paracellular pathway may represent a defensive mechanism, which flushes out microorganisms so contributing to the innate immune response of the host against bacterial colonization of the small intestine.

Besides bacterial exposure, we have shown that also gliadin, a storage protein present in wheat and that triggers CD in genetically susceptible individuals, also affects the intestinal barrier function by releasing zonulin.[29] This effect of gliadin is polarized, i.e., gliadin increases intestinal permeability only when administered on the luminal side of the intestinal tissue.[29] This observation led us to the identification of the chemokine receptor CXCR3 as the target intestinal receptor for gliadin.[30] Our data demonstrate that in the intestinal epithelium, CXCR3 is expressed at the luminal level, is overexpressed in CD

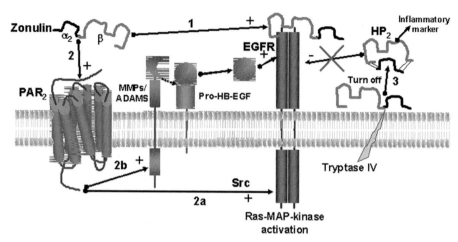

Figure 2. Zonulin can activate EGFR through direct binding (1) and/or through PAR2 transactivation (2). This second mechanism can be mediated by either Src signaling (2a) or by the release of MMPs and/or ADAMS that in turn will activate Pro-HB-EGF. When cell tryptase IV cleaves zonulin in its two subunits (so eliminating one of the three required disulfide bridges necessary for EGF activity), the molecule is not able to bind to EGFR (3), while it acquires a different function (Hb binding) and becomes an inflammatory marker.

patients, colocalizes with gliadin, and that this interaction coincides with recruitment of the adapter protein, MyD88, to the receptor.[28] We also demonstrated that binding of gliadin to CXCR3 is crucial for the release of zonulin and subsequent increase of intestinal permeability, since CXCR3-deficient mice failed to respond to gliadin challenge in terms of zonulin release and TJ disassembly.[30] Using an α-gliadin synthetic peptide library, we identified two α-gliadin 20mers (QVLQQSTYQLLQELCC-QHLW and QQQQQQQQQQQQQILQQILQQ) that bind to CXCR3 and release zonulin.[30]

Role of zonulin in autoimmune diseases

Celiac disease
CD is an immune-mediated chronic enteropathy with a wide range of presenting manifestations of variable severity. It is triggered by the ingestion of gliadin fraction of wheat gluten and similar alcohol-soluble proteins (prolamines) of barley and rye in genetically susceptible subjects with subsequent immune reaction leading to small bowel inflammation and normalization of the villous architecture in response to a gluten-free diet.[31] CD is a unique model of autoimmunity in which, in contrast to most other autoimmune diseases, a close genetic association with HLA genes, a highly specific humoral autoimmune response against tissue transglutaminase autoantigen, and, most importantly, the triggering environmental factor (gliadin), are all known. It is the interplay between genes (both HLA

and non-HLA associated) and environment (i.e., gluten) that leads to the intestinal damage typical of the disease.[32] Under physiological circumstances, this interplay is prevented by competent intercellular TJs. Early in CD, TJs are opened[33–36] and severe intestinal damage ensues.

The repertoire of gluten peptides involved in the disease pathogenesis is greater than appreciated previously and may differ between children and adult patients.[37] There are at least 50 toxic epitopes in gluten peptides exerting cytotoxic, immunomodulatory, and gut-permeating activities.[38] The effect of the permeating gliadin peptides *in vivo* was confirmed by the analysis of intestinal tissues from patients with active CD and non-CD controls probed for zonulin expression.[23] Quantitative immunoblotting of intestinal tissue lysates from active CD patients confirmed the increase in zonulin protein compared to control tissues.[23] Zonulin upregulation during the acute phase of CD was confirmed by measuring zonulin concentration in sera of 189 CD patients using a sandwich ELISA (Fig. 3A).[13]

Type 1 diabetes
The trigger of the autoimmune destruction of pancreatic beta cells in T1D is unknown. T1D has the same pathogenic challenges as other autoimmune diseases: what are the environmental triggers, and how do these triggers cross the intestinal barrier to interact with the immune system?[39,40] Certain HLA class II alleles account for 40% of the genetic

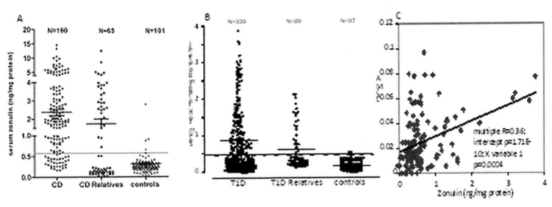

Figure 3. (A) CD patients showed higher serum zonulin levels compared to both their relatives and controls. (B) Similar results were obtained in T1D patients. (C) Serum zonulin correlated with intestinal permeability evaluated by the LA/MA test. The percentage of CD patients (81%) and their relatives (50%) with elevated serum zonulin levels was higher compared to T1D patients (42%) and their relatives (29%), respectively.

susceptibility to T1D in Caucasians;[41] however, the majority of individuals with these HLA alleles do not develop T1D. This supports the concept that reaction to some environmental products triggers autoimmune destruction of beta cells and leads to T1D. Gastrointestinal symptoms in T1D have been generally ascribed to altered intestinal motility secondary to autonomic neuropathy.[42] However, more recent studies have shown that altered intestinal permeability occurs in T1D prior to the onset of complications,[43] which is not the case in type 2 diabetes. This has led to the suggestion that an increased intestinal permeability due to alteration in intestinal TJs is responsible for the onset of T1D.[43,44] This hypothesis is supported by studies performed in an animal model that develops T1D spontaneously and showed an increased permeability of the small intestine (but not of the colon) of the BioBreeding diabetic-prone (BBDP) rats that preceded the onset of diabetes by at least a month.[45] Further, histological evidence of pancreatic islet destruction was absent at the time of increased permeability but clearly present at a later time.[45] Therefore, these studies provided evidence that increased gut permeability occurred before either histological or overt manifestation of diabetes in this animal model. We confirmed these data by reporting in the same rat model that zonulin-dependent increase in intestinal permeability precedes the onset of T1D by 2–3 weeks.[16]

Several reports have linked gliadin (the environmental trigger of CD autoimmunity that also causes zonulin release from the gut, see Refs. 23 and 29) to T1D autoimmunity both in animal models and in human studies. Findings from studies using nonobese diabetic (NOD) mice and BBDP rats have implicated wheat gliadin as a dietary diabetogen.[46,47] We have recently reported a direct link between antibodies to Glo-3a (a wheat-related protein), zonulin upregulation, and islet autoimmunity (IA) in children at increased risk for T1D.[48]

Association of HP genotypes with serum zonulin expression and clinical severity in human disease states

We have recently discovered that human zonulin is identical to pre-Hp2 (see above). Two co-dominant allele variants, termed *HP1* and *HP2*, the latter unique to the human species, are variously distributed in the general population, resulting in three phenotypes: Hp1-1 (~20% in Western populations), Hp2-1 (~50%), and Hp2-2 (~30%).[49] Several studies have suggested that the presence of the *HP2* allele correlates with higher risk to develop immune-mediated diseases[50,51] and that *HP2* homozygosis is associated with poor prognosis[51] and decreased longevity.[52]

Zonulin phenotype in CD and T1D and its correlation with intestinal permeability. Using a serum zonulin ELISA, we measured serum zonulin levels in both CD and T1D patients, their relatives, and age- and sex-matched healthy controls. CD patients showed statistically higher serum zonulin levels (2.37 ± 0.17 ng/mg protein) compared to both their relatives (1.75 ± 0.27 ng/mg protein, $P = 0.05$) and control subjects (0.31 ± 0.03 ng/mg protein, $P < 0.00001$) (Fig. 3A). A total of 81% (154/190) of CD patients and 50% of their first-degree relatives (33/65) had serum zonulin levels that were 2 SD above the mean zonulin levels detected in age-matched healthy controls. Only 4.9% (5/101) of controls had zonulin levels 2 SD above the mean ($P < 0.01$). Serum zonulin was higher in CD compared to their relatives ($P < 0.00001$). Similar results were obtained in T1D patients in which we detected increased serum zonulin levels (0.83 ± 0.05 ng/mg protein) compared to both their relatives (0.62 ± 0.07 ng/mg protein, $P = 0.011$) and control subjects (0.21 ± 0.02 ng/mg protein, $P < 0.00001$; Fig. 3B).[15] A total of 42% (141/339) of T1D patients and 29% of their first-degree relatives (26/89) had serum zonulin levels that were 2 SD above the mean zonulin levels detected in age-matched healthy controls. Only 4% (4/97) of controls had zonulin levels 2 SD above the mean ($P < 0.01$). Serum zonulin was higher in T1D compared to their relatives ($P = 0.01$). To establish whether serum zonulin levels correlated with intestinal permeability, lactulose (LA)/mannitol (MA) urine ratio was determined in both a subset of T1D subjects with documented zonulin upregulation (N = 36) and their relatives (N = 56). Intestinal permeability correlated with serum zonulin (Fig. 3C).[15]

HP2 (alias zonulin) allele is overrepresented in immune-mediated diseases. To establish the distribution of *HP1* and *HP2* genes among CD patients and matched controls, we developed a single-step RT-PCR protocol using specific primers in exon 2 and exon 5 of *HP1* corresponding to exons 2 and 7 of *HP2*. After PCR, the amplicons were run on a 1% agarose gel and read under a UV bulb. The *HP1* genotype ran at the predicted size of 2.5 kb while the *HP2* ran at 4.3 kb. Our results showed that in CD patients HP1-1 genotype (0 copies of the zonulin gene) was decreased, while the HP2-2 genotype (2 copies of zonulin gene) was increased compared to healthy controls (Table 1). Interestingly, the percentage of HP 1-1 CD patients (0 copies of the zonulin gene) was in the same range of the percentage of CD patients that tested negative by zonulin ELISA (see above). Similar distribution of the HP genes have been reported by other investigators in other immune-mediated diseases, including Crohn's disease,[53] schizophrenia,[54] and chronic kidney disease (CKD)[55] (Table 1).

Proof of concept for the pathogenic role of zonulin-dependent intestinal barrier defect in CD and T1D

CD and T1D autoimmune models suggest that when the finely tuned trafficking of macromolecules is deregulated in genetically susceptible individuals, autoimmune disorders can occur.[56] This new paradigm subverts traditional theories underlying the development of autoimmunity, which are based on molecular mimicry and/or the bystander effect, and suggests that the autoimmune process can be arrested if the interplay between genes and environmental triggers is prevented by reestablishing the intestinal barrier function. To challenge this hypothesis, zonulin inhibitor AT1001 was used with encouraging results in the BBDP rat model of autoimmunity.[16] Besides preventing the loss of intestinal barrier function, the appearance of autoantibodies, and the onset of disease, pretreatment with AT1001 protected against the insult of pancreatic islets and, therefore, of the insulitis, is responsible for the onset of T1D.

This proof of concept in an animal model of autoimmunity provided the rationale to design human clinical trials in which AT1001 was initially tested in an inpatient, double-blind, randomized placebo controlled trial to determine its safety, tolerability, and preliminary efficacy.[57] No increase in adverse events was recorded among patients exposed to AT-1001 compared to placebo. Following acute gluten exposure, a 70% increase in intestinal permeability was detected in the placebo group, while no changes were seen in the AT-1001 group.[57] After gluten exposure, IFN-γ levels increased in four out of seven patients (57.1%) of the placebo-group, but only in 4 out of 14 patients (28.6%) of the AT-1001 group. Gastrointestinal symptoms were significantly more frequent among patients of the placebo group compared to the AT-1001 group.[57] Together, these data suggest that AT-1001 is well tolerated and appears to reduce pro-inflammatory cytokine production and gastrointestinal symptoms in CD patients. AT1001 has now been tested in approximately 500 subjects with an excellent safety profile and promising efficacy in protecting against symptoms caused by gluten exposure in CD patients.[58]

Table 1. HP genotype distribution in several immune-mediated diseases

Genotype	CD[19]		Crohn's disease[56]		Schizophrenia[57]		CKD[58]	
	Cntr	Pts	Cntr	Pts	Cntr	Pts	Cntr	Pts
HP 1-1	20.6	7.1	23.9	10.1	14.1	9.2	9.4	3.8
HP 1-2	43.5	35.7	44.0	46.2	46.9	38.8	46.5	43.7
HP 2-2	35.9	57.2	32.1	43.7	39.0	52.0	44.1	52.6

Concluding remarks

The gastrointestinal tract has been extensively studied for its digestive and absorptive functions. A more attentive analysis of its anatomo-functional characteristics, however, clearly indicates that its functions go well beyond the handling of nutrients and electrolytes. The exquisite regional-specific anatomical arrangements of cell subtypes and the finely regulated cross talk between epithelial, neuroendocrine, and immune cells highlight other less-studied, yet extremely important functions of the gastrointestinal tract. Of particular interest is the regulation of antigen trafficking by the zonulin-dependent paracellular pathway and its activation by intestinal mucosa–microbiota/gluten interactions. These functions dictate the switch from tolerance to immunity, and are likely integral mechanisms involved in the pathogenesis of inflammatory and autoimmune processes.

The classical paradigm of inflammatory pathogenesis involving specific genetic makeup and exposure to environmental triggers has been challenged recently by the addition of a third element, the loss of intestinal barrier function. Genetic predisposition, miscommunication between innate and adaptive immunity, exposure to environmental triggers, and loss of intestinal barrier function secondary to the activation of the zonulin pathway by food-derived environmental triggers or changes in gut microbiota all seem to be key ingredients involved in the pathogenesis of inflammation, autoimmunity, and cancer. This new theory implies that once the pathologic process is activated, it is not auto-perpetuating. Rather, it can be modulated or even reversed by preventing the continuous interplay between genes and the environment. Since zonulin-dependent TJ dysfunction allows such interactions, new therapeutic strategies aimed at reestablishing the intestinal barrier function by downregulating the zonulin pathway offer innovative and not yet-explored approaches for the management of these debilitating chronic diseases.

Acknowledgments

Work presented in this review was supported in part by the National Institutes of Health Grants DK-48373 and DK-078699 to AF.

Conflict of interest

Dr. Fasano is a stock holder of Alba Therapeutics.

References

1. Okada, H., C. Kuhn & H. Feillet. 2010. The 'hygiene hypothesis' for autoimmune and allergic diseases: an update. *Clin. Exp. Immunol.* 2010: 1–9.
2. Progress in Autoimmune Diseases Research, Report to Congress, National Institutes of Health, The Autoimmune Diseases Coordinating Committee, March 2005.
3. National Institutes of Health Autoimmune Diseases research Plan 2002.
4. Tao, B., M. Pietropaolo, M. Atkinson, *et al.* 2010. Estimating the cost of type 1 diabetes in the USA propensity score matching method. *PLoS One* 5: 1–11.
5. Arrieta, M.C., L. Bistritz & J.B. Meddings. 2006. Alterations in intestinal permeability. *Gut* 55: 1512–1520.
6. Fasano, A. & T. Shea-Donohue. 2005. Mechanisms of disease: the role of intestinal barrier function in the pathogenesis of gastrointestinal autoimmune diseases. *Nat. Clin. Pract. Gastroenterol. Hepatol.* 2: 416–422.
7. Wapenaar, M.C., A.J. Monsuur, A.A. van Bodegraven, *et al.* 2008. Associations with tight junction genes PARD3 and MAGI2 in Dutch patients point to a common barrier defect for coeliac disease and ulcerative colitis. *Gut* 57: 463–467.
8. Monsuur, A.J., P.I. de Bakker, B.Z. Alizadeh, *et al.* 2005. Myosin IXB variant increases the risk of celiac disease and points toward a primary intestinal barrier defect. *Nat. Genet.* 37: 1341–1344.
9. Ménard, S., N. Cerf-Bensussan & M. Heyman. 2010. Multiple facets of intestinal permeability and epithelial handling of dietary antigens. *Mucosal. Immunol.* 3: 247–259.

10. Chieppa, M., M. Rescigno, A.Y. Huang & R.N. Germain. 2006. Dynamic imaging of dendritic cell extension into the small bowel lumen in response to epithelial cell TLR engagement. *J. Exp. Med.* **203:** 2841–2852.

11. Mowat, A.M., O.R. Millington & F.G. Chirdo. 2004. Anatomical and cellular basis of immunity and tolerance in the intestine. *J. Pediatr. Gastroenterol. Nutr.* **39:** S723–S724.

12. Wang, W., S. Uzzau, S.E. Goldblum & A. Fasano. 2000. Human zonulin, a potential modulator of intestinal tight junctions. *J. Cell Sci.* **113:** 4435–4440.

13. Fasano, A., T. Not, W. Wang, *et al.* 2000. Zonulin, a newly discovered modulator of intestinal permeability, and its expression in coeliac disease. *Lancet* **355:** 1518–1519.

14. Fasano, A. 2000. Regulation of intercellular tight junctions by zonula occludens toxin and its eukaryotic analogue zonulin. *Ann. N. Y. Acad. Sci.* **915:** 214–222.

15. Sapone, A., L. de Magistris, M. Pietzak, *et al.* 2006. Zonulin upregulation is associated with increased gut permeability in subjects with type 1 diabetes and their relatives. *Diabetes* **55:** 1443–1449.

16. Watts, T., I. Berti, A. Sapone, *et al.* 2005. Role of the intestinal tight junction modulator zonulin in the pathogenesis of type I diabetes in BB diabetic-prone rats. *Proc. Natl. Acad. Sci. USA* **102:** 2916–2921.

17. Paterson, B.M., K.M. Lammers, M.C. Arrieta, *et al.* 2007. The safety, tolerance, pharmacokinetic and pharmacodynamic effects of single doses of AT-1001 in coeliac disease subjects: a proof of concept study. *Aliment. Pharmacol. Ther.* **26:** 757–766.

18. Fasano, A. 2011. Zonulin and its regulation of intestinal barrier function: the biological door to inflammation, autoimmunity, and cancer. *Physiol. Rev.* **91:** 151–175.

19. Tripathi, A., K.M. Lammers, S. Goldblum, *et al.* 2009. Identification of human zonulin, a physiological modulator of tight junctions, as prehaptoglobin-2. *Proc. Natl. Acad. Sci. USA* **106:** 16799–16804.

20. Bjorkman, P.J., M.A. Saper, B. Samraoui, *et al.* 1987. Structure of the human class I histocompatibility antigen, HLA-A2. *Nature* **329:** 506–512.

21. Asleh, R., S. Marsh, M. Shilkrut, *et al.* 2003. Genetically determined heterogeneity in hemoglobin scavenging and susceptibility to diabetic cardiovascular disease. *Circ. Res.* **92:** 1193–1200.

22. Wicher, K.B. & E. Fries. 2004. Prohaptoglobin is proteolytically cleaved in the endoplasmic reticulum by the complement C1r-like protein. *Proc. Natl. Acad. Sci. USA* **101:** 14390–14395.

23. Drago, S., A.R. El, P.M. Di, *et al.* 2006. Gliadin, zonulin and gut permeability: effects on celiac and non-celiac intestinal mucosa and intestinal cell lines. *Scand. J. Gastroenterol.* **41:** 408–419.

24. Hollande, F., E.M. Blanc, J.P. Bali, *et al.* 2001. HGF regulates tight junctions in new nontumorigenic gastric epithelial cell line. *Am. J. Physiol. Gastrointest. Liver Physiol.* **280:** G910–G921.

25. Jin, M., E. Barron, S. He, *et al.* 2002. Regulation of RPE intercellular junction integrity and function by hepatocyte growth factor. *Invest. Ophthalmol. Vis. Sci.* **43:** 2782–2790.

26. van der Merwe, J.Q., M.D. Hollenberg & W.K. MacNaughton. 2008. EGF receptor transactivation and MAP kinase mediate proteinase-activated receptor-2-induced chloride secretion in intestinal epithelial cells. *Am. J. Physiol. Gastrointest. Liver Physiol.* **294:** G441–G451.

27. Goldblum, S.E., U. Rai, A. Tripathi, *et al.* 2011. The active Zot domain (aa 288–293) increases ZO-1 and myosin 1C serine/threonine phosphorylation, alters interaction between ZO-1 and its binding partners, and induces tight junction disassembly through proteinase activated receptor 2 activation. *FASEB J.* **25:** 144–158.

28. El Asmar, R., P. Panigrahi, P. Bamford, *et al.* 2002. Host-dependent activation of the Zonulin system is involved in the impairment of the gut barrier function following bacterial colonization. *Gastroenterology* **123:** 1607–1615.

29. Clemente, M.G., S. De Virgiliis, J.S. Kang, *et al.* 2003. Early effects of gliadin on enterocyte intracellular signaling involved in intestinal barrier function. *Gut* **52:** 218–223.

30. Lammers, K.M., R. Lu, J. Brownley, *et al.* 2008. Gliadin induces an increase in intestinal permeability and zonulin release by binding to the chemokine receptor CXCR3. *Gastroenterology* **135:** 194–204.

31. Branski, D., A. Fasano & R. Troncone. 2006. Latest developments in the pathogenesis and treatment of celiac disease. *J. Pediatr.* **149:** 295–300

32. Plenge, R.M. 2010. Unlocking the pathogenesis of celiac disease. *Nat. Genet.* **42:** 281–282.

33. Madara, J.L. & J.S. Trier. 1980. Structural abnormalities of jejunal epithelial cell membranes in celiac sprue. *Lab. Inves.* **43:** 254–261.

34. Wolters, V.M., B.Z. Alizadeh, M.E. Weijerman, *et al.* 2010. Intestinal barrier gene variants may not explain the increased levels of antigliadin antibodies, suggesting other mechanisms than altered permeability. *Hum. Immunol.* **71:** 392–396.

35. Szakál, D.N., H. Gyorffy, A. Arató, *et al.* 2010. Mucosal expression of claudins 2, 3 and 4 in proximal and distal part of duodenum in children with coeliac disease. *Virchows Arch.* **456:** 245–250.

36. Schumann, M., D. Günzel, N. Buergel, *et al.* 2012. Cell polarity-determining proteins Par-3 and PP-1 are involved in epithelial tight junction defects in coeliac disease. *Gut* **61:** 220–228.

37. Arentz-Hansen, H., S. McAdam, O. Molberg, *et al.* 2003. Celiac lesion T cells recognized epitopes that cluster in regions of gliadin rich in proline residues. *Gastroenterology* **123:**803–809.

38. Nikulina, M. *et al.* 2004. Wheat gluten causes dendritic cell maturation and chemokine secretion. *J. Immunol.* **173:** 1925–1933.

39. Fasano, A. 2001. Pathological and therapeutical implications of macromolecule passage through the tight junction. In *Tight Junctions.* 697–722. CRC Press, Inc., Boca Raton, FL.

40. Fasano, A. 2008. Physiological, pathological, and therapeutic implications of zonulin-mediated intestinal barrier modulation: living life on the edge of the wall. *Am. J. Pathol.* **173:** 1243–1252.

41. Brorsson, C., N. Tue Hansen, R. Bergholdt, *et al.* 2010. The type 1 diabetes—HLA susceptibility interactome—identification of HLA genotype-specific disease genes for type 1 diabetes. *PLoS One.* **5:** e9576.

42. Feldman, M. & L.R. Schiller. 1983. Disorders of gastrointestinal motility associated with diabetes mellitus. *Ann. Intern. Med.* **98:** 378–384.

43. De Magistris, L., M. Secondulfo, D. Iafusco, *et al.* 1996. Altered mannitol absorption in diabetic children. *Ital. J. Gastroenterol.* **28:** 367.

44. Mooradian, A.D., J.E. Morley, A.S. Levine, *et al.* 1996. Abnormal intestinal permeability to sugars in diabetes mellitus. *Diabetologia* **29:** 221–224.

45. Meddings, J.B., J. Jarand, S.J. Urbanski, *et al.* 1999. Increased gastrointestinal permeability is an early lesion in the spontaneously diabetic BB rat. *Am. J. Physiol.* **276:** G951–G957.

46. Funda, D.P. & A. Kaas, H. Tlaskalová-Hogenová & K. Buschard. 2008. Gluten-free but also gluten-enriched (gluten+) diet prevent diabetes in NOD mice; the gluten enigma in type 1 diabetes. *Diabetes Metab. Res. Rev.* **24:** 59–63.

47. Visser, J.T., K. Lammers, A. Hoogendijk, *et al.* 2010. Restoration of impaired intestinal barrier function by the hydrolysed casein diet contributes to the prevention of type 1 diabetes in the diabetes-prone BioBreeding rat. *Diabetologia.* **53:** 2621–2628.

48. Simpson, M., M. Mojibian, K. Barriga, *et al.* 2009. An exploration of Glo-3A antibody levels in children at increased risk for type 1 diabetes mellitus. *Pediatr. Diabetes* **10:** 563–572.

49. Melamed-Frank, M., O. Lache, B.I. Enav, *et al.* 2001. Structure-function analysis of the antioxidant properties of haptoglobin. *Blood* **98:** 3693–3698.

50. Blum, S., U. Milman, C. Shapira & A.P. Levy. 2008. Pharmacogenomic application of the haptoglobin genotype in the prevention of diabetic cardiovascular disease. *Pharmacogenomics* **9:** 989–991.

51. Papp, M., I. Foldi, E. Nemes, *et al.* 2008. Haptoglobin polymorphism: a novel genetic risk factor for celiac disease development and its clinical manifestations. *Clin. Chem.* **54:** 697–704.

52. Napolioni, V., P. Giannì, F.M. Carpi, *et al.* 2011. Haptoglobin (HP) polymorphisms and human longevity: a cross-sectional association study in a Central Italy population. *Clin. Chim. Acta.* **412:** 574–577.

53. Márquez, L. *et al.* 2012. Effects of haptoglobin polymorphisms and deficiency on susceptibility to inflammatory bowel disease and on severity of murine colitis. *Gut* **61:** 528–534.

54. Wan, C. *et al.* 2007. Abnormal changes of plasma acute phase proteins in schizophrenia and the relation between schizophrenia and haptoglobin (Hp) gene. *Amino. Acids* **32:** 101–108.

55. Chen, Y. C. *et al.* 2011. Haptoglobin polymorphism as a risk factor for chronic kidney disease: a case-control study. *Am. J. Nephrol.* **33:** 510–514.

56. Fasano, A. 2009. Surprises from celiac disease. *Sci. Am.* **301:** 54–61.

57. Paterson, B.M., K.M. Lammers, M.C. Arrieta, *et al.* 2007. The safety, tolerance, pharmacokinetic and pharmacodynamic effects of single doses of AT-1001 in celiac disease subjects: a proof of concept study. *Aliment. Pharmacol. Ther.* **26:** 757–766.

58. Kelly, C.P., P.H. Green, J.A. Murray, *et al.* 2009. Safety, tolerability and effects on intestinal permeability of larazotide acetate in celiac disease: results of a phase IIB 6-week gluten-challenge clinical trial. *Gastro.* **136,5:** A-474.

Ann. N.Y. Acad. Sci. ISSN 0077-8923

ANNALS OF THE NEW YORK ACADEMY OF SCIENCES

Issue: *Barriers and Channels Formed by Tight Junction Proteins*

Myosin light chain kinase: pulling the strings of epithelial tight junction function

Kevin E. Cunningham and Jerrold R. Turner

Department of Pathology, the University of Chicago, Chicago, Illinois

Address for correspondence: Jerrold R. Turner, M.D., Ph.D., Department of Pathology, the University of Chicago, 5841 South Maryland, MC 1089, Chicago, IL 60637. jturner@bsd.uchicago.edu

Dynamic regulation of paracellular permeability is essential for physiological epithelial function, while dysregulated permeability is common in disease. The recent elucidation of the molecular composition of the epithelial tight junction complex has been accompanied by characterization of diverse intracellular mediators of paracellular permeabiltiy. Myosin light chain kinase (MLCK), which induces contraction of the perijunctional actomyosin ring through myosin II regulatory light chain phosphorylation, has emerged as a key regulator of tight junction permeability. Examination of the regulation and role of MLCK in tight junction dysfunction has helped to define pathological processes and characterize the role of barrier loss in disease pathogenesis, and may provide future therapeutic targets to treat intestinal disease.

Keywords: tight junction; myosin light chain kinase; TNF-α; inflammatory bowel disease

Introduction

Epithelial cells serve an essential role in homeostasis by providing and regulating the physical barrier between tissue compartments.[1] In the intestine, the epithelium forms a barrier between the sterile tissues and the harsh environment of the lumen. This barrier must be selectively permeable to allow the absorption and secretion of large volumes of solutes and fluids while preventing flux of pathogens and xenobiotic materials. In addition, it is now clear that barrier function is dynamic and can be regulated by a variety of physiological and pathophysiological stimuli.[2] The apical and basal plasma membranes of mechanically linked epithelial cells form the majority of the barrier surface and constitute the transcellular barrier, but the space between adjacent cells, that is, the paracellular space, must also be sealed to maintain the continuity of the barrier. Throughout the gastrointestinal tract, the *transcellular* and *paracellular* routes are used for transport of solutes and fluids through and between epithelial cells. Specific pumps, transporters, and channels on the apical and basolateral plasma membranes facilitate transcellular transport. In contrast,

paracellular transport is passive, occurring down electrochemical gradients established either by the activity of the transcellular transporters or by external events, such as ingestion and luminal digestion of nutrients.

Paracellular transport is regulated by an intercellular seal formed by two main protein complexes located at the apical-most part of the lateral membrane, collectively termed the apical junctional complex (Fig. 1).[3] The adherens junction, which contains cadherin and catenin proteins, is linked to the dense ring of perijunctional actin and myosin that underlies the apical junction complex. While critical for structural integrity, the adherens junction does not contribute to sealing of the paracellular space. The tight junction, which is located just apical to the adherens junction, is the essential determinant of paracellular flux.[4,5] Tight junction–associated proteins, including the claudin family and occludin,[6] and the specialized lipid composition of local membranes[7] form a seal that limits paracellular flux. As with the adherens junction, the tight junction protein complex is bound to the perijunctional actomyosin ring[8] by direct and indirect protein–protein interactions.[9,10] As discussed

doi: 10.1111/j.1749-6632.2012.06526.x

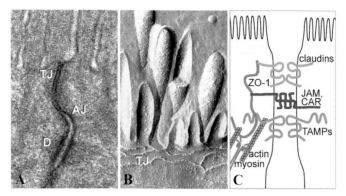

Figure 1. The apical junctional complex. (A) Transmission electron micrograph shows the tight junction (TJ), located most apically; the adherens junction (AJ), located below; and desmosomes, D, located basolaterally. (B) Freeze-fracture electron microscopy shows that the TJ is composed of particulate intramembranous strands. (C) Diagrammatic representation of the apical junctional complex. Myosin and actin interact with the tight junction through plaque proteins, such as ZO-1. Integral membrane proteins, including junctional adhesion molecules (JAM), Coxsackie adenovirus receptor (CAR), and TJ-associated MARVEL proteins (TAMPs), such as occludin, bridge the intercellular space. Figure reproduced from Shen *et al.*[72] with permission.

below, these interactions are critical to tight junction structure and function.[11–13]

Myosin light chain kinase

Original studies assumed that the tight junction was a static barrier. However, demonstration of rapid modulation of structure and barrier function by plant cytokinins, compounds derived from purines, suggested the possibility of physiological tight junction regulation. This was subsequently described as a consequence of Na^+-nutrient cotransport in mammalian small intestine.[14–19] However, the contribution of the paracellular pathway to overall nutrient and water absorption was considered controversial.[20–25] Many reasons exist for the failure of some studies to demonstrate paracellular water and nutrient absorption, most of which reflect technical issues and, potentially, the species studied.[24,25] The clearest *in vivo* example of the role of physiologic tight junction regulation in paracellular glucose transport is provided by a perfusion study of rat jejunum.[26] This work showed that D-glucose absorption over a wide range of perfusate concentrations could only be explained as the sum of active (i.e., transcellular) and passive (i.e., paracellular) transport.[26] As discussed below, *in vivo* perfusion studies of mouse jejunum have confirmed the critical contributions of tight junction regulation and paracellular transport to overall nutrient and water absorption in health and disease.

While the study of intact tissue, either *in vivo* or *ex vivo*, was critical to the initial demonstration of tight junction regulation, it was not suitable for investigation of the underlying mechanisms. However, the study of intact tissue did provide one essential clue. Ultrastructural analyses showed that Na^+-glucose cotransport–induced tight junction regulation was associated with perijunctional actomyosin condensation.[15,18,27] This led to the hypothesis that perijunctional actomyosin contraction might regulate the tight junction barrier function. Development of a model for the study of Na^+-glucose cotransport-induced tight junction regulation in cultured epithelial monolayers demonstrated that permeability was affected in a size-selective manner, that is, permeability to small, but not large, molecules was increased.[28] This model also allowed biochemical analysis, which revealed increased myosin II regulatory light chain (MLC) phosphorylation, a biochemical marker of actomyosin contraction, following Na^+-glucose cotransport activation.[28] Further, pharmacological inhibition of myosin light chain kinase (MLCK) prevented Na^+-glucose cotransport–induced tight junction regulation, both in cultured monolayers and isolated rodent mucosae.[28] Subsequent *ex vivo* analyses confirmed Na^+-glucose cotransport–induced tight junction regulation in human jejunum, and quantitative fluorescence microscopy demonstrated MLC phosphorylation within the perijunctional actomyosin ring of absorptive enterocytes within these tissues. Thus, MLCK-dependent MLC phosphorylation is

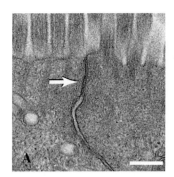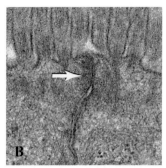

Figure 2. T cell activation induces perijunctional actomyosin condensation. The tight junctions (arrows) of jejunal villous enterocytes within control (A) and anti-CD3–treated (B) mice. Note the perijunctional cytoskeletal condensation induced by T cell activation. Figure reproduced from Clayburgh et al.[33] with permission.

an essential intermediate in physiological tight junction regulation.

MLCK is necessary for TNF-α–induced barrier loss

A role for the cytoskeleton in pathophysiological tight junction regulation was first suggested by the observation that MLC phosphorylation is markedly increased following infection with enteropathogenic *Escherichia coli*.[29] Shortly thereafter, a role for MLCK in tight junction dysfunction induced by tumor necrosis factor (TNF-α) was defined.[30] Analyses of cultured monolayers demonstrated TNF-α–induced barrier loss could be corrected acutely using a highly specific pseudosubstrate peptide MLCK inhibitor.[31] Further study demonstrated that TNF-α activated MLCK by at least two separate mechanisms: increased transcription and increased enzymatic function.[32]

While some details of TNF-α–induced barrier loss are best identified using highly manipulable *in vitro* models, determination of the effect on transepithelial transport requires use of *ex vivo* or *in vivo* models. Thus, an *in vivo* perfusion system was developed to allow quantitative analysis of barrier function and fluid transport in neurovascularly intact mouse jejunum.[33] Mice were treated with anti-CD3 to activate T cells systemically.[34] This induced jejunal barrier loss and reversal of net fluid movement, from absorption to secretion, in a TNF-α–dependent manner.[33,35] Ultrastructural examination revealed that this was associated with perijunctional actomyosin contraction (Fig. 2) similar to that induced by Na$^+$-glucose cotransport, and phosphorylation of perijunctional MLC.[33] Further-

more, mice lacking long MLCK, the intestinal epithelial MLC isoform,[36] or mice treated with the peptide inhibitor of MLCK failed to phosphorylate intestinal epithelial MLC and were protected from both barrier loss and fluid secretion.[33] Thus, MLCK is critical effector of pathological barrier dysfunction *in vivo*.

MLCK regulation in the gastrointestinal tract

The emergence of MLCK as a critical regulator of epithelial paracellular permeability has provided an opportunity to prevent intestinal barrier dysfunction in experimental models and examine the potential therapeutic benefit of this intervention. As noted above, TNF regulates MLCK transcription and enzymatic activity.[31,32,37] Moreover, MLCK is expressed in intestinal epithelia as two splice variants. Short, or smooth muscle, MLCK[38] is not expressed in intestinal epithelia.[36] Long MLCK is derived from the same gene as short MLCK, but uses an upstream promoter that gives rise to 5′ transcriptional and translational start sites and additional amino terminal protein sequence.[38] Two long MLCK isoforms, MLCK1, or full-length long MLCK, and MLCK2, which lacks a single exon within the unique, long MLCK upstream sequences, are expressed in intestinal epithelia. These splice variants have distinct subcellular localizations and functions,[36] and their expression is differentially regulated during epithelial differentiation. MLCK1 is predominantly expressed in villous epithelium, while MLCK2 is expressed throughout the crypt villus axis. Moreover, MLCK1 is concentrated at the perijunctional actomyosin ring and specific MLCK1 knockdown increases barrier function.[36,39] Finally, the ability of cultured intestinal epithelial monolayers to regulate barrier function after initiation of Na^+-glucose cotransport, which develops during enterocyte differentiation, coincides with the onset of MLCK1 expression.[36] Given that MLCK participates in multiple cellular processes, MLCK1 may be the preferred molecular target for therapeutic MLCK inhibition.

Despite the unique role of MLCK1 in tight junction regulation, MLCK1 and MLCK2 transcription appear to be activated similarly by TNF-α.[37] There has been debate regarding the signaling events that lead to increased MLCK transcription. The first study examining this found that inhibitors of NF-κB could block TNF-α–induced MLCK upregulation at extremely low concentrations.[32] In contrast, NF-κB inhibition required use of these agents at significantly greater doses that actually enhanced TNF-α–induced barrier loss.[32] A subsequent study suggested that NF-κB was critical to TNF-α–induced MLCK upregulation.[40,41] Although both of these studies used the Caco-2 intestinal epithelial cell line, which is derived from a human colonic adenocarcinoma, the first used the well-differentiated, absorptive (surface) enterocyte-like BBe subclone,[31,32,42] while the second study used the less well-differentiated parent line.[40,41] A detailed analysis of the human long MLCK promoter showed that this discrepancy likely explains the difference in mechanism of transcriptional regulation.[37] While the promoter was responsive to both AP-1 and NF-κB, the data show that TNF preferentially activates NF-κB in poorly differentiated monolayers and AP-1 in well-differentiated monolayers.[37] While activation of MLCK transcription by TNF has been demonstrated in vivo,[37,39] the mechanism of transcriptional regulation has not been defined. However, transgenic mice expressing an epithelial-specific IκBα mutant, which functions as an NF-κB super repressor, were protected from anti-CD3–induced tight junction regulation.[43]

A plethora of studies have identified distinct mechanisms to control MLCK expression and activity, which appear to be altered in disease states. While numerous groups have shown that MLCK phosphorylates MLC at Ser19, the regulation of MLCK expression in disease is less well defined. Perhaps with greatest relevance to inflammatory bowel disease (IBD), numerous studies have established the ability of inflammatory cytokines TNF-α and interferon (IFN)-γ, which are elevated in Crohn's disease, to induce barrier loss in vitro. Critically, TNF-α and IFN-γ inhibition can reverse barrier loss and substantially reduce inflammation in patients and animal models. In vitro monolayers primed with IFN-γ respond to TNF-α by decreasing barrier function and increasing MLC phosphorylation. Barrier function was restored by specifically inhibiting MLCK, suggesting that MLCK activity was responsible for the loss of barrier function.[31] The molecular mechanism leading to increased MLCK activity was subsequently studied in this model and it was found that MLCK protein[32] and gene[37] expression was also increased by IFN-γ/TNF-α, and corresponded with increased MLC phosphorylation. Thus, MLCK was found to be inducible by TNF-α, uncovering a novel mechanism of epithelial barrier regulation by the cytokine. Crucially, this observation is supported by patient data. In intestinal resections and biopsies, MLCK expression was slightly increased in ileal epithelia of patients with inactive Crohn's disease, and this increased further in active disease, correlating with histological disease activity.[39] MLC phosphorylation was also increased in colonocytes of patients with active disease (Fig. 3). Taken together, these studies provide key insight into the regulation of MLCK by inflammatory cytokines and the role of MLCK in IBD-associated epithelial barrier loss.

Mechanism of MLCK-dependent barrier regulation

While the data above demonstrate that MLCK is a critical mediator of tight junction barrier function, the downstream events activated by MLCK are only beginning to be defined. First, while Na^+-glucose cotransport and TNF-α both regulate tight junctions by MLCK-dependent processes, each has a distinct impact on barrier function. Na^+-glucose cotransport induces a size-selective increase in permeability that is limited to small molecules.[25,27,28] In contrast, TNF-α–induced increases in paracellular flux of small and large molecules.[32,33,45]

Inducible expression of constitutively active MLCK in cultured intestinal epithelia caused a size-selective increase in permeability similar to that following initiation of Na^+-glucose cotransport.[46] Detailed analysis of the tight junction structure in these monolayers demonstrated MLCK-dependent reorganization of perijunctional actin, occludin,

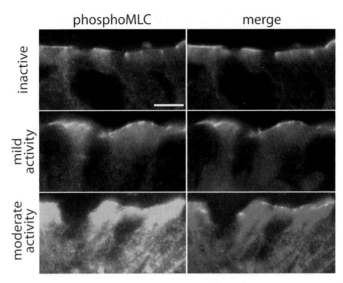

Figure 3. Myosin light chain (MLC) phosphorylation is correlated with inflammatory activity in IBD. Phosphorylated MLC (red) is primarily detected at the perijunctional actomyosin ring (green) in biopsies without active disease. While still predominantly within the perijunctional actomyosin ring, the intensity of phosphorylated MLC detection is markedly enhanced with increasing disease activity. Matched exposures are shown. Bar = 5 μm. Figure reproduced from Blair *et al.*[39] with permission.

and ZO-1.[46] The normally smooth, arc-like tight junction profiles viewed *en face* were modified to irregularly undulating profiles after MLCK activation.[46] This resulted in a nearly 20% increase in tight junction length that, by virtue of an increase in potential paracellular channels, could partially explain the MLCK-induced increase in paracellular flux. This may have also been associated with a change in lipid composition of tight junction membrane microdomains, as occludin was redistributed to a higher density population of glycoplipid- and cholesterol-rich membranes.[46] However, given the critical role of claudins in defining paracellular permeability,[47–51] it is notable that neither the *en face* profiles nor density of membranes containing claudin-1 and claudin-2 were affected by MLCK activation.

To better understand the mechanism of physiological MLCK-dependent barrier regulation, the dynamic behaviors of occludin, ZO-1, claudin-1, and actin were examined in monolayers with active Na$^+$-glucose cotransport.[52,53] MLCK inhibition markedly reduced ZO-1 exchange between tight junction and cytosolic pools, but did not affect dynamic behavior of other tight junction proteins or perijunctional actin.[53] Exchange of a ZO-1 mutant lacking the actin binding region was not af-

fected by MLCK inhibition, thereby demonstrating that this domain mediated the observed increased in ZO-1 anchoring.[53] Further, either ZO-1 knockdown or expression of the free actin binding region, as a dominant negative inhibitor, prevented MLCK-dependent barrier regulation.[53] Thus, physiological MLCK-dependent barrier regulation occurs via a ZO-1–dependent process. The role of occludin in this form of tight junction barrier regulation has not been established.

Similar to physiological tight junction regulation, pathophysiological MLCK-dependent barrier loss is associated with increased undulation of *en face* ZO-1 profiles.[33] However, TNF-α–induced tight junction regulation is also accompanied by occludin internalization.[33] This caveolin-1–dependent endocytosis is prevented by MLCK inhibition.[33,54] Further, *in vivo* occludin overexpression reduced TNF-α–induced barrier dysfunction by ~50%.[54] Similar data have demonstrated a critical role for occludin during *in vitro* TNF-α–induced barrier loss.[55] Thus, occludin endocytosis is a critical intermediate in pathophysiological tight junction regulation, both *in vitro* and *in vivo*. While not reported, it may be that this occludin endocytosis is responsible for the loss of size selectivity in pathophysiological, relative to physiological, MLCK-dependent barrier loss.

The role of MLCK-dependent barrier regulation in disease initiation and progression

Identification of MLCK as a central regulator of intestinal epithelial tight junctions has enabled further characterization of its role in disease pathogenesis as well as the wider role of barrier loss in disease initiation and progression. Patient data supporting the critical role of barrier function include the increased risk of relapse from remission in Crohn's disease patients with increased intestinal permeability.[56] The observed permeability increases in a subset of healthy first degree relatives of Crohn's disease patients[57,58] also suggests a role for tight junction dysregulation in disease initiation, while, simultaneously, demonstrating that barrier dysfunction alone is insufficient to cause disease.

To determine the contribution of MLCK to initiation and development of disease, transgenic mice that express constitutively active MLCK within the intestinal epithelium were developed.[59] As expected, these mice displayed increased intestinal epithelial MLC phosphorylation and paracellular permeability. While mucosal immune activation, including increased TNF-α, IFN-γ, IL-10, and IL-13, as well as increased numbers of lamina propria T cells were observed,[47,59] these mice did not develop spontaneous disease. Thus, the mice may be similar to healthy first-degree relatives of Crohn's disease patients. However, when immunodeficient mice expressing constitutively active MLCK were challenged with adoptive transfer of naive T cells, they developed colitis more rapidly than nontransgenic littermates.[59] In addition, disease in the transgenic mice was more severe in terms of cytokine production, histopathology, and overall survival.[59] Thus, while insufficient to initiate disease, epithelial tight junction dysregulation can accelerate disease progression and enhance disease severity. Conversely, delayed disease onset and reduced severity have been reported after adoptive transfer of naive T cells into long MLCK knockout mice.[60] Thus, targeted MLCK inhibition could be of therapeutic benefit.

Potential of MLCK inhibition as a therapeutic approach

The prospect of preventing initial development of Crohn's disease, maintaining remission, and reduc-ing severity of active flares by inhibition of epithelial long MLCK is compelling. The safety of such an approach is supported by the observation that long MLCK knockout mice are healthy[61] and are at least partially protected from many stressors.[33,61–64] However, established pharmacological inhibitors such as ML-7 and ML-9 are not useful, as they inhibit many kinases at concentrations necessary to block MLCK.[65] Although inhibitors with greater specificity are available,[61,66,67] these are also unsuitable, as they cannot distinguish between long and short MLCK, whose catalytic domains are derived from a single gene and are, therefore, identical.[38] Thus, a pharmacological approach targeting epithelial long MLCK enzymatic activity will also inhibit the short MLCK expressed in smooth muscle. Toxicities that would follow are demonstrated by the perinatal death of genetically modified mice lacking the MLCK catalytic domain,[68] and the hypotension, gut dysmotility, and viscus obstruction of mice with smooth muscle–specific MLCK deletion.[69,70] Thus, targeting of MLCK enzymatic activity may not be a viable approach to therapy when the risk of adverse effects is considered. One possible future direction may involve targeting of specific MLCK isoforms. Due to its expression within well-differentiated intestinal epithelia, specific inhibition of long MLCK1 might prevent tight junction dysfunction without systemic toxicities. The presence of a unique IgCAM domain that contains src phosphorylation sites could represent an alternative targets to inhibit MLCK1.[71] However, available data are limited and further investigation is needed.

Conclusions

In summary, understanding the contribution of epithelial paracellular permeability to physiological processes in the intestine, and its dysregulation in disease, has provided invaluable insight into disease mechanisms. While not yet practical, isoform-specific MLCK inhibition may ultimately provide a viable approach to restoring tight junction barrier function and preventing or treating intestinal disease.

Acknowledgments

We are grateful to past and present members of our research group, our collaborators, and many others in the field for the insights they have provided into tight junction structure and function. We also apologize to colleagues whose outstanding work was not cited due to length restrictions. Our work is supported by the National Institutes of Health (R01DK61931, R01DK68271, P01DK067887), the Department of Defense (W81XWH-09-1-0341), the Broad Medical Research Foundation (IBD-022), the University of Chicago Cancer Center (P30CA14599), and the University of Chicago Institute for Translational Medicine (UL1RR024999).

Conflicts of interest

The authors have no conflicts of interest.

References

1. Marchiando, A.M., W.V. Graham & J.R. Turner. 2010. Epithelial barriers in homeostasis and disease. *Annu. Rev. Pathol.* **5:** 119–144.
2. Nusrat, A., J.R. Turner & J.L. Madara. 2000. Regulation of tight junctions by extracellular stimuli: nutrients, cytokines, and immune cells. *Am. J. Physiol.–Gastrointest. Liver Physiol.* **279:** G851–G857.
3. Farquhar, M. & G. Palade. 1963. Junctional complexes in various epithelia. *J. Cell Biol.* **17:** 375–412.
4. Martinez-Palomo, A. & D. Erlij. 1975. Structure of tight junctions in epithelia with different permeability. *Proc. Natl. Acad. Sci. USA* **72:** 4487–4491.
5. Bentzel, C.J., B. Hainau, A. Edelman, T. Anagnostopoulos & E.L. Benedetti. 1976. Effect of plant cytokinins on microfilaments and tight junction permeability. *Nature* **264:** 666–668.
6. Tsukita, S. & M. Furuse. 1999. Occludin and claudins in tight-junction strands: leading or supporting players? *Trends Cell Biol.* **9:** 268–273.
7. Nusrat, A., C.A. Parkos, P. Verkade, *et al.* 2000. Tight junctions are membrane microdomains. *J. Cell Sci.* **113:** 1771–1781.
8. Madara, J.L. 1987. Intestinal absorptive cell tight junctions are linked to cytoskeleton. *AJP–Cell Physiol.* **253:** C171–C175.
9. Fanning, A.S., B.J. Jameson, L.A. Jesaitis & J.M. Anderson. 1998. The tight junction protein zo-1 establishes a link between the transmembrane protein occludin and the actin cytoskeleton. *J. Biol. Chem.* **273:** 29745–29753.
10. Fanning, A.S., T.Y. Ma & J.M. Anderson. 2002. Isolation and functional characterization of the actin binding region in the tight junction protein zo-1. *FASEB J.* **16:** 1835–1837.
11. Bentzel, C.J., B. Hainau, S. Ho, *et al.* 1980. Cytoplasmic regulation of tight-junction permeability: effect of plant cytokinins. *Am. J. Physiol.* **239:** C75–C89.
12. Madara, J.L., D. Barenberg & S. Carlson. 1986. Effects of cytochalasin d on occluding junctions of intestinal absorptive cells: further evidence that the cytoskeleton may influence paracellular permeability and junctional charge selectivity. *J. Cell Biol.* **102:** 2125–2136.
13. Shen, L. & J.R. Turner. 2005. Actin depolymerization disrupts tight junctions via caveolae-mediated endocytosis. *Mol. Biol. Cell.* **16:** 3919–3936.
14. Madara, J.L. & S. Carlson. 1991. Supraphysiologic l-tryptophan elicits cytoskeletal and macromolecular permeability alterations in hamster small intestinal epithelium *in vitro. J. Clin. Invest.* **87:** 454–462.
15. Madara, J.L. & J.R. Pappenheimer. 1987. Structural basis for physiological regulation of paracellular pathways in intestinal epithelia. *J. Membr. Biol.* **100:** 149–164.
16. Pappenheimer, J.R. 1987. Physiological regulation of transepithelial impedance in the intestinal mucosa of rats and hamsters. *J. Membr. Biol.* **100:** 137–148.
17. Pappenheimer, J.R. & K.Z. Reiss. 1987. Contribution of solvent drag through intercellular junctions to absorption of nutrients by the small intestine of the rat. *J. Membr. Biol.* **100:** 123–136.
18. Atisook, K., S. Carlson & J.L. Madara. 1990. Effects of phlorizin and sodium on glucose-elicited alterations of cell junctions in intestinal epithelia. *Am. J. Physiol.* **258:** C77–C85.
19. Turner, J.R., D.E. Cohen, R.J. Mrsny & J.L. Madara. 2000. Noninvasive *in vivo* analysis of human small intestinal paracellular absorption: regulation by na$^+$-glucose cotransport. *Dig. Dis. Sci.* **45:** 2122–2126.
20. Madara, J.L. 1994. Sodium-glucose cotransport and epithelial permeability. *Gastroenterology* **107:** 319–320.
21. Turner, J.R. & J.L. Madara. 1995. Physiological regulation of intestinal epithelial tight junctions as a consequence of na$^+$-coupled nutrient transport. *Gastroenterology* **109:** 1391–1396.
22. Fine, K.D., C.A. Santa Ana, J.L. Porter & J.S. Fordtran. 1993. Effect of d-glucose on intestinal permeability and its passive absorption in human small intestine *in vivo. Gastroenterology* **105:** 1117–1125.
23. Fine, K.D., C.A. Santa Ana, J.L. Porter & J.S. Fordtran. 1994. Mechanism by which glucose stimulates the passive absorption of small solutes by the human jejunum *in vivo. Gastroenterology* **107:** 389–395.
24. Lane, J.S., E.E. Whang, D.A. Rigberg, *et al.* 1999. Paracellular glucose transport plays a minor role in the unanesthetized dog. *Am. J. Physiol.* **276:** G789–G794.
25. Fihn, B.M., A. Sjoqvist & M. Jodal. 2000. Permeability of the rat small intestinal epithelium along the villus- crypt axis: effects of glucose transport. *Gastroenterology* **119:** 1029–1036.
26. Meddings, J.B. & H. Westergaard. 1989. Intestinal glucose transport using perfused rat jejunum *in vivo*: model analysis and derivation of corrected kinetic constants. *Clin. Sci.* **76:** 403–413.
27. Atisook, K. & J.L. Madara. 1991. An oligopeptide permeates intestinal tight junctions at glucose-elicited dilatations. *Gastroenterology* **100:** 719–724.
28. Turner, J.R., B.K. Rill, S.L. Carlson, *et al.* 1997. Physiological regulation of epithelial tight junctions is associated with myosin light-chain phosphorylation. *Am. J. Physiol.* **273:** C1378–C1385.
29. Yuhan, R., A. Koutsouris, S.D. Savkovic & G. Hecht. 1997. Enteropathogenic escherichia coli-induced myosin light chain phosphorylation alters intestinal epithelial permeability. *Gastroenterology* **113:** 1873–1882.
30. Taylor, C.T., A.L. Dzus & S.P. Colgan. 1998. Autocrine regulation of epithelial permeability by hypoxia: role for polarized

release of tumor necrosis factor alpha. *Gastroenterology* **114:** 657–668.

31. Zolotarevsky, Y., G. Hecht, A. Koutsouris, *et al*. 2002. A membrane-permeant peptide that inhibits mlc kinase restores barrier function in *in vitro* models of intestinal disease. *Gastroenterology* **123:** 163–172.

32. Wang, F., W.V. Graham, Y. Wang, *et al*. 2005. Interferon-gamma and tumor necrosis factor-alpha synergize to induce intestinal epithelial barrier dysfunction by up-regulating myosin light chain kinase expression. *Am. J. Pathol.* **166:** 409–419.

33. Clayburgh, D.R., T.A. Barrett, Y. Tang, *et al*. 2005. Epithelial myosin light chain kinase-dependent barrier dysfunction mediates T cell activation-induced diarrhea *in vivo*. *J. Clin. Invest.* **115:** 2702–2715.

34. Ferran, C., M. Dy, K. Sheehan, *et al*. 1991. Inter-mouse strain differences in the *in vivo* anti-cd3 induced cytokine release. *Clin. Exp. Immunol.* **86:** 537–543.

35. Musch, M.W., L.L. Clarke, D. Mamah, *et al*. 2002. T cell activation causes diarrhea by increasing intestinal permeability and inhibiting epithelial na+/k+-atpase. *J. Clin. Invest* **110:** 1739–1747.

36. Clayburgh, D.R., S. Rosen, E.D. Witkowski, *et al*. 2004. A differentiation-dependent splice variant of myosin light chain kinase, mlck1, regulates epithelial tight junction permeability. *J. Biol. Chem.* **279:** 55506–55513.

37. Graham, W.V., F. Wang, D.R. Clayburgh, *et al*. 2006. Tumor necrosis factor-induced long myosin light chain kinase transcription is regulated by differentiation-dependent signaling events. Characterization of the human long myosin light chain kinase promoter. *J. Biol. Chem.* **281:** 26205–26215.

38. Kamm, K.E. & J.T. Stull. 2001. Dedicated myosin light chain kinases with diverse cellular functions. *J. Biol. Chem.* **276:** 4527–4530.

39. Blair, S.A., S.V. Kane, D.R. Clayburgh & J.R. Turner. 2006. Epithelial myosin light chain kinase expression and activity are upregulated in inflammatory bowel disease. *Lab. Invest.* **86:** 191–201.

40. Ma, T.Y., M.A. Boivin, D. Ye, *et al*. 2005. Mechanism of tnf-alpha modulation of caco-2 intestinal epithelial tight junction barrier: role of myosin light-chain kinase protein expression. *Am. J. Physiol.–Gastrointest. Liver Physiol.* **288:** G422–G430.

41. Ma, T.Y., G.K. Iwamoto, N.T. Hoa, *et al*. 2004. Tnf-alpha-induced increase in intestinal epithelial tight junction permeability requires nf-kappa b activation. *Am. J. Physiol.–Gastrointest. Liver Physiol.* **286:** G367–G376.

42. Peterson, M.D. & M.S. Mooseker. 1992. Characterization of the enterocyte-like brush border cytoskeleton of the c2bbe clones of the human intestinal cell line, caco-2. *J. Cell Sci.* **102**(Pt 3): 581–600.

43. Tang, Y., D.R. Clayburgh, N. Mittal, *et al*. 2010. Epithelial nf-kappab enhances transmucosal fluid movement by altering tight junction protein composition after T cell activation. *Am. J. Pathol.* **176:** 158–167.

44. Bruewer, M., A. Luegering, T. Kucharzik, *et al*. 2003. Proinflammatory cytokines disrupt epithelial barrier function by apoptosis-independent mechanisms. *J. Immunol.* **171:** 6164–6172.

45. Clayburgh, D.R., M.W. Musch, M. Leitges, *et al*. 2006. Co-ordinated epithelial nhe3 inhibition and barrier dysfunction are required for tnf-mediated diarrhea *in vivo*. *J. Clin. Invest.* **116:** 2682–2694.

46. Shen, L., E.D. Black, E.D. Witkowski, *et al*. 2006. Myosin light chain phosphorylation regulates barrier function by remodeling tight junction structure. *J. Cell Sci.* **119:** 2095–2106.

47. Weber, C.R., D.R. Raleigh, L. Su, *et al*. 2010. Epithelial myosin light chain kinase activation induces mucosal interleukin-13 expression to alter tight junction ion selectivity. *J. Biol. Chem.* **285:** 12037–12046.

48. Furuse, M., K. Fujita, T. Hiiragi, *et al*. 1998. Claudin-1 and -2: novel integral membrane proteins localizing at tight junctions with no sequence similarity to occludin. *J. Cell Biol.* **141:** 1539–1550.

49. Simon, D.B., Y. Lu, K.A. Choate, *et al*. 1999. Paracellin-1, a renal tight junction protein required for paracellular mg2+ resorption. *Science* **285:** 103–106.

50. Van Itallie, C., C. Rahner & J.M. Anderson. 2001. Regulated expression of claudin-4 decreases paracellular conductance through a selective decrease in sodium permeability. *J. Clin. Invest.* **107:** 1319–1327.

51. Amasheh, S., N. Meiri, A.H. Gitter, *et al*. 2002. Claudin-2 expression induces cation-selective channels in tight junctions of epithelial cells. *J. Cell Sci.* **115:** 4969–4976.

52. Shen, L., C.R. Weber & J.R. Turner. 2008. The tight junction protein complex undergoes rapid and continuous molecular remodeling at steady state. *J. Cell Biol.* **181:** 683–695.

53. Yu, D., A.M. Marchiando, C.R. Weber, *et al*. 2010. Mlck-dependent exchange and actin binding region-dependent anchoring of zo-1 regulate tight junction barrier function. *Proc. Natl. Acad. Sci. USA* **107:** 8237–8241.

54. Marchiando, A.M., L. Shen, W.V. Graham, *et al*. 2010. Caveolin-1-dependent occludin endocytosis is required for tnf-induced tight junction regulation *in vivo*. *J. Cell Biol.* **189:** 111–126.

55. Van Itallie, C.M., A.S. Fanning, J. Holmes & J.M. Anderson. 2010. Occludin is required for cytokine-induced regulation of tight junction barriers. *J. Cell Sci.* 2844–2852.

56. Wyatt, J., H. Vogelsang, W. Hubl, T. Waldhoer & H. Lochs. 1993. Intestinal permeability and the prediction of relapse in crohn's disease. *Lancet* **341:** 1437–1439.

57. Hollander, D., C.M. Vadheim, E. Brettholz, *et al*. 1986. Increased intestinal permeability in patients with Crohn's disease and their relatives. A possible etiologic factor. *Ann. Intern. Med.* **105:** 883–885.

58. Buhner, S., C. Buning, J. Genschel, *et al*. 2006. Genetic basis for increased intestinal permeability in families with crohn's disease: role of card15 3020insc mutation? *Gut* **55:** 342–347.

59. Su, L., L. Shen, D.R. Clayburgh, *et al*. 2009. Targeted epithelial tight junction dysfunction causes immune activation and contributes to development of experimental colitis. *Gastroenterology* **136:** 551–563.

60. Su, L., S.C. Nalle, E.A. Sullivan, *et al*. 2009. Genetic ablation of myosin light chain kinase limits epithelial barrier dysfunction and attenuates experimental inflammatory bowel disease. *Gastroenterology* **136:** A81 (Abstract).

61. Wainwright, M.S., J. Rossi, J. Schavocky, *et al.* 2003. Protein kinase involved in lung injury susceptibility: evidence from enzyme isoform genetic knockout and *in vivo* inhibitor treatment. *Proc. Natl. Acad. Sci. USA* **100:** 6233–6238.

62. Reynoso, R., R.M. Perrin, J.W. Breslin, *et al.* 2007. A role for long chain myosin light chain kinase (mlck-210) in microvascular hyperpermeability during severe burns. *Shock* **28:** 589–595.

63. Rossi, J.L., A.V. Velentza, D.M. Steinhorn, *et al.* 2007. Mlck210 gene knockout or kinase inhibition preserves lung function following endotoxin-induced lung injury in mice. *Am. J. Physiol.–Lung Cell Mol. Physiol.* **292:** L1327–L1334.

64. Xu, J., X.P. Gao, R. Ramchandran, *et al.* 2008. Nonmuscle myosin light-chain kinase mediates neutrophil transmigration in sepsis-induced lung inflammation by activating beta2 integrins. *Nat. Immunol.* **9:** 880–886.

65. Bain, J., H. McLauchlan, M. Elliott & P. Cohen. 2003. The specificities of protein kinase inhibitors: an update. *Biochem. J.* **371:** 199–204.

66. Behanna, H.A., D.M. Watterson & H.R. Ranaivo. 2006. Development of a novel bioavailable inhibitor of the calmodulin-regulated protein kinase mlck: a lead compound that attenuates vascular leak. *Biochim. Biophys. Acta.* **1763:** 1266–1274.

67. Owens, S.E., W.V. Graham, D. Siccardi, J.R. Turner & R.J. Mrsny. 2005. A strategy to identify stable membrane-permeant peptide inhibitors of myosin light chain kinase. *Pharm. Res.* **22:** 703–709.

68. Somlyo, A.V., H. Wang, N. Choudhury, *et al.* 2004. Myosin light chain kinase knockout. *J. Muscle Res. Cell Motil.* **25:** 241–242.

69. He, W.Q., Y.N. Qiao, C.H. Zhang, *et al.* 2011. Role of myosin light chain kinase in regulation of basal blood pressure and maintenance of salt-induced hypertension. *Am. J. Physiol.–Heart Circ. Physiol.* **301:** H584–H591.

70. He, W.Q., Y.J. Peng, W.C. Zhang, *et al.* 2008. Myosin light chain kinase is central to smooth muscle contraction and required for gastrointestinal motility in mice. *Gastroenterology* **135:** 610–620.

71. Birukov, K.G., C. Csortos, L. Marzilli, *et al.* 2000. Differential regulation of alternatively spliced endothelial cell myosin light chain kinase isoforms by p60src. *J. Biol. Chem.* **276:** 8567–8573.

72. Shen, L., C.R. Weber, D.R. Raleigh, *et al.* 2011. Tight junction pore and leak pathways: a dynamic duo. *Annu. Rev. Physiol.* **73:** 283–309.

Ann. N.Y. Acad. Sci. ISSN 0077-8923

ANNALS OF THE NEW YORK ACADEMY OF SCIENCES
Issue: *Barriers and Channels Formed by Tight Junction Proteins*

Defective tight junctions in refractory celiac disease

Michael Schumann,[1] Sarah Kamel,[1] Marie-Luise Pahlitzsch,[1] Lydia Lebenheim,[1] Claudia May,[1] Michael Krauss,[2] Michael Hummel,[3] Severin Daum,[1] Michael Fromm,[4] and Jörg-Dieter Schulzke[1,5]

[1]Department of Gastroenterology, Infectious Diseases and Rheumatology, Berlin, Germany. [2]Laboratory for Membrane Biochemistry and Molecular Cell Biology, Institute of Chemistry and Biochemistry, Freie Universität Berlin, Berlin, Germany. [3]Institute of Pathology, Campus Benjamin Franklin, Charité—Universitätsmedizin Berlin, Berlin, Germany. [4]Institute of Clinical Physiology, Berlin, Germany. [5]Division of Nutritional Medicine, Berlin, Germany

Address for correspondence: Dr. Michael Schumann, Charité—Universitätsmedizin Berlin, Campus Benjamin Franklin, Medizinische Klinik für Gastroenterologie, Infektiologie und Rheumatologie, 12200 Berlin, Germany. michael.schumann@charite.de

In celiac disease, the gut-associated immune system is activated in response to the ingestion of gluten, causing an atrophy of the small intestinal mucosa. Although this condition is, in most cases, responsive to a gluten-free diet, celiac disease refractory to treatment occurs in a small percentage of celiacs. An epithelial barrier defect is known to be an integral part of celiac pathophysiology. However, the mucosa in refractory celiac disease underlies a constant inflammatory process. The epithelial barrier has not been addressed in this condition so far. Herein, the tight junction-associated barrier in refractory celiac disease is investigated functionally and structurally. Although normally expressed in celiac disease, claudin-4 is shown to be downregulated in refractory cases, presumably by two mechanisms, reduced protein expression and increased claudin endocytosis. Furthermore, the tightening claudin-5 is downregulated and the pore-forming claudin-2 is upregulated.

Keywords: celiac disease; refractory celiac disease; epithelial barrier; tight junction

Introduction

In celiac disease, ingestion of gluten-containing grains triggers a reaction of the adaptive as well as the innate mucosa-associated immune system leading to severe alterations of small intestinal mucosal architecture, which causes a malabsorption syndrome of nutrients and vitamins. Duodenal histology hallmarks of celiac pathology include an atrophy of intestinal villi, crypt hyperplasia, and an infiltration of the epithelial layer by CD3+ T lymphocytes (so-called intraepithelial lymphocytes, IEL).[1] The immune response responsible for these drastic changes involves the activation of both, adaptive and innate immune system and finally targets the epithelial compartment. The adaptive response is built up around the presentation of so-called immune-dominant gliadin fragments by antigen-presenting cells to T helper cells generating a specific cytokine response, including TNF-α

and interferon-γ, which damage epithelial cells either directly or by activation of matrix metalloproteinases.[2,3] Innate immunity is activated by a second subset of gliadin fragments, that trigger IL-15 release and thereby upregulate atypical MHC proteins including MIC-A on epithelial cells. These cells are then recognized by NKG2D+ IELs triggering epithelial cytolysis.[4]

In over 90% of celiac cases, these immunopathologies are reversible after implementation of a gluten-free diet (GFD). However, about 3% of the patients do not improve with a GFD, a situation referred to as refractory celiac disease (RCD).[5] Dependent on the presence of aberrant IELs in the mucosae of these patients, they are subclassified as RCD II (with aberrant IELs) or RCD I (aberrant IELs absent).[6] RCD II is known to be a precursor lesion for an enteropathy-associated T cell lymphoma (EATL), a most often fatal malignancy.[7]

doi: 10.1111/j.1749-6632.2012.06565.x

Previous studies have revealed a defective epithelial barrier in celiac disease, presumably secondary to the inflammatory process. However, since the polarity protein partitioning-defective 3 (Par-3) was shown to be genetically associated with celiac disease, and the Par-3/aPKC complex was shown to be associated with structural TJ defects in CD, a primary barrier dysfunction in celiac disease cannot be ruled out.[8] In RCD, the mucosal immune activation is autonomous, that is, it does not require further activation by gliadin peptides, which is believed to involve high interleukin-15 (IL-15) secretion and possibly a reduced number of mucosal $\gamma\delta$ T cells.[6] The role of the epithelial barrier or the tight junction network in RCD is unknown. In the following study, mucosa samples from RCD small intestine were analyzed for barrier integrity and compared to celiac and control mucosae.

Methods

Patient recruitment, biopsy collection, and TCR monoclonality

Four patient groups were recruited: (1) Control: healthy controls without any CD-typical clinical symptoms at the time of the biopsy and a normal duodenal mucosal architecture under normal (gluten-containing) diet; (2) Celiac: celiac disease patients who were under a gluten-containing diet who had villus atrophy on duodenal biopsy (Marsh II–III) and a positive tissue-transglutaminase IgA or endomysium-IgA serology; (3) GFD: celiac disease patients who were under a GFD with recovery of mucosal architecture (Marsh 0–I) and negativity for tissue-transglutaminase IgA (tTg-IgA); and (4) RCD: patients with RCD, which is defined as lack of response to a GFD after at least 6 months of GFD. Biopsy specimens were sampled during upper GI endoscopy (either conventional gastroscopy or double ballon enteroscopy). For subclassification in RCS I or RCD II, the presence of clonally rearranged T cell receptor gamma (TCRG) and beta (TCRB) chain genes was examined by polymerase chain reaction (PCR). For this purpose, DNA was extracted from paraffin sections using the QIAamp DNA Mini Kit (QIAGEN) and a TCRG-PCR and TCRB-PCR was performed according to the BioMed-2 protocols.[9] All fluorescence-labeled PCR products were analyzed by capillary electrophoresis (GeneScan; 3130 Genetic Analyser, Applied Biosystems), and only cases with identical dominant PCR products in two independent PCR runs were regarded to harbor T cell populations.

The study was approved by the ethics committee (#227-44) and written consent was obtained from all patients in our study.

One-path impedance spectroscopy

Endoscopic biopsies of human duodenal mucosa were mounted into Ussing-type chambers and analyzed for one-path impedance spectroscopy as described earlier.[8,10] The frequency-dependent electrical characteristics of the capacitor allowed to obtain transmural resistance (R^t) at low frequencies and subepithelial resistance (R^{sub}) at high frequencies. The epithelial resistance (R^{epi}) was calculated from $R^{epi} = R^t - R^{sub}$.

Western blotting

TJ protein expression was determined from membrane extracts of duodenal biopsies as described previously.[8] Primary antibodies were either rabbit or mouse and were purchased at Life Technology/Invitrogen (Carlsbad, CA), secondary antirabbit IgG and antimouse IgG were POD-conjugated.

Cell culture and endocytosis inhibition

Caco-2 cells were maintained in culture and plated on PCF filter inserts as previously published.[8] On days 10–14 after plating, Caco-2 cells on filters were mounted in Ussing chambers to measure transepithelial resistance. After 15 min of equilibration, the endocytosis blocker dynasore was added to both chamber compartments (80 μM, 30 min, 37 °C). Transepithelial resistance was measured continously. After completion filters were unmounted, fixed with paraformaldehyde 2% (15 min, RT), and processed for staining as described below.

Immunostaining and confocal microscopy

Staining and LSM visualization was performed as reported earlier.[8] Endoscopic biopsies of human duodenal mucosa were fixed using paraformaldehyde 1% (60 min, RT), followed by glycine treatment (25 mM, 5 min, RT), and stepwise dehydration of tissue (sucrose 10%, 60 min → 20%, 60 min → 30% overnight, all at 4 °C). After tissue was embedded in TissueTek, cryo sections (5 μm) were accomplished using a Leica CM 1900 cryostat. Immunostaining was carried out using rabbit polyclonal or mouse monoclonal anticlaudin antibodies from Life Technology/Invitrogen.

Table 1. Patient characteristics

Patient #	Age (years)	Sex	Histology (Marsh)	IELs/ 100 EC	CD8+ IELs	TCR mono- clonality	tTG-IgA	HLA-DQ2
Control								
#1	68	F	0	19	ND	ND	−	−
#2	59	M	0	<40	ND	ND	−	−
#3	51	M	0	<40	ND	ND	−	+
#4	59	M	0	<40	ND	ND	ND	ND
#5	51	M	0	<40	ND	ND	−	−
#6	68	M	0	<40	ND	ND	ND	ND
#7	62	M	0	<40	ND	ND	−	+
#8	15	M	0	<40	ND	ND	−	−
#9	22	F	0	22	ND	ND	−	−
Celiac								
#10	76	M	IIIb	20–30	ND	ND	−[a]	+
#11	29	F	IIIa-b	91	+	ND	+	+
#12	36	F	IIIb-c	75	−	ND	+	+
#13	60	M	IIIb	100	−	ND	+	+
#14	34	F	IIIa-b	45	+	ND	+	+
#15	40	F	IIIb	70	+	ND	+	+
#16	75	F	IIIb	100	ND	ND	+	+
#17	28	F	IIIa	110	ND	ND	+	+
#18	70	M	IIIa-IIIb	54	−	ND	+	+
GFD								
#19	65	F	0	30	ND	ND	−	+
#20	24	F	I-IIIa	48	+	ND	−	+
#21	29	F	I	78	+	ND	−	+
#22	49	F	0	13	+	ND	−	−
#23	68	F	0	15	+	ND	−	+
#24	60	F	0	<40	+	ND	−	+
RCD								
#25	70	M	IIIa	80	+	+	+	+
#26	56	F	IIIc	30–50	−	+	−	+
#27	66	F	IIIa-b	80	+	+	+	+
#28[b]	48	F	IIIc	85	+	ND	+	+
#29	60	M	IIIb	35	−	+	+	+
#30[b]	57	F	IIIb	80	+	−	+	+
#31	40	M	IIIb	60	−	−	−	+

Clinical characteristics of the patients involved in the study. Histology of duodenal mucosa was graded according to the Marsh-Oberhuber classification. IELs were counted per 100 enterocytes (EC). Tg-IgA, serology for Transglutaminase-IgA.

[a]Patient with selective IgA deficiency, Tg-IgG-positive.

[b]Patients with RCD I.

Secondary antibodies (antirabbit- or antimouse-IgG) from Life Technology/Molecular Probes were labeled with AlexaFluor594 and AlexaFluor488 dyes.

Statistics

All data were means ± SEM. The statistical evaluation is based on two-sided Student's *t*-test for unpaired data or Mann–Whitney

U-test as appropriate. $P < 0.05$ was considered significant.

Results

Clinical characteristics of patients

As expected, patients with celiac disease had a Marsh IIIa-c duodenal mucosa with a significantly increased number of IELs compared to control mucosae (control, 28 ± 1; celiac disease 74 ± 10 IELs/100 enterocytes, $P < 0.001$, Table 1). Celiac patients on a GFD had Marsh 0 or I mucosa (with one exception) with an only modest increase of IELs, which was significantly lower compared to the celiac IEL number (GFD 36 ± 10, $P < 0.01$). Interestingly, in RCD IEL numbers had a similar level as in untreated celiac disease (RCD 66 ± 8). RCD II, as identified by monoclonality of the T cell receptor-γ and -β gene and/or by >50% loss of CD8 staining in CD3$^+$ IELs, was present in five cases (total RCD: seven cases). The remaining RCD cases were classified as RCD I.

Impedance spectroscopy

A defective epithelial barrier in celiac disease is well known from clinical studies as well as Ussing chamber experiments.[8,11,12] To confirm this defect in our studies, endoscopic biopsies from duodenal mucosa were mounted in Ussing chambers and analyzed by one-path impedance spectroscopy. As detailed in Table 2, an epithelial resistance (R^{epi}) of 25 Ω cm^2 was found in control duodenal mucosa, similar to what has been published previously.[8] R^{epi} was strongly decreased in both, celiac disease and RCD by 48% and 40%, respectively. Interestingly, reduction in R^{epi} was partially reversed in celiac patients

on a GFD with a R^{epi} higher than in the group of untreated patients ($P < 0.05$), implying that barrier defects might at least in part be secondary to the celiac inflammatory process.

Western blotting of claudins

As previously shown, claudin protein level is changed in celiac disease. However, the molecular structure of the TJ in RCD is unknown. Therefore, we analyzed the claudin protein level by Western blotting in RCD patients (Fig. 1). Comparisons were made to untreated CD patients and CD patients on a GFD. Interestingly, the increase of claudin-2 in RCD was even more pronounced than in CD (3.9-fold compared to 2.6-fold in controls, see Fig. 1B). Secondly, the previously published reduction in claudin-3 level in celiac disease was not found in RCD.[8] Moreover, claudin-4 was found to be decreased to $60 \pm 9\%$ of control in RCD, whereas it was unchanged in untreated and treated CD. In contrast, reduction in claudin-5 level was similar in untreated CD and RCD, whereas it approached a normal level in patients on GFD.

Subcellular localization of claudins

Since claudin-2 was strongly upregulated in refractory celiac disease, RCD small intestinal mucosa was analyzed with confocal microscopy. Immunostained cryosections revealed a strong upregulation of claudin-2 in small intestinal crypts (Fig. 2, detail below), whereas in surface epithelia the claudin-2 signal remained undetectable. Furthermore, analysis of claudin-4 revealed a punctuated intracellular pattern in RCD mucosa (Fig. 3A) that was not found in celiac disease or control mucosae (Fig. 3B and C). This finding was in contrast to claudin-3, which is localized to the lateral membrane of intestinal

Table 2. One-path impedance spectroscopy

Group	R^{epi} Ω cm^2	P	R^{sub} Ω cm^2	P	R^t Ω cm^2	P	n
Control	25 ± 2		36 ± 2		61 ± 3		7
Celiacs	13 ± 2	<0.01	23 ± 3	<0.01	36 ± 4	<0.001	6
GFD	21 ± 3	NS	28 ± 2	<0.05	49 ± 5	NS	6
RCD	15 ± 1	<0.05	23 ± 3	<0.01	38 ± 4	<0.01	3

One-path impedance spectroscopy subdivides transmural electrical resistance (R^t) into epithelial (R^{epi}) and subepithelial resistance (R^{sub}). *R* values are means $\pm$ SEM. *P* values are calculated with Student's *t*-test versus the control group. NS, not significant.

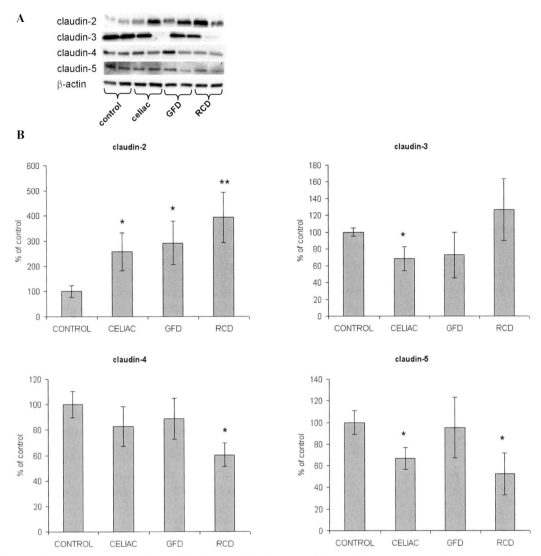

Figure 1. Western blotting for claudins. Duodenal biopsies of patients with celiac disease, patients on a GFD, and RCD patients were compared to controls. (A) Representative Western blots for claudins. β-actin was used as a loading control. (B) Densitometric analysis of the TJ protein level by Western blotting. GFD, gluten-free diet; RCD, refractory celiac disease. Significances refer to control mucosae. $^*P < 0.05$, $^{**}P < 0.01$, $n = 6$.

epithelial cells and is strongly concentrated to the TJ. Thus, subcellular localization of claudin-4 revealed a claudin-4-depleted TJ in RCD.

Inhibition of claudin-4 endocytosis in Caco-2 cells

To investigate the functional relevance of the intracellular claudin-4 compartment in RCD intestinal epithelial cells a Caco-2 model was used. Untreated Caco-2 cells were compared to cells treated with the endocytosis blocker dynasore (80 μM, 30 minutes). Transepithelial resistance was significantly higher in the dynasore group ($596 \pm 4\ \Omega\ cm^2$, compared to $417 \pm 5\ \Omega\ cm^2$ in control cells, $P < 0.01$, $n = 5$). In untreated cells, immunostaining for claudin-3 and claudin-4 revealed claudin-4-positive vesicles close to the TJ (Fig. 4A). Claudin-3 was mostly confined to the TJ. In dynasore-treated Caco-2 cells both, claudin-4 and claudin-3 were mostly localized to the TJ (Fig. 4B).

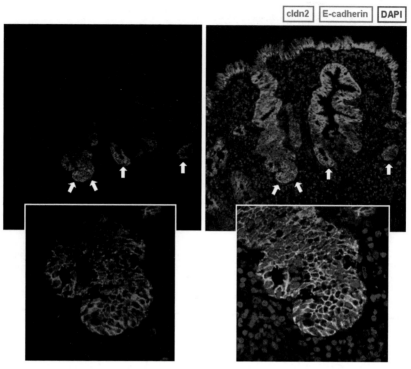

Figure 2. Immunostaining of claudin-2. Cryosections of small intestinal mucosa of RCD patients were immunostained for claudin-2 (red) and E-cadherin (green). Nuclei were stained with DAPI (blue). For a larger field-of-view, a lower magnification (20× objective) is shown in the upper panels. To demonstrate details of the TJ-network in the lower crypt, images were generated using the 40× objective in the lower panel.

Discussion

In celiac disease, the immune system specifically activates after ingestion of gluten targets small intestinal epithelial cells causing a severe defect of the epithelial barrier. This involves various effector mechanisms including induction of epithelial cell apoptosis, downregulation of tight junction function, as well as an increased rate of macromolecular transcytosis.[8,11,13–15] However, data on the epithelial barrier in RCD are sparse. RCD is a condition derived from celiac disease, where responsiveness to a GFD has been vanished. So far, it is known that IL-15, supposedly secreted by epithelia stimulated by a gliadin-19mer, induces epithelial expression of NK cell receptors ligands.[4,16] One of them, MICA, triggers IEL-mediated cytolysis of the MICA-expressing epithelial cell by interaction with IELs bearing the NK cell receptor NKG2D. Another NK cell receptor associated with epithelial cytolysis in RCD is DNAM-1, being strongly expressed on CD3-negative IELs.[17] Thus, published scientific

evidence on barrier defects in RCD focuses on IEL-mediated epithelial cytotoxicity involving specific interactions of NK cell receptors and their ligands, so far. The current knowledge on alternative mechanisms is negligible.

Work provided herein complements existing knowledge, mostly with regard to the role of tight junctions in the epithelial barrier in RCD. Firstly, we present data on a decreased epithelial resistance in RCD, that is, an increased conductance for small ions. In our previous work on celiac mucosae, this finding was paralleled with an increased flux for mannitol, a marker for the paracellular route.[8] An increased paracellular flux of smaller and medium-sized molecules can be secondary to either an increased epithelial apoptotic rate or defective tight junctions, both of which have been shown for celiac disease.[18]

Herein, we provide evidence for the TJ to be strongly altered in RCD, namely by a dysregulation of several claudins, which is the second result of this work. Of interest, two findings resemble the

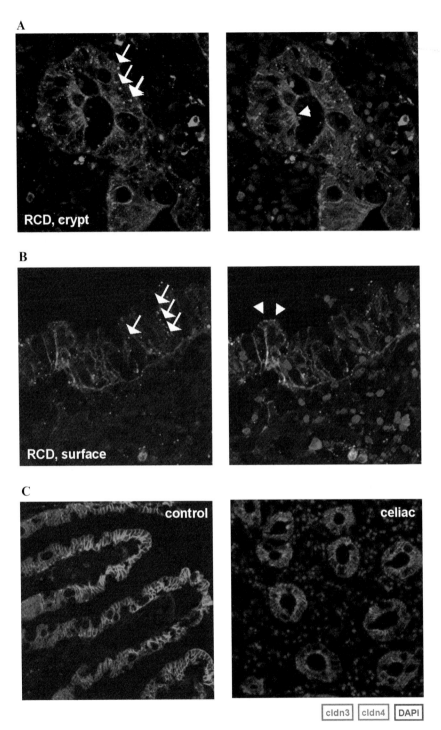

Figure 3. Immunostaining of claudin-4. Cryosections of small intestinal mucosa of RCD patients were immunostained for claudin-3 (red) and claudin-4 (green). Nuclei were stained with DAPI (blue). (A) and (B) depict the crypt and surface pattern of claudin-3 and -4. Claudin-4 staining is shown to the left, while merged images of claudin-3, claudin-4, and DAPI are shown to the right. Note that the lateral membrane and TJ staining of claudin-3 (arrow heads), which is contrasted by the intracellular, punctuated staining of claudin-4 (arrows). (C) In controls and celiac disease, claudin-4 localizes to the lateral membrane.

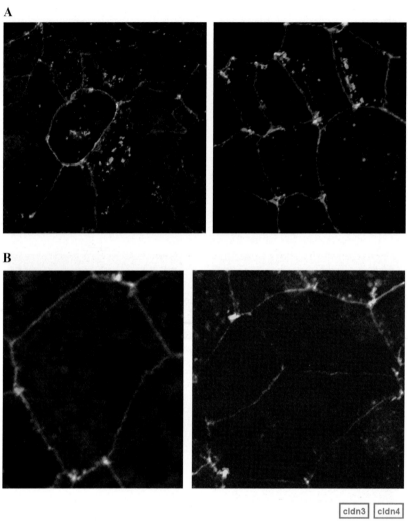

cldn3 cldn4

Figure 4. Endocytosis of claudin-4 in Caco-2 cells. Caco-2 cells grown on PCF filter inserts were treated with dynasore (80 μM, 30 min) and immunostained for claudin-3 (red) and claudin-4 (green). (A) Untreated cells reveal intracellular claudin-4 positive vesicles, while claudin-3 is confined to the membrane. (B) In dynasore-treated cells, both, claudin-4 and claudin-3 are localized to the membrane. Two representative high-resolution images are shown per condition.

pattern of changes published before for celiac disease, whereas two other findings are in contrast to what has been found previously: The fact that the cation pore-forming claudin-2 is strongly upregulated within the lower crypts especially and that claudin-5 is downregulated is not surprising, as it mirrors the situation found in celiac disease. More interesting is the fact that claudin-4, a tightening claudin being clearly unaltered in celiac disease, is reduced in its protein level as well as localized off the TJ. Reasoning about this discrepancy is purely speculative and includes the effect of a prolonged chronic mucosal inflammation as it is found in RCD or a difference in cytokine levels found in RCD (e.g., IL-15). Similar to this result is the unaltered claudin-3 level as this was found reduced in celiac disease.

As briefly mentioned before, such alterations at the apical junctional complex generally can be primary or secondary in nature. Par-3, a cell polarization protein being genetically linked to celiac disease and having a role in the TJ assembly, can be raised as an argument in favor for a primary barrier defect.[8,19] However, the significant epithelial barrier improvement (as well as the normalization of the claudin-5 protein level) after GFD implementation suggests that this is rather secondary to the reduction of mucosal inflammation.

A third conclusion that can be drawn from this study is the fact that claudin turnover includes a relevant baseline endocytosis, which can be deduced from the endocytosis blocking experiments with Caco-2 cells: The epithelial barrier of untreated cells was significantly strengthened by "merely" blocking endocytosis. Since the increase in R^t was accompanied by an altered claudin-4 trafficking, one can presume that the intracellular claudin-4 found in RCD is secondary to claudin-4 endocytosis.

In summary, this study reveals an impaired epithelia barrier in refractory celiac disease that is at least partially explained by an altered claudin-2, -4, and -5 expression and a claudin-4 localization of the TJ.

Acknowledgment

This work was supported by Deutsche Forschungsgemeinschaft (Schu 2389/1-1 and FOR721/2 TP2).

Conflicts of interest

The authors declare no conflicts of interest.

References

1. Sollid, L.M. & K.E. Lundin. 2009. Diagnosis and treatment of celiac disease. *Mucosal Immunol.* **2:** 3–7.
2. Daum, S. *et al.* 1999. Increased expression of mRNA for matrix metalloproteinases-1 and -3 and tissue inhibitor of metalloproteinases-1 in intestinal biopsy specimens from patients with coeliac disease. *Gut* **44:** 17–25.
3. Schmitz, H. *et al.* 1999. Tumor necrosis factor-alpha (TNFalpha) regulates the epithelial barrier in the human intestinal cell line HT-29/B6. *J. Cell Sci.* **112**(Pt 1)**:** 137–146.
4. Hue, S. *et al.* 2004. A direct role for NKG2D/MICA interaction in villous atrophy during celiac disease. *Immunity* **21:** 367–377.
5. Tjon, J.M., J. van Bergen & F. Koning. Celiac disease: how complicated can it get? *Immunogenetics* **62:** 641–651.
6. Verbeek, W. *et al.* 2008. Novel approaches in the management of refractory celiac disease. *Expert Rev. Clin. Immunol.* **4:** 205–219.
7. Daum, S., C. Cellier & C.J. Mulder. 2005. Refractory coeliac disease. *Best Pract. Res. Clin. Gastroenterol.* **19:** 413–424.
8. Schumann, M. *et al.* 2011. Cell polarity-determining proteins Par-3 and PP-1 are involved in epithelial tight junction defects in coeliac disease. *Gut* **61:** 220–228.
9. van Dongen, J.J. *et al.* 2003. Design and standardization of PCR primers and protocols for detection of clonal immunoglobulin and T-cell receptor gene recombinations in suspect lymphoproliferations: report of the BIOMED-2 Concerted Action BMH4-CT98–3936. *Leukemia* **17:** 2257–2317.
10. Fromm, M. *et al.* 2009. High-resolution analysis of barrier function. *Ann. N. Y. Acad. Sci.* **1165:** 74–81.
11. Schulzke, J.D. *et al.* 1998. Epithelial tight junction structure in the jejunum of children with acute and treated celiac sprue. *Pediatr. Res.* **43:** 435–441.
12. Vogelsang, H. *et al.* 2001. In vivo and in vitro permeability in coeliac disease. *Aliment Pharmacol. Ther.* **15:** 1417–1425.
13. Matysiak-Budnik, T. *et al.* 2008. Secretory IgA mediates retrotranscytosis of intact gliadin peptides via the transferrin receptor in celiac disease. *J. Exp. Med.* **205:** 143–154.
14. Moss, S.F. *et al.* 1996. Increased small intestinal apoptosis in coeliac disease. *Gut* **39:** 811–817.
15. Schumann, M. *et al.* 2008. Mechanisms of epithelial translocation of the alpha(2)-gliadin-33mer in coeliac sprue. *Gut* **57:** 747–754.
16. Mention, J.J. *et al.* 2003. Interleukin 15: a key to disrupted intraepithelial lymphocyte homeostasis and lymphomagenesis in celiac disease. *Gastroenterology* **125:** 730–745.
17. Tjon, J.M. *et al.* 2011. DNAM-1 mediates epithelial cell-specific cytotoxicity of aberrant intraepithelial lymphocyte lines from refractory celiac disease type II patients. *J. Immunol.* **186:** 6304–6312.
18. Mankertz, J. & J.D. Schulzke. 2007. Altered permeability in inflammatory bowel disease: pathophysiology and clinical implications. *Curr. Opin. Gastroenterol.* **23:** 379–383.
19. Wapenaar, M.C. *et al.* 2008. Associations with tight junction genes PARD3 and MAGI2 in Dutch patients point to a common barrier defect for coeliac disease and ulcerative colitis. *Gut* **57:** 463–467.

Ann. N.Y. Acad. Sci. ISSN 0077-8923

ANNALS OF THE NEW YORK ACADEMY OF SCIENCES
Issue: *Barriers and Channels Formed by Tight Junction Proteins*

Microbial butyrate and its role for barrier function in the gastrointestinal tract

Svenja Plöger,[1] Friederike Stumpff,[1] Gregory B. Penner,[2] Jörg-Dieter Schulzke,[3] Gotthold Gäbel,[4] Holger Martens,[1] Zanming Shen,[5] Dorothee Günzel,[6] and Joerg R. Aschenbach[1]

[1]Institute of Veterinary Physiology, Free University of Berlin, Berlin, Germany. [2]Department of Animal and Poultry Science, University of Saskatchewan, Saskatoon, Canada. [3]Department of Gastroenterology, Division of Nutritional Medicine, Charité, Berlin, Germany. [4]Institute of Veterinary Physiology, University of Leipzig, Leipzig, Germany. [5]College of Animal Medicine, Nanjing Agricultural University, Nanjing, China. [6]Institute of Clinical Physiology, Charité, Campus Benjamin Franklin, Berlin, Germany

Address for correspondence: Dr. Jörg R. Aschenbach, Institute of Veterinary Physiology, Free University of Berlin, Oertzenweg 19b, D-14163 Berlin, Germany. joerg.aschenbach@fu-berlin.de

Butyrate production in the large intestine and ruminant forestomach depends on bacterial butyryl-CoA/acetate-CoA transferase activity and is highest when fermentable fiber and nonstructural carbohydrates are balanced. Gastrointestinal epithelia seem to use butyrate and butyrate-induced endocrine signals to adapt proliferation, apoptosis, and differentiation to the growth of the bacterial community. Butyrate has a potential clinical application in the treatment of inflammatory bowel disease (IBD; ulcerative colitis). Via inhibited release of tumor necrosis factor α and interleukin 13 and inhibition of histone deacetylase, butyrate may contribute to the restoration of the tight junction barrier in IBD by affecting the expression of claudin-2, occludin, cingulin, and zonula occludens poteins (ZO-1, ZO-2). Further evaluation of the molecular events that link butyrate to an improved tight junction structure will allow for the elucidation of the cofactors affecting the reliability of butyrate as a clinical treatment tool.

Keywords: butyrate; colon; inflammatory bowel disease; rumen; tight junction

Introduction

Acetate, propionate, and butyrate are the three major short-chain fatty acids (SCFA) produced during bacterial carbohydrate fermentation.[1,2] Butyrate is the least abundant but also the most dynamic of these three acids, varying from ~5% to more than 20% of total fermentation acids.[3] Maximum butyrate fermentation is achieved when degradable fiber and degradable starch coincide in a balanced way (see next section), whereby butyrate might be considered a signal molecule for balanced bacterial growth. The butyrate signal is apparently received and utilized by the mammalian host to adapt gastrointestinal functions to the growth of the bacterial community (Fig. 1).[4,5]

Some of the gastrointestinal effects of butyrate have clinical implications. For example, butyrate induces epithelial proliferation in normal intestinal tissue[6,7] but decreases cell proliferation, increases apoptosis, and stimulates cell differentiation in colonic cancer cells,[8–12] which may minimize the incidence and progression of colon cancer.[4,9,13] Butyrate also stimulates NaCl absorption in the rat distal colon[14,15] and inhibits the prosecretory action of several cAMP-generating secretagogs, which can be beneficial in the treatment of diarrheal disorders.[8,14] Finally, butyrate may improve the barrier function of gastrointestinal epithelia[16,17] and thus ameliorate those diarrheal disorders that are sustained by barrier failure, for example, inflammatory bowel disease (IBD).[8,18] Due to the great importance of IBD, this last aspect of butyrate action is receiving increasing attention and will be the focus of this review. The main intention is to analyze the current knowledge on the effect of butyrate on the paracellular tight junction (TJ) barrier and the molecular events that link butyrate generation by luminal microbes to an improved TJ structure.

doi: 10.1111/j.1749-6632.2012.06553.x

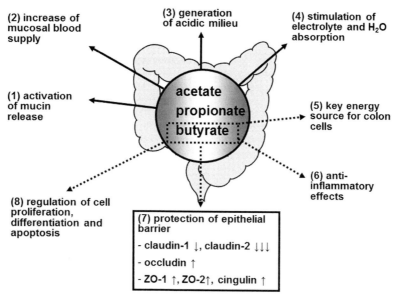

Figure 1. Effects of butyrate in the gastrointestinal tract. Acetate, propionate, and butyrate are produced by microbes in fermentative organs. All three acids have beneficial effects for gut health by stimulating mucin release in mucosecretory organs (1), increasing mucosal blood flow (2), the generation of an acidic milieu (3), and also by stimulation of electrolyte and water absorption (4). While butyrate is already more potent than acetate and propionate for some of these effects (e.g., 2 and 4), it has additional beneficial effects that are either attributed to the specific utilization of butyrate for epithelial cell metabolism (5) or to butyrate-inhibited histone deacetylation. The latter effect is involved in the regulation of cell proliferation, differentiation, and apoptosis (8) that is critical in the prevention of colon cancer, as well as the anti-inflammatory (6) and barrier-preserving actions (7) that are critical in the prevention and therapy of IBD.

In an attempt at interspecies comparison, the analysis not only includes butyrate effects in the large intestine but also butyrate generation and effects in the forestomach of ruminants. The rumen is interesting for comparative research because partly different concepts of butyrate action have been postulated for this gastrointestinal compartment.

Fermentation conditions favoring microbial butyrate production

Microbial and biochemical presuppositions of butyrate fermentation

Acetate is the dominating SCFA under almost all fermentation conditions and is always present in the highest concentration relative to other SCFA in ruminal fluid[19,20] and intestinal contents.[19,21] The percentage of acetate is highest when the rate of carbohydrate degradation is slow.[19] In situations where there is abundant supply of rapidly fermentable carbohydrates, such as starch or sugars, fermentation end-product accumulation shifts toward propionate, and if such excessive fermentation conditions persist, including increased

acidity, lactate production and accumulation is promoted.[22]

Butyrate production and accumulation, on the other hand, seems to rise when high-fiber degradability and high availability of nonstructural carbohydrates coincide. This may be related to the use of acetate as a precursor for butyrate synthesis under conditions where lactate fermentation could occur.[23] Under those conditions, lactate-producing bacteria like *Butyrivibrio fibrosolvens* may utilize acetate directly for butyrate production via butyryl-CoA/acetate-CoA transferase rather than through the conversion of two acetyl-CoA molecules to acetoacetyl-CoA.[24] The replenishing of NAD^+ via the butyryl-CoA/acetate-CoA pathway has significant energetic benefits for the involved bacteria.[25] These bacteria are strictly anaerobic, Gram-positive firmicutes and include *Faecalibacterium prausnitzii*, *Butyrivibrio fibrosolvens*, *Eubacterium rectale*, *Roseburia faecis*, and *Eubacterium hallii*.[25–27]

Nutritional modulation of butyrate fermentation

While the biochemistry of butyrate production is rather well understood,[25] achieving elevated butyrate concentrations in digesta contents is not as easily achieved. In this regard, the rumen is a comparatively simple model because nutrients are directly introduced into this compartment without prior modulation by small intestinal absorption. An analysis of a large data set for ruminal metabolism showed positive correlations of butyrate production with the sugar and fermentable neutral detergent fiber contents of the diet.[28] Dietary lactose addition appears to be a particularly efficient measure to increase the molar proportion of butyrate in ruminal fluid.[29,30]

In monogastric species, dietary sugars usually do not reach the large intestine due to small intestinal digestion and absorption. Alternatively, soluble fiber in combination with slowly degradable starch (e.g., corn starch) appears to promote large intestinal butyrate production. Rehman *et al.* fed chicks a basal diet (control) primarily consisting of corn and soya combined with either additional inulin, sucrose, or the combination of inulin and sucrose. Their study revealed that providing inulin or inulin with sucrose increased cecal butyrate concentration relative to the control, while sucrose alone failed to increase butyrate concentration.[31] Similarly, Metzler-Zebeli *et al.* observed increasing butyrate concentrations in the stomach, caecum, and colon of young pigs when supplementing a corn starch–based diet with high amounts of β-glucan.[32] These results underline that the coincidence of easily fermentable fiber and nonstructural carbohydrates are a common prerequisite for high rates of butyrate fermentation. However, the nutritional strategy to achieve such conditions has to consider the site of digesta fermentation.

Role of butyrate for barrier function of mucosecretory gastrointestinal mucosa

Butyrate and IBD
Beneficial effects of butyrate have been suggested in acute gastroenteritis, cholera, congenital chloride diarrhea, and, most often, in IBD.[8] The special consideration of butyrate in IBD could be linked to the prominent involvement of TJ lesions in this disease complex. The two types of IBD, Crohn's disease (CD) and ulcerative colitis (UC), are characterized by leak flux diarrhea with loss of plasma/interstitial fluid into the gut lumen due to an insufficient TJ. The main feature of UC is a reduction in the number of TJ strands,[33] while CD additionally involves TJ breaks.[34]

The ongoing inflammation has been suggested to be a crucial factor triggering TJ leakiness in IBD.[35] On the other hand, it is well known that butyrate enemas can alleviate inflammation in patients with UC.[36] In an attempt to verify the positive effects of butyrate on the colonic barrier in IBD, Venkatraman *et al.* harvested colonic epithelia from the rat model of IBD, dextran sulfate sodium (DSS)–induced colitis. Butyrate treatment of those isolated colonic sheets led to a recovery in transepithelial resistance (R^t), which was ascribed to a preservation of TJ integrity and inhibition of TNF-α release.[37] Other studies supported this assumption by demonstrating that butyrate inhibits the release of TNF-α[38] and IL-13, the latter being a key effector cytokine of UC.[39] Common to both TNF-α and IL-13 is their ability to upregulate the expression of the TJ molecule claudin-2.[40] Claudin-2 is a cation-selective pore[41] that is also upregulated in patients with IBD, where it explains at least part of the disturbances in barrier function.[35] One major effect of butyrate is to downregulate claudin-2 expression, as shown by microarray analysis in butyrate-treated colonic epithelial cells.[42] This suggests a relationship between the alleviation of inflammation by butyrate, the downregulation of claudin-2, and the improvement of barrier function.

Another study demonstrated a decrease of butyrate absorption in DSS mice and patients with IBD. This was mediated by a reduced expression of the butyrate transporter, monocarboxylate transporter (MCT-)1. A similar reduction in MCT-1 expression could be induced by TNF-α in HT-29 cells.[43] These findings support an alternative concept in which inflammation during IBD initially targets on MCT-1 to decrease butyrate uptake. The low intracellular availability of butyrate would, subsequently, result in a loss of the regulatory depression of claudin-2 expression by butyrate. However, further experiments are needed to clearly delineate the sequence of events that underlie the effects of butyrate on the TJs, especially on claudin-2 expression.

Influence of butyrate on barrier function and TJ integrity

While many studies see the merit of butyrate treatment on TJ integrity mainly in its immune-modulatory and anti-inflammatory action, *in vitro* studies showed that barrier function of intestinal epithelial cells can also be enhanced by butyrate in the absence of concurrent immune stimulation. Nevertheless, the results are equivocal. While butyrate promoted a higher R^t and a reduced mannitol permeability in Caco-2 cells,[16] a study by Suzuki *et al.* showed a short-term increase in R^t induced by a mix of SCFA, but not of butyrate alone.[44] Those discrepancies may depend on other modulating factors and on the dose of butyrate. With regard to the latter, Peng *et al.* observed an increased R^t in Caco-2 cells after treatment with 2 mmol/L butyrate and attributed this effect to a reorganization of the TJ molecules ZO-1 and occludin via activation of the AMP-activated protein kinase (AMPK).[17] However, higher concentrations of butyrate (8 mmol/L) decreased R^t, which was associated with a marked increase in the apoptosis level of Caco-2 cells.[45]

The ability of butyrate to modulate TJ protein expression was also demonstrated in noncolonic cell models. For example, Harten *et al.*[46] showed a restored epithelial barrier after sodium butyrate treatment in a tumor suppressor gene *VHL*-defective renal epithelial cell line that normally shows disrupted TJ structure. The protein concentrations of E-cadherin, occluding, and claudin-1 were increased after butyrate treatment, and TJ labeling for ZO-1, occludin, and claudin-1 was restored. These effects were assumed to be linked to a decreased HIF-1α expression and an ability of butyrate to act as histone deacetylase inhibitor (HDACi).[46] Furthermore, butyrate also upregulated cingulin, ZO-1, and ZO-2 in Rat-1 fibroblasts, cingulin in COS-7 cells, and cingulin and occludin in HeLa cells.[47] The authors also attributed the results to an HDACi action of butyrate.[47]

Regulation of cell processes by butyrate acting as HDACi

As already indicated in the previous section, butyrate may influence cell processes based on its action as an HDACi.[12,47,48] In many cases, this leads to hyperacetylation of histones that is followed by increased gene expression. On the other hand, a decrease in histone acetylation by butyrate, especially in downregulated genes, could also be demonstrated.[49] Either way, gene regulation by butyrate is very often the result of changes in histone acetylation. For example, in the gastric cancer cell line HGC-27, butyrate and the known HDACi trichostatin A (TSA) both enhanced vitamin D–induced

apoptosis by upregulating the gene PTEN. This was based on an increased histone acetylation level at the PTEN-promoter, which promoted the binding of the transcription factor Egr-1.[50]

Transcriptional regulation of gene expression by butyrate also plays an important role in cell differentiation. Gaschott *et al.* showed a synergism between butyrate and vitamin D receptor (VDR) activation on cell differentiation in the colon cancer cell line Caco-2.[51,52] Cell differentiation implies the expression of TJ proteins in intestinal epithelial cells.[53,54] So far, however, no direct link has been established between the expression of the vitamin D receptor and the expression of TJ proteins in the intestine. Nonetheless, butyrate action as HDACi seems to play a role in the regulation of TJ protein expression. For example, claudin-1 overexpression induced by increased histone deacetylation is linked to dedifferentiation and increased invasion in colon cancer. Claudin-1 overexpression can be reversed by butyrate or TSA, which decrease claudin-1 mRNA half-life time through their HDACi action.[55] Moreover, the downregulation of claudin-2 by butyrate mentioned earlier in this review[42] depends on a reduced binding affinity of transcription factors within the claudin-2 promoter.[56] It has been recently shown that TSA and butyrate similarly decreased claudin-2 expression in an intestinal cell line, which points to the possibiliy that this butyrate effect may also be dependent on altered histone acetylation.[56]

Role of butyrate for barrier function of cutaneous gastrointestinal mucosa

The forestomach as a model

Although current research on the interaction of butyrate with epithelia is focused largely on the colon, far larger quantities of SCFA are produced in certain species that ferment forages in a forestomach, with production of butyrate alone estimated at approximately 5–10 mol/day in high-yielding dairy cows.[2,3,19] Accordingly, both the transepithelial absorption of SCFA and the intraepithelial metabolism of butyrate were first discovered in the forestomachs of ruminants.[2,57,58] Comparative research on forestomach epithelia may thus lead to valuable clues in understanding the general principles of butyrate action.

In contrast to the colon, the forestomachs are covered with multilayered stratified squamous epithelia, the proliferation of which leads to increases in papillae length and in the number and composition of cell layers.[59–61] The influence of butyrate on epithelial growth and differentiation can be observed in milk-fed ruminants, who are born with a rumen that is nonfunctional and initially bypassed until transition to a grain or forage based diet occurs. Butyrate was identified as a key stimulus to initiate growth and differentiation of the rumen into the mature absorptive organ;[62–64] however, a number of hormonal mediators (glucagon-like peptide 2, insulin, insulin-like growth factor 1) appear to be involved as cofactors.[60,65–69] Interestingly, the mediation of butyrate effects via hormonal mediators has been intensively discussed for the rumen but has rarely been considered for the large intestine.[7] *In vitro* and without any hormonal stimulus, butyrate itself has antiproliferative action on primary cultures of ruminal epithelial cells.[65,66,68]

Butyrate and the forestomach barrier

A current limitation on attempts to delineate the effects of butyrate on the forestomach barrier is that knowledge of its functional organization is far less advanced than that concerning monolayered gastrointestinal epithelia. Due to the multilayered structure of forestomach epithelia, a separation into a clearly distinct apical and basolateral domain appears too simplistic.[70] Instead, at least two functional barriers emerge. The first barrier is that of the uppermost stratum corneum, the keratinocytes of which are thought to form a loose physical boundary that may serve as an apical microclimate,[71] with cells possibly loosely interconnected by claudin-7.[72] The second barrier is that of the stratum granulosum, which clearly shows the highest incidence in morphological correlates of occluding junctions,[70,73,74] while staining for claudin-1, claudin-4, ZO-1, and occludin can be observed.[70,72] Staining intensity for claudin-4 drops sharply within the stratum spinosum toward the basal layer, whereas claudin-1 and occludin persist down to the stratum basale. Notably, and in contrast to the findings in monolayered epithelia, attempts to demonstrate the presence of the leaky claudin-2 via immunohistochemical staining, PCR, or Western blot in tissues and cultured cells of the rumen and other forestomach compartments have so far been unsuccessful.[72] Possibly both due to a lack of claudin-2 and due to the peculiarities of their multilayered organization,

forestomach epithelia appear to be relatively tight, allowing the absorption of ions against formidable gradients.[75–77] There is no indication that SCFA in general,[78–80] or butyrate in particular (Penner *et al.*, unpublished data), can upregulate what is already a formidable barrier, although they may enhance the absorptive capacity of the epithelium[79,81,82] and its ability to withstand osmotic challenges[78] and to metabolize butyrate.[80,83,84]

Instead, and in conjunction with decreasing luminal pH, rising concentrations of butyrate are suspected to be a detrimental factor for the forestomach barrier in a condition known as ruminal acidosis.[81] Clinicians expect symptoms ranging from decreased food intake to severe ruminal inflammation with liver abscesses and immunological manifestations.[85,86] Histologically, rapid structural changes of the epithelium are observed with a decline in cellular junctions, sloughing of the stratum corneum and a thinning of all underlying epithelial layers.[87] Chronic exposure leads to a thickening of the stratum corneum at the expense of the stratum granulosum,[88,89] suggesting profound alterations in barrier structure in this pathological situation. While butyrate is thus essential for the differentiation and maturation of the forestomachs into organs of impressive absorptive capacity, there is considerable reason to assume that excessive amounts of butyrate may be toxic, especially when coinciding with low luminal pH.

Future perspectives

Butyrate is an important metabolic signal in the gastrointestinal tract with a proven role for the tightness of the epithelial barrier. The latter makes it a promising tool in the treatment of gut disorders like IBD. However, butyrate effects are variable and may depend on the butyrate concentration, pH, the differentiation state of affected cells and on confounding indirect effects of butyrate due to altered profiles of (or sensitivity to) hormones, growth factors, and inflammatory mediators. A better understanding of these complex interactions and the related molecular events, as well as a better understanding of the conditions favoring microbial butyrate production, are necessary to improve the therapeutic usability of butyrate. To this end, comparative research in the forestomach of ruminants is suggested to be of value for delineating primary butyrate actions from cofactor-dependent effects of butyrate.

Acknowledgments

We wish to thank the German Research Foundation (FOR 721 to J.D.S. and D.G.; Ga 329/6-1 and 7-1 to G.G.), the Canadian Natural Sciences and Engineering Research Council (NSERC Discovery Grant to G.B.P.), the German Federal Ministry for Education and Research (BMBF-Fugato-plus to H.M.), and the National Nature Science Foundation of China (No.31010103020 to Z.S.) for financial support.

Conflicts of interest

The authors declare no conflicts of interest.

References

1. Bugaut, M. 1987. Occurrence, absorption and metabolism of short chain fatty acids in the digestive tract of mammals. *Comp. Biochem. Physiol. B.* **86:** 439–472.

2. Bergman, E.N. 1990. Energy contributions of volatile fatty acids from the gastrointestinal tract in various species. *Physiol. Rev.* **70:** 567–590.

3. Aschenbach, J.R., G.B. Penner, F. Stumpff & G. Gabel. 2011. Ruminant Nutrition Symposium: role of fermentation acid absorption in the regulation of ruminal pH. *J. Anim. Sci.* **89:** 1092–1107.

4. Bugaut, M. & M. Bentejac. 1993. Biological effects of short-chain fatty acids in nonruminant mammals. *Annu. Rev. Nutr.* **13:** 217–241.

5. Gill, R.K. & P.K. Dudeja. 2011. A novel facet to consider for the effects of butyrate on its target cells. Focus on "The short-chain fatty acid butyrate is a substrate of breast cancer resistance protein". *Am. J. Physiol. Cell Physiol.* **301:** C977–C979.

6. Kripke, S.A., A.D. Fox, J.M. Berman, *et al.* 1989. Stimulation of intestinal mucosal growth with intracolonic infusion of short-chain fatty acids. *JPEN J. Parenter. Enteral. Nutr.* **13:** 109–116.

7. Sakata, T. & W. von Engelhardt. 1983. Stimulatory effect of short chain fatty acids on the epithelial cell proliferation in rat large intestine. *Comp. Biochem. Physiol. A. Comp. Physiol.* **74:** 459–462.

8. Canani, R.B., M.D. Costanzo, L. Leone, *et al.* 2011. Potential beneficial effects of butyrate in intestinal and extraintestinal diseases. *World J. Gastroenterol.* **17:** 1519–1528.

9. Clarke, J.M., G.P. Young, D.L. Topping, *et al.* 2012. Butyrate delivered by butyrylated starch increases distal colonic epithelial apoptosis in carcinogen-treated rats. *Carcinogenesis* **33:** 197–202.

10. Gibson, P.R., I. Moeller, O. Kagelari, *et al.* 1992. Contrasting effects of butyrate on the expression of phenotypic markers of differentiation in neoplastic and non-neoplastic colonic epithelial cells in vitro. *J. Gastroenterol. Hepatol.* **7:** 165–172.

11. Hague, A. & C. Paraskeva. 1995. The short-chain fatty acid butyrate induces apoptosis in colorectal tumor cell lines. *Eur. J. Cancer Prev.* **4:** 359–364.

12. Siavoshian, S., J.P. Segain, M. Kornprobst, *et al.* 2000. Butyrate and trichostatin A effects on the proliferation/differentiation of human intestinal epithelial cells: induction of cyclin D3 and p21 expression. *Gut.* **46:** 507–514.

13. Comalada, M., E. Bailon, O. de Haro, *et al.* 2006. The effects of short-chain fatty acids on colon epithelial proliferation and survival depend on the cellular phenotype. *J. Cancer Res. Clin. Oncol.* **132:** 487–497.

14. Binder, H.J. 2010. Role of colonic short-chain fatty acid transport in diarrhea. *Annu. Rev. Physiol.* **72:** 297–313.

15. Binder, H.J. & P. Mehta. 1989. Short-chain fatty acids stimulate active sodium and chloride absorption in vitro in the rat distal colon. *Gastroenterology* **96:** 989–996.

16. Mariadason, J.M., D.H. Barkla & P.R. Gibson. 1997. Effect of short-chain fatty acids on paracellular permeability in Caco-2 intestinal epithelium model. *Am. J. Physiol.* **272:** 705–712.

17. Peng, L., Z.R. Li, R.S. Green, *et al.* 2009. Butyrate enhances the intestinal barrier by facilitating tight junction assembly via activation of AMP-activated protein kinase in Caco-2 cell monolayers. *J. Nutr.* **139:** 1619–1625.

18. Lewis, K., F. Lutgendorff, V. Phan, *et al.* 2010. Enhanced translocation of bacteria across metabolically stressed epithelia is reduced by butyrate. *Inflamm. Bowel. Dis.* **16:** 1138–1148.

19. Sutton, J.D., M.S. Dhanoa, S.V. Morant, *et al.* 2003. Rates of production of acetate, propionate, and butyrate in the rumen of lactating dairy cows given normal and low-roughage diets. *J. Dairy Sci.* **86:** 3620–3633.

20. Penner, G.B., L.L. Guan & M. Oba. 2009. Effects of feeding Fermenten on ruminal fermentation in lactating Holstein cows fed two dietary sugar concentrations. *J. Dairy Sci.* **92:** 1725–1733.

21. Penner, G.B. & M. Oba. 2009. Increasing dietary sugar concentration may improve dry matter intake, ruminal fermentation, and productivity of dairy cows in the postpartum phase of the transition period. *J. Dairy Sci.* **92:** 3341–3353.

22. Owens, F.N., D.S. Secrist, W.J. Hill & D.R. Gill. 1998. Acidosis in cattle: a review. *J. Anim. Sci.* **76:** 275–286.

23. Pryde, S.E., S.H. Duncan, G.L. Hold, *et al.* 2002. The microbiology of butyrate formation in the human colon. *FEMS Microbiol. Lett.* **217:** 133–139.

24. Diez-Gonzalez, F., D.R. Bond, E. Jennings & J.B. Russell. 1999. Alternative schemes of butyrate production in Butyrivibrio fibrisolvens and their relationship to acetate utilization, lactate production, and phylogeny. *Arch. Microbiol.* **171:** 324–330.

25. Louis, P. & H.J. Flint. 2009. Diversity, metabolism and microbial ecology of butyrate-producing bacteria from the human large intestine. *FEMS Microbiol. Lett.* **294:** 1–8.

26. Kopecny, J., M. Zorec, J. Mrazek, *et al.* 2003. Butyrivibrio hungatei sp. nov. and Pseudobutyrivibrio xylanivorans sp. nov., butyrate-producing bacteria from the rumen. *Int. J. Syst. Evol. Microbiol.* **53:** 201–209.

27. Louis, P., P. Young, G. Holtrop & H.J. Flint. 2010. Diversity of human colonic butyrate-producing bacteria revealed by analysis of the butyryl-CoA: acetate CoA-transferase gene. *Environ. Microbiol.* **12:** 304–314.

28. Morvay, Y., A. Bannink, J. France, *et al.* 2011. Evaluation of models to predict the stoichiometry of volatile fatty acid

profiles in rumen fluid of lactating Holstein cows. *J. Dairy Sci.* **94:** 3063–3080.

29. DeFrain, J.M., A.R. Hippen, K.F. Kalscheur & D.J. Schingoethe. 2004. Feeding lactose increases ruminal butyrate and plasma beta-hydroxybutyrate in lactating dairy cows. *J. Dairy Sci.* **87:** 2486–2494.

30. DeFrain, J.M., A.R. Hippen, K.F. Kalscheur & D.J. Schingoethe. 2006. Feeding lactose to increase ruminal butyrate and the metabolic status of transition dairy cows. *J. Dairy Sci.* **89:** 267–276.

31. Rehman, H., J. Bohm & J. Zentek. 2008. Effects of differentially fermentable carbohydrates on the microbial fermentation profile of the gastrointestinal tract of broilers. *J. Anim. Physiol. Anim. Nutr. (Berl)* **92:** 471–480.

32. Metzler-Zebeli, B.U., R.T. Zijlstra, R. Mosenthin & M.G. Ganzle. 2011. Dietary calcium phosphate content and oat beta-glucan influence gastrointestinal microbiota, butyrate-producing bacteria and butyrate fermentation in weaned pigs. *FEMS Microbiol. Ecol.* **75:** 402–413.

33. Schmitz, H., C. Barmeyer, M. Fromm, *et al.* 1999. Altered tight junction structure contributes to the impaired epithelial barrier function in ulcerative colitis. *Gastroenterology* **116:** 301–309.

34. Marin, M.L., A.J. Greenstein, S.A. Geller, *et al.* 1983. A freeze fracture study of Crohn's disease of the terminal ileum: changes in epithelial tight junction organization. *Am. J. Gastroenterol.* **78:** 537–547.

35. Zeissig, S., N. Burgel, D. Gunzel, *et al.* 2007. Changes in expression and distribution of claudin 2, 5 and 8 lead to discontinuous tight junctions and barrier dysfunction in active Crohn's disease. *Gut* **56:** 61–72.

36. Steinhart, A.H., A. Brzezinski & J.P. Baker. 1994. Treatment of refractory ulcerative proctosigmoiditis with butyrate enemas. *Am. J. Gastroenterol.* **89:** 179–183.

37. Venkatraman, A., B.S. Ramakrishna, A.B. Pulimood, *et al.* 2000. Increased permeability in dextran sulphate colitis in rats: time course of development and effect of butyrate. *Scand J. Gastroenterol.* **35:** 1053–1059.

38. Inan, M.S., R.J. Rasoulpour, L. Yin, *et al.* 2000. The luminal short-chain fatty acid butyrate modulates NF-kappaB activity in a human colonic epithelial cell line. *Gastroenterology* **118:** 724–734.

39. Nancey, S., J. Bienvenu, B. Coffin, *et al.* 2002. Butyrate strongly inhibits in vitro stimulated release of cytokines in blood. *Dig. Dis. Sci.* **47:** 921–928.

40. Heller, F., P. Florian, C. Bojarski, *et al.* 2005. Interleukin-13 is the key effector Th2 cytokine in ulcerative colitis that affects epithelial tight junctions, apoptosis, and cell restitution. *Gastroenterology* **129:** 550–564.

41. Amasheh, S., N. Meiri, A.H. Gitter, *et al.* 2002. Claudin-2 expression induces cation-selective channels in tight junctions of epithelial cells. *J. Cell. Sci.* **115:** 4969–4976.

42. Daly, K. & S.P. Shirazi-Beechey. 2006. Microarray analysis of butyrate regulated genes in colonic epithelial cells. *DNA Cell Biol.* **25:** 49–62.

43. Thibault, R., F. Blachier, B. Darcy-Vrillon, *et al.* 2010. Butyrate utilization by the colonic mucosa in inflammatory bowel diseases: a transport deficiency. *Inflamm. Bowel. Dis.* **16:** 684–695.

44. Suzuki, T., S. Yoshida & H. Hara. 2008. Physiological concentrations of short-chain fatty acids immediately suppress colonic epithelial permeability. *Br. J. Nutr.* **100:** 297–305.

45. Peng, L., Z. He, W. Chen, *et al.* 2007. Effects of butyrate on intestinal barrier function in a Caco-2 cell monolayer model of intestinal barrier. *Pediatr. Res.* **61:** 37–41.

46. Harten, S.K., D. Shukla, R. Barod, *et al.* 2009. Regulation of renal epithelial tight junctions by the von Hippel-Lindau tumor suppressor gene involves occludin and claudin 1 and is independent of E-cadherin. *Mol. Biol. Cell* **20:** 1089–1101.

47. Bordin, M., F. D'Atri, L. Guillemot & S. Citi. 2004. Histone deacetylase inhibitors upregulate the expression of tight junction proteins. *Mol. Cancer Res.* **2:** 692–701.

48. Boffa, L.C., G. Vidali, R.S. Mann & V.G. Allfrey. 1978. Suppression of histone deacetylation in vivo and in vitro by sodium butyrate. *J. Biol. Chem.* **253:** 3364–3366.

49. Rada-Iglesias, A., S. Enroth, A. Ameur, *et al.* 2007. Butyrate mediates decrease of histone acetylation centered on transcription start sites and down-regulation of associated genes. *Genome Res.* **17:** 708–719.

50. Pan, L., A.F. Matloob, J. Du, *et al.* 2010. Vitamin D stimulates apoptosis in gastric cancer cells in synergy with trichostatin A/sodium butyrate-induced and 5-aza-2'-deoxycytidine-induced PTEN upregulation. *FEBS J.* **277:** 989–999.

51. Gaschott, T., A. Wachtershauser, D. Steinhilber & J. Stein. 2001. 1,25-Dihydroxycholecalciferol enhances butyrate-induced p21(Waf1/Cip1) expression. *Biochem. Biophys. Res. Commun.* **283:** 80–85.

52. Gaschott, T., O. Werz, A. Steinmeyer, *et al.* 2001. Butyrate-induced differentiation of Caco-2 cells is mediated by vitamin D receptor. *Biochem. Biophys. Res. Commun.* **288:** 690–696.

53. Schulzke, J.D., A.H. Gitter, J. Mankertz, *et al.* 2005. Epithelial transport and barrier function in occludin-deficient mice. *Biochim. Biophys. Acta.* **1669:** 34–42.

54. Balda, M.S. & K. Matter. 2009. Tight junctions and the regulation of gene expression. *Biochim. Biophys. Acta.* **1788:** 761–767.

55. Krishnan, M., A.B. Singh, J.J. Smith, *et al.* 2010. HDAC inhibitors regulate claudin-1 expression in colon cancer cells through modulation of mRNA stability. *Oncogene* **29:** 305–312.

56. Plöger, S. 2010. Effect of sodium butyrate on barrier function of HT-29/B6/GR-MR cells in conjunction with down-regulation of claudin-2. *Genes. Nutr.* **5**(Suppl. 1): S28 (abstract)

57. Phillipson, A.T. & R.A. McAnnally. 1942. Studies on the fate of carbohydrates in the rumen of sheep. *J. Exp. Biol.* **19:** 199.

58. Masson, M.J. & A.T. Phillipson. 1951. The absorption of acetate, propionate and butyrate from the rumen of sheep. *J. Physiol.* **113:** 189–206.

59. Lesmeister, K.E. & A.J. Heinrichs. 2004. Effects of corn processing on growth characteristics, rumen development, and rumen parameters in neonatal dairy calves. *J. Dairy Sci.* **87:** 3439–3450.

60. Shen, Z., H.M. Seyfert, B. Lohrke, *et al.* 2004. An energy-rich diet causes rumen papillae proliferation associated with more IGF type 1 receptors and increased plasma IGF-1 concentrations in young goats. *J. Nutr.* **134:** 11–17.

61. Zitnan, R., J. Voigt, U. Schonhusen, *et al.* 1998. Influence of dietary concentrate to forage ratio on the development of rumen mucosa in calves. *Arch. Tierernahr.* **51:** 279–291.

62. Flatt, W.P., R.G. Warner & J.K. Loosli. 1958. Influence of purified materials on the development of the ruminant stomach. *J. Dairy Sci.* **41:** 1593–1600.

63. Mentschel, J., R. Leiser, C. Mulling, *et al.* 2001. Butyric acid stimulates rumen mucosa development in the calf mainly by a reduction of apoptosis. *Arch. Tierernahr.* **55:** 85–102.

64. Sander, E.G., R.G. Warner, H.N. Harrison & J.K. Loosli. 1959. The stimulatory effect of sodium butyrate and sodium propionate on the development of rumen Mucosa in the young calf. *J. Dairy Sci.* **42:** 1600–1605.

65. Baldwin, R.L. 1999. The proliferative actions of insulin, insulin-like growth factor-1, epidermal growth factor, butyrate and propionate on ruminal epithelial cells in vitro. *Small Rumin. Res.* **32:** 261–268.

66. Galfi, P., G. Gäbel & H. Martens. 1993. Influences of extracellular matrix components on the growth and differentiation of ruminal epithelial cells in primary culture. *Res. Vet. Sci.* **54:** 102–109.

67. Gorka, P., Z.M. Kowalski, P. Pietrzak, *et al.* 2011. Effect of method of delivery of sodium butyrate on rumen development in newborn calves. *J. Dairy Sci.* **94:** 5578–5588.

68. Neogrady, S., P. Galfi & F. Kutas. 1989. Effects of butyrate and insulin and their interaction on the DNA synthesis of rumen epithelial cells in culture. *Experientia* **45:** 94–96.

69. Sakata, T., K. Hikosaka, Y. Shiomura & H. Tamate. 1980. Stimulatory effect of insulin on ruminal epithelium cell mitosis in adult sheep. *Br. J. Nutr.* **44:** 325–331.

70. Graham, C. & N.L. Simmons. 2005. Functional organization of the bovine rumen epithelium. *Am. J. Physiol. Regul. Integr. Comp. Physiol.* **288:** 173–181.

71. Abdoun, K., F. Stumpff, I. Rabbani & H. Martens. 2010. Modulation of urea transport across sheep rumen epithelium in vitro by SCFA and CO2. *Am. J. Physiol. Gastrointest. Liver Physiol.* **298:** 190–202.

72. Stumpff, F., M.I. Georgi, L. Mundhenk, *et al.* 2011. Sheep rumen and omasum primary cultures and source epithelia: barrier function aligns with expression of tight junction proteins. *J. Exp. Biol.* **214:** 2871–2882.

73. Schnorr, B. & K.H. Wille. 1972. Zonulae occludentes in the goat ruminal epithelium. *Z. Zellforsch Mikrosk. Anat.* **124:** 39–43.

74. Scott, A., I.C. Gardner, D.R. Fulton & G.B. McInroy. 1972. Tight junctions in the stratum basale of ruminal epithelium. *Z. Zellforsch. Mikrosk. Anat.* **131:** 199–203.

75. Dobson, A. & A.T. Phillipson. 1958. The absorption of chloride ions from the retie-ulo-rumen sac. *J. Physiol.* **140:** 94–104.

76. Sellers, A.F. & A. Dobson. 1960. Studies on reticulo-rumen sodium and potassium concentration and electrical potentials in sheep. *Res. Vet. Sci.* **1:** 95–102.

77. Sperber, I. & S. Hyden. 1952. Transport of chloride through the ruminal mucosa. *Nature* **169:** 587.

78. Lodemann, U. & H. Martens. 2006. Effects of diet and osmotic pressure on Na+ transport and tissue conductance of sheep isolated rumen epithelium. *Exp. Physiol.* **91:** 539–550.

79. Uppal, S.K., K. Wolf, S.S. Khahra & H. Martens. 2003. Modulation of Na +transport across isolated rumen epithelium by short-chain fatty acids in hay- and concentrate-fed sheep. *J. Anim. Physiol. Anim. Nutr. (Berl)* **87:** 380–388.

80. Penner, G.B., M.A. Steele, J.R. Aschenbach & B.W. McBride. 2011. Ruminant Nutrition Symposium: molecular adaptation of ruminal epithelia to highly fermentable diets. *J. Anim. Sci.* **89:** 1108–1119.

81. Gäbel, G., J.R. Aschenbach & F. Müller. 2002. Transfer of energy substrates across the ruminal epithelium: implications and limitations. *Anim. Health Res. Rev.* **3:** 15–30.

82. Rabbani, I., C. Siegling-Vlitakis, B. Noci & H. Martens. 2011. Evidence for NHE3-mediated Na transport in sheep and bovine forestomach. *Am. J. Physiol. Regul. Integr. Comp. Physiol.* **301:** 313–319.

83. Weigand, E., J.W. Young & A.D. McGilliard. 1972. Extent of butyrate metabolism by bovine ruminoreticulum epithelium and the relationship to absorption rate1. *J. Dairy. Sci.* **55:** 589–597.

84. Krehbiel, C.R., D.L. Harmon & J.E. Schneider. 1992. Effect of increasing ruminal butyrate on portal and hepatic nutrient flux in steers. *J. Anim. Sci.* **70:** 904–914.

85. Kleen, J.L., G.A. Hooijer, J. Rehage & J.P. Noordhuizen. 2003. Subacute ruminal acidosis (SARA): a review. *J. Vet. Med. A. Physiol. Pathol. Clin. Med.* **50:** 406–414.

86. Plaizier, J.C., D.O. Krause, G.N. Gozho & B.W. McBride. 2008. Subacute ruminal acidosis in dairy cows: the physiological causes, incidence and consequences. *Vet. J.* **176:** 21–31.

87. Steele, M.A., J. Croom, M. Kahler, *et al.* 2011. Bovine rumen epithelium undergoes rapid structural adaptations during grain-induced subacute ruminal acidosis. *Am. J. Physiol. Regul. Integr. Comp. Physiol.* **300:** 1515–1523.

88. Gilliland, R.L., L.J. Bush & J.D. Friend. 1962. Relation of ration composition to rumen development in early-weaned dairy calves with observations on ruminal parakeratosis1. *J. Dairy Sci.* **45:** 1211–1217.

89. Hinders, R.G. & F.G. Owen. 1965. Relation of Ruminal parakeratosis development to volatile fatty acid absorption1. *J. Dairy Sci.* **48:** 1069–1073.

Ann. N.Y. Acad. Sci. ISSN 0077-8923

ANNALS OF THE NEW YORK ACADEMY OF SCIENCES

Issue: *Barriers and Channels Formed by Tight Junction Proteins*

SUMOylation of claudin-2

Christina M. Van Itallie,[1] Laura L. Mitic,[2] and James M. Anderson[1]

[1]National Heart Lung and Blood Institute, National Institutes of Health, Bethesda, Maryland. [2]University of California at San Francisco, San Francisco, California

Address for correspondence: Christina M. Van Itallie, National Institutes of Health, NHLBI, Bldg 50, Rm 4525 Bethesda, MD 20892. Christina.VanItallie@nih.gov

The *C*-terminal cytoplasmic tails of claudins are likely sites for interaction with proteins that regulate their function. We performed a yeast two-hybrid screen with the tail of human claudin-2 against a human kidney cDNA library and identified interactions with the PDZ3 domain of ZO-2 as well as ubiquitin-conjugating enzyme E2I (SUMO ligase-1) and E3 SUMO-protein ligase PIAS; the first is a predicted interaction, while the latter two are novel and suggest that claudin-2 is a substrate for SUMOylation. Using an *in vitro* SUMOylation assay, we identified K218 as a conjugation site on claudin-2; mutation of that lysine to arginine blocked SUMOylation. Stable expression of inducible GFP-SUMO-1 in MDCK cells resulted in decreased levels of claudin-2 protein by immunoblot and decreased claudin-2 membrane expression by immunofluorescence microscopy. We conclude that the cellular levels of claudin-2 may be modulated by SUMOylation, warranting further investigation of cellular pathways that regulate this modification *in vivo*.

Keywords: claudin; claudin-2; MDCK; SUMO-1; tight junction; yeast two-hybrid

Introduction

Claudins form a large family of tetraspanning membrane proteins that create the variable permselective barrier properties of tight junctions.[1] It is presumed that their function, including barrier properties, assembly, trafficking, and half-lives, might be regulated by posttranslational modifications or through binding other proteins to their C-terminal cytoplasmic sequences. Currently, it is known that most claudins bind to the PDZ domains of the ZO-1, -2, and -3 MAGUK proteins through PDZ binding motifs on their extreme C-termini,[2] and several claudins are known to be phosphorylated or palmitoylated with functional consequences.[3,4] Claudin-1 provides the single example of covalent modification by ubiquitin, a modification that enhances delivery to and destruction by the proteosome.[5,6] We performed yeast two-hybrid (Y2H) screening with the C-terminal tail of claudin-2 in an effort to identify novel binding proteins, which could provide further insight into the regulation of claudin function. Our results demonstrate that claudin-2 has the ca-

pacity for modification on lysine-218 by SUMO-1 (small ubiquitin-like modifier-1), a modification-like ubiquitination that in other proteins is known to regulate a range of protein functions, including protein–protein interactions, subcellular localization, and trafficking.[7–12] Like ubiquitination, the covalent attachment of SUMO proteins to lysine residues on target proteins requires a series of conjugation factors that recognize sequences surrounding the target lysine;[13–19] interaction with SUMO ligases in Y2H screen provided the initial indication that numerous proteins were SUMOylated.

Materials and methods

Y2H screening was performed as previously described in our laboratory using the L40 yeast strain as described in Niethammer *et al.*[20,21] The bait consisted of the cytoplasmic C-terminal sequence of human claudin-2 (residues 185–230) subcloned in frame with the lexA DNA-binding domain into vector pBHA5. The bait was used to screen a human kidney cDNA library constructed in pGAD5 (Clontech Laboratories, Inc., Mountain View, CA). DNA

doi: 10.1111/j.1749-6632.2012.06541.x

 Ann. N.Y. Acad. Sci. 1258 (2012) 60–64 © 2012 New York Academy of Sciences.

from positive interacting clones, as assayed by beta-galactosidase staining, was rescreened against the claudin-2–containing vector in binary assays and autoactivators eliminated by unitary transformation assays.

For *in vitro* SUMOylation assays, the carboxyl-terminal tail of canine claudin-2 (amino acids 189–230) was amplified from MDCK II cell mRNA and cloned into pCR TOPO. The sequence was verified and the insert subcloned into pGEX4T (GE Healthcare, Port Washington, NY). For use in some studies, K218 was mutated to arginine by site-directed mutagenesis (Quik Change Site-Directed Mutagenesis Kit, Agilent Technologies, Santa Clara, CA); sequences were verified and proteins expressed in *Escherichia coli*, as described previously by our laboratory.[22] *In vitro* SUMOylation was performed using a Active Motif SUMOlink SUMO-1 kit (Active Motif, Carlsbad, CA); this kit includes a mutant SUMO-1 that lacks SUMOylation function.

GFP-SUMO-1 was constructed from a human SUMO-1 cDNA (Open Biosystems, Lafayette, CO) cloned into the pTRE vector (Clontech) that had been modified to include EGFP-human SUMO-1 was cloned downstream of the EGFP-coding region. This vector was cotransfected into MDCK II Tet-off cells (Clontech) with pSVZeo; stable cell lines were selected with 1 mg/mL Zeocin (InVivogen, San Diego, CA). Transfected cells were maintained without GFP-SUMO-1 expression by addition of 50 ng/mL doxycycline and protein induction was performed by removal of doxycycline from media. GFP-SUMO-1 expression was verified by fluorescence microscopy and immunoblot analysis. MDCK cell culture, protein induction, immunoblots, and immunofluorescence microscopy were performed as described elsewhere.[23] All antibodies were purchased from Invitrogen (Carlsbad, CA).

Results and discussion

Y2H screening
A human kidney cDNA library in the pGAD10 Y2H vector was screened using a sequence encoding the entire 45 residue C-terminal cytoplasmic tail sequence of human claudin-2 in pBHA5. Of 99 positive sequences, 65 encoded ubiquitin-conjugating enzyme E2I (also called SUMO-1 ligase) (Genbank NM˙003345), 17 encoded the E3 SUMO ligase protein inhibitor of activated STAT-2 (PIAS2, Genbank NM˙173206), 5 included the third PDZ domain of

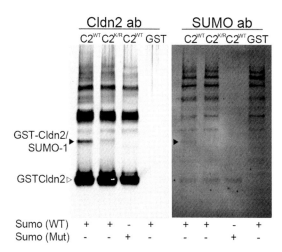

Figure 1. *In vitro* assay reveals that claudin-2 can be SUMOylated on K218. Wild-type (lanes 1 and 3) and mutant K218R (lane 2) claudin-2 tails were expressed as GST fusion proteins and purified by glutathione-affinity chromatography; GST alone was used as a nonspecific control (lane 4). GST proteins were mixed with SUMOylation reagents according to the manufacturer's directions; SUMO-1 mutant that cannot be complexed was used as a negative control (lane 3). Immunoblot with claudin-2 antibody (left immunoblot) reveals that GST-claudin-2 (≈30 kDa) and a single unique band at approximately 42 kDa representing claudin-2/SUMO-1; this band is also immunoreactive with a SUMO-1 antibody (right immunoblot). This 42 kDa band is not present in the reaction mixture containing either claudin-2 K218R or the mutant SUMO-1. Other bands are present in all lanes and thus are unlikely to represent claudin-2/-1 forms. These are the same blot probed (at the same time) with anti-mouse claudin-2 primary antibody and IR700 antimouse secondary antibody and SUMO-1 rabbit primary antibody and IR800 antirabbit secondary antibody. The secondary antibodies do not cross react, but the faint staining at the site of the GST-cldn2 signal with the SUMO-1 primary antibody is probably a nonspecific protein: protein interaction due to the large amount of fusion protein.

ZO-2, and the other 12 were out of coding frame. An interaction with the PDZ3 domain of ZO-2 has not been reported but is not unexpected, since the PDZ1 and PDZ2 domains of ZO-1 and ZO-2 are known to bind PDZ motifs of claudins;[2] however, those with proteins involved in SUMOylation was unexpected and novel. Regardless of length, all E3 SUMO ligase PIAS2 clones included sequences encoding the SP-RING domain.[8] This domain is required for binding to both the SUMO donors like UbcE2I and the acceptor recognition region on the targets for E3 SUMO ligase, suggesting that the tail of claudin-2 might be a substrate for the SUMO ligases. Further, as expected, the C-terminal three residues of claudin-2 were required in Y2H assays for interaction with ZO-2 but not to bind UbcE2I or SUM0-ligase PIAS2. Of note, claudin-2 is the only claudin with a potential lysine acceptor residue (K218) within a SUMOylation recognition motif (ΨKXE) at VKSEFNSYSLTGYV, although it is possible that other claudins could be SUMOylated on cytoplasmic lysines. Lysine 218 is positioned 13 residues preceding the C-terminal PDZ binding motif suggesting the possibility that conjugation of a SUMO protein at K218 might sterically inhibit binding to ZO proteins; however, this was not tested.

Claudin-2 can be SUMOylated on K218

Next we tested whether the tail of claudin-2 purified as GST-fusion protein from *E. coli* was a substrate for SUMO-1 modification in a standard *in vitro* test assay and whether the predicted acceptor residue at K218 was the site of conjugation. The latter was tested by mutating K218 to arginine, which lacks the epsilon-amino conjugation nitrogen. Immunoblot analysis of purified GST-claudin-2 tail after incubation in the SUMOylation assay (including E1 activating enzyme, E2 conjugating enzyme, and SUMO-1 protein) with GST-claudin-2 tail reveals a band at the expected size for GST-claudin-2 (approximately 30 kDa, GSTCldn2) and a higher molecular weight (MW) band corresponding to a GST-claudin-2-SUMO-1 complex (approximately 42 kDa, GSTCldn2). This higher MW band is also detected with a SUMO-1 antibody (Fig. 1, left-most lane of both immunoblots), verifying that it is the conjugated GST-claudin-2 tail. This band is not present in incubation mix containing the mutated GST-claudin-2 K218R tail (Fig. 1, middle lanes) or in incubation mix containing wild-type GST-claudin-2 tail but with a mutated control SUMO-1 that cannot be conjugated (Fig. 1, third lanes). These data demonstrate that *in vitro*, K218 can be SUMOylated.

Expression of SUMO-1 in MDCK cells reduces the level of claudin-2

We were unable to detect baseline SUMOlyation of claudin-2 in cultured MDCK epithelial monolayers by immunoblotting of cell lysates or after immunoprecipitating of claudin-2. In the absence of knowledge about how to stimulate physiologic conjugation, we overexpressed SUMO-1 to drive conjugation of all substrates and assayed for po-

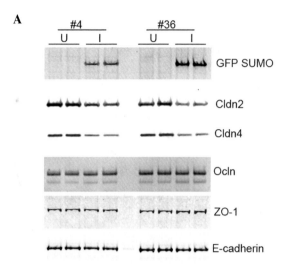

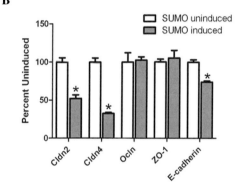

Figure 2. Immunoblot analysis of MDCK cells stably expressing GFP-SUMO-1. Two separate MDCK cell lines were not induced (U) or induced (I) to express GFP-SUMO-1 for seven days; wells were plated in duplicate. Cells were processed for immunoblot analysis (A); both clones expressed GFP-SUMO-1 only when induced. Both claudin-2 and claudin-4 levels were decreased in cells expressing GFP-SUMO-1 compared with uninduced cells while there was no change in occludin or ZO-1 protein levels and only a small change in E-cadherin expression. The changes in expression levels (average of the two clones) are quantified in (B) (mean ± SEM), revealing a 50% decrease in protein expression for claudin-2 and -4 and a 20% decrease in E-cadherin levels, $^*P < 0.05$ by untailed Student's *t*-test.

tential changes in the levels or location of claudin-2. SUMO-1 was expressed in a tet-inducible system fused to GFP so that changes in claudin-2 could be detected in the same clone before and after induction and so that the expression of SUMO-1 could be detected by fluorescence microscopy. Induction of GFP-SUMO-1 in two separate clones of MDCK II tet-off cells (immunoblot, Fig. 2A; quantified in Fig. 2B) resulted in significant decreases in the levels

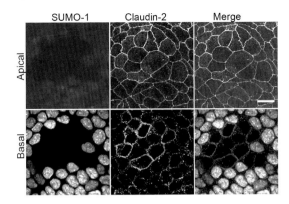

Figure 3. Confocal microscopy analysis of a cocultured mixture of nontransfected MDCK cells and cells expressing GFP-SUMO-1 reveals that GFP-SUMO-1 is concentrated in nuclei (left panel). Cells were imaged below the TJ at the level of the mid-lateral members and revealed that claudin-2 staining is reduced in cells expressing GFP-SUMO-1, middle panel and merge, right panel. Bar = 10 μm.

of claudin-2 (to 50%), claudin-4 (to 35%), and to a lesser extent E-cadherin, but had no effect on the levels of ZO-1, ZO-2, or occludin.

Changes in claudin-2 were also detected by immunofluorescence microscopy after induction of SUMO-1. A cocultured mixture of untransfected MDCK II cell and GFP-SUMO-1–expressing cells reveals that cells expressing GFP-SUMO-1 have reduced levels of claudin-2 on their lateral membranes (Fig. 3). The inverse correlation between induced expression of SUMO-1 and decreased lateral claudin-2, was obvious in two different GFP-SUMO-1–expressing cell lines. In contrast, immunofluorescent claudin-2 expression at the level of the tight junction was apparently identical in all cells, regardless of GFP-SUMO-1 expression (not shown). This observation leads to speculation that SUMOylation may target lateral but not tight junction claudin-2 for removal and degradation, but this conclusion would require verification.

The mechanistic relationship between expression of GFP-SUMO-1 and the changes in claudin-2 is unclear. Although claudin-2 can be SUMOylated *in vitro*, we have so far been unable to demonstrate *in vivo* SUMOylation. However, the correlation between increased GFP-SUMO-1 and decreased claudin-2 levels, demonstrated both by immunoblotting and immunofluorescence microscopy, suggests that SUMO-1 acts to regulate claudin-2 level. This could occur through altered

trafficking or endocytosis and degradation, either by direct tagging of claudin-2 or by altering an indirect pathway, which alters claudin-2 levels. The observation that claudin-4 levels (and cadherin levels) are reduced in GFP-SUMO-1–expressing cells is likely an indirect effect, since claudin-4 does not contain a consensus peri-lysine sequence for SUMOylation. However, the lack of effect of GFP-SUMO-1 expression on other tight junction proteins argues that the effects on claudin levels are not due to a global increase in degradation, but may represent a specific, physiologically relevant regulatory mechanism. There is considerable interest in how cells might regulate tight junction barrier properties through differential regulation of specific claudin levels. There is considerable knowledge about how claudins are differentially regulated at a transcriptional level;[24] however, regulation at a posttranscriptional level by SUMO-1 conjugation is novel and deserves further study.

Conflicts of interest

The authors declare no conflicts of interest.

References

1. Furuse, M. & S. Tsukita. 2006. Claudins in occluding junctions of humans and flies. *Trends Cell Biol.* **16:** 181–188
2. Itoh, M. *et al.* 1999. Direct binding of three tight junction-associated MAGUKs, ZO-1, ZO-2, and ZO-3, with the COOH termini of claudins. *J. Cell Biol.* **147:** 1351–1363.
3. Koval, M. 2009. Tight junctions, but not too tight: fine control of lung permeability by claudins. *Am. J. Physiol. Lung Cell Mol. Physio.* **297:** L217–L218.
4. Van Itallie, C. *et al.* 2005. Palmitoylation of claudins is required for efficient tight-junction localization. *J. Cell Sci.* **118:** 1427–1436.
5. Takahashi, S. *et al.* 2009. The E3 ubiquitin ligase LNX1p80 promotes the removal of claudins from tight junctions in MDCK cells. *J. Cell Sci.* **122:** 985–994.
6. Dukes, J.D. *et al.* 2011. Functional ESCRT machinery is required for constitutive recycling of claudin-1 and maintenance of polarity in vertebrate epithelial cells. *Mol. Biol. Cell* **22:** 3192–3205.
7. Anckar, J. & L. Sistonen. 2007. SUMO: getting it on. *Biochem. Soc. Trans.* **35:** 1409–1413.
8. Geiss-Friedlander, R. & F. Melchior. Concepts in sumoylation: a decade on. *Nat. Rev. Mol. Cell Biol.* **8:** 947–956.
9. Mukhopadhyay, D. & M. Dasso. 2007. Modification in reverse: the SUMO proteases. *Trends Biochem. Sci.* **32:** 286–295.
10. Ulrich, H.D. 2008. The fast-growing business of SUMO chains. *Mol. Cell* **32:** 301–305.
11. Yeh, E.T. 2009. SUMOylation and De-SUMOylation: wrestling with life's processes. *J. Biol. Chem.* **284:** 8223–8227.

12. Zhao, J. 2007. Sumoylation regulates diverse biological processes. *Cell Mol. Life Sci.* **64:** 3017–3033.

13. Desterro, J.M. *et al.* 1997. Ubch9 conjugates SUMO but not ubiquitin. *FEBS Lett.* **417:** 297–300.

14. Johnson, E.S. & G. Blobel. 1997. Ubc9p is the conjugating enzyme for the ubiquitin-like protein Smt3p. *J. Biol. Chem.* **272:** 26799-26802.

15. Sampson, D.A. *et al.* 2001. The small ubiquitin-like modifier-1 (SUMO-1) consensus sequence mediates Ubc9 binding and is essential for SUMO-1 modification. *J. Biol. Chem.* **276:** 21664–21669.

16. Johnson, E.S. & A.A. Gupta. 2001. An E3-like factor that promotes SUMO conjugation to the yeast septins. *Cell* **106:** 735–744.

17. Kagey, M.H. *et al.* 2003. The polycomb protein Pc2 is a SUMO E3. *Cell* **113:** 127–137

18. Kahyo, T. *et al.* 2001. Involvement of PIAS1 in the sumoylation of tumor suppressor p53. *Mol. Cell* **8:** 713–718.

19. Pichler, A. & F. Melchior. 2002. Ubiquitin-related modifier SUMO1 and nucleocytoplasmic transport. *Traffic* **3:** 381–387.

20. Cohen, A.R. *et al.* 1998. Human CASK/LIN-2 binds syndecan-2 and protein 4.1 and localizes to the basolateral membrane of epithelial cells. *J. Cell Biol.* **142:** 129–138.

21. Niethammer, M., E. Kim & M. Sheng. 1996. Interaction between the C terminus of NMDA receptor subunits and multiple members of the PSD-95 family of membrane-associated guanylate kinases. *J. Neurosci.* **16:** 2157–2163.

22. Fanning, A.S. *et al.* 2007. The unique-5 and -6 motifs of ZO-1 regulate tight junction strand localization and scaffolding properties. *Mol. Biol. Cell* **18:** 721–731.

23. Van Itallie, C.M. *et al.* 2009. ZO-1 stabilizes the tight junction solute barrier through coupling to the perijunctional cytoskeleton. *Mol. Biol. Cell* **20:** 3930–3940.

24. Van Itallie, C.M. & J.M. Anderson. 2006. Claudins and epithelial paracellular transport. *Annu. Rev. Physiol.* **68:** 403–429.

Ann. N.Y. Acad. Sci. ISSN 0077-8923

ANNALS OF THE NEW YORK ACADEMY OF SCIENCES

Issue: *Barriers and Channels Formed by Tight Junction Proteins*

Proof of concept for claudin-targeted drug development

Hidehiko Suzuki, Masuo Kondoh, Azusa Takahashi, and Kiyohito Yagi

Laboratory of Bio-Functional Molecular Chemistry, Graduate School of Pharmaceutical Sciences, Osaka University, Suita, Japan

Address for correspondence: Masuo Kondoh, Ph.D., Laboratory of Bio-Functional Molecular Chemistry, Graduate School of Pharmaceutical Sciences, Osaka University, Suita, Osaka 565-0871, Japan. masuo@phs.osaka-u.ac.jp

Claudins (CLs) are a family of tetra-integral membrane proteins that are a key structural and functional component of tight junctions. CLs are overexpressed in some malignant tumors. Claudin-4 is highly expressed in the epithelial cells covering mucosal immune tissues. CLs may therefore be a potential target for improving drug absorption, treating cancer, and developing mucosal vaccine. Research using *Clostridium perfringens* enterotoxin has resulted in proofs of concept for CL-targeted drug development. A platform for the creation of CL binders, such as immunization of CL and preparation of CL proteins, is now beginning to be established.

Keywords: *Clostridium perfringens* enterotoxin; claudin; mucosal absorption; cancer-targeting; vaccination

Introduction

Epithelium plays a pivotal role in separating the inside and the outside of the body. Drugs must pass across the epithelium to be absorbed; most malignant tumors derive from epithelium, and most pathogens invade the body via the epithelium. Epithelium is therefore a potent target for improving drug absorption, treating cancer, and curing infectious diseases.

Tight junctions (TJs) are intercellular sealing complexes between epithelial cells. TJs prevent the free movement of solutes through the intercellular spaces.[1] Modulation of TJ seals is a popular strategy for improving drug absorption. TJs compartmentalize the apical and basal membrane domains of epithelial cells, leading to the formation of cellular polarity. Loss of cell–cell interaction and cellular polarity, which often occurs in cancer cells during carcinogenesis, leads to exposure of TJ components on the cellular surface.[2] Therefore, TJ components could be potential targets for TJ-targeted drug development.

TJs are composed of biochemical complexes including occludin, claudins (CLs), and junctional adhesion molecules.[3] Treatment with a CL-binding molecule enhanced jejunal absorption of dextran (MW: 4 kDa) 400-fold more than did a currently used clinical absorption enhancer.[4] Expression of CLs is deregulated in malignant tumors.[5,6] CL-expressing epithelium covers mucosal immune tissues.[7] Therefore, CL, a member of a family of tetra-transmembrane proteins, is a promising candidate for TJ-targeted drug development. However, the low immunogenicity and hydrophobicity of CLs make it difficult to determine their structures and prepare CL-binding reagents, including antibodies, thereby delaying CL-targeted drug development.

Claudin-3 (CL-3) and CL-4 are receptors for *Clostridium perfringens* enterotoxin (CPE),[8,9] which causes the symptoms associated with *C. perfringens* type A food poisoning in humans.[10] Half of the C-terminal fragment of CPE (C-CPE) is a receptor-binding domain, and C-CPE is the most frequently used CL-binding molecule.[10] Proof of concept for CL-targeted drug development has been established by using C-CPE as a model CL binder.

Proof of concept for CL-targeted mucosal absorption

Modulation of the epithelial barrier for mucosal absorption of drugs was first explored for TJ-targeted drug development approximately a half

doi: 10.1111/j.1749-6632.2012.06503.x

century ago.[11] Due to the identification of a mucosal absorption-enhancing magnesium and calcium chelator EDTA, modulation of TJs has been a popular strategy for improving drug absorption.[12] However, loosening TJs increases the risk of inducing the nonspecific influx of solutes other than the intended drug. Most researchers therefore think that clinical application is difficult or impossible.

Epithelial cells maintain the intratissue environment not only by preventing the nonspecific influx of solutes through the paracellular space, but also by permitting selective paracellular transport of solutes.[13] Control of the epithelial transport system may reduce the nonspecific influx of solutes during paracellular drug delivery.

There are at least 27 members of the CL family, whose TJ-sealing properties differ depending on the tissues. CL-1 deficiency results in disruption of the epidermal barrier, leading to leakage of small molecules (<600 Da).[14] CL-5–deficient mice exhibited a disrupted blood–brain barrier, which resulted in the influx of solutes (~800 Da).[15] Modulation of CL was proposed as a novel transepithelial drug delivery system by Tsukita and Furuse in the first paper identifying CLs.[16]

C-CPE, a CL-3– and CL-4–binding molecule, is a modulator of TJs.[10] Treatment of jejunum with C-CPE dose dependently enhanced jejunal absorption of dextran (MW: 4 kDa). The absorption-enhancing effect was 400-fold greater than that of sodium caprate, a mucosal absorption enhancer used clinically.[4] Deletion of the CL-binding domains of C-CPE attenuated the absorption-enhancing effect. Sodium caprate also enhanced absorption in the colon, whereas C-CPE did not. C-CPE did not result in any histological injury.[4] Therefore, CL may be a potent target for the development of tissue-specific paracellular transport.

Thirty percent of newly developed drugs are biologics, such as peptides, proteins, and nucleic acids. Because of their low permeability in epithelium, most biologics must be injected into patients. C-CPE has also been shown to enhance nasal, pulmonary, and jejunal absorption of the biologic human parathyroid hormone derivative hPTH(1–34).[17] Taken together, these findings indicate that CLs may be a promising target for improving mucosal absorption of drugs.

Development of a broadly specific CL-binding molecule

CLs form homo- and hetero-type strands on the lateral membrane of the adjacent epithelial cells. Adjacent strands associate with each other resulting in the formation of TJs.[18] The combination of CLs determines the tissue- and solute-specificity of TJs.[14,18] Using CL-binding agents with optimal CL specificity for particular applications likely will reduce the influx of solutes other than the intended drug.

Although CL proteins are needed to screen CL-binding agents, it is difficult to prepare recombinant CL proteins. To our knowledge, CL-4 is the only CL whose recombinant protein is able to be prepared. Display of protein on a viral surface is one method of membrane protein preparation.[19,20] Budded baculovirus (BV) display large amounts of membrane proteins on the surface in their active form.[21,22] CL-4 is displayed on the BV membrane (Fig. 1).[23] When a mixture of CL-4–bound phages and negative

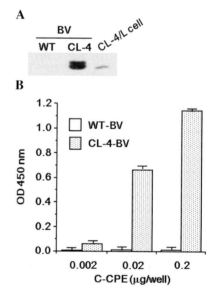

Figure 1. Preparation of CL-4–displaying BV. (A) Western blot analysis. Wild-type BV (WT-BV) and CL-4–BV (0.1 µg/lane) were subjected to SDS-PAGE followed by immunoblot analysis with anti–CL-4 antibody. The lysate of CL-4–expressing L (CL-4/L) cells was used as the positive control. (B) Interaction of a CL-4 binder with CL-4–BV. Immunotubes were coated with WT-BV or CL-4–BV, and C-CPE was added to the BV-coated immunotubes at the indicated concentration. C-CPE bound to the BV-coated tubes was detected by using ELISA with an anti-his-tag antibody. Data are means ± SD (*n* = 3).

control phages is added to the tube coated with CL-4–displaying BV, CL-bound phages were enriched (data not shown). Therefore, CL-displaying BV may be a screening reagent for obtaining candidate CL-binding molecules.

Functional domain mapping of C-CPE revealed that the 16 amino acids of the C-terminal are critical for the binding of C-CPE to CL-4 and that alanine-substituted C-CPE mutants showed either increased or decreased affinity to CL-4.[24,25] We prepared a C-CPE mutant library in which the affinity-increasing residues were randomly replaced by 20 amino acids. CL-1–binding mutants were selected by using CL-1–displaying BV. Interestingly, one of the CL-1–binding mutants also bound to CL-2, CL-4, and CL-5 (data not shown). The broadly specific CL binder was a more potent TJ modulator and jejunal absorption enhancer than was C-CPE (data not shown).

CL-targeted cancer therapy

Abnormal expression of CLs often occurs in human malignancies, such as breast, prostate, ovarian, gastric, and pancreatic cancers.[6,26] Moreover, deregulation of CL expression is associated with metastasis.[27,28] Intratumoral administration of a CL-targeting toxin, CPE, suppresses tumor growth in mice inoculated with CL-4–expressing human pancreatic cancer cell line Panc-1.[26,29] Intracranial administration of CPE inhibited tumor growth in murine brain metastasis model using the human breast cancer cell line MDA-MB-468 or the murine breast cancer cell line NT2.5 without any apparent local or systemic toxicity.[5] Therefore, CL may be a potential target for cancer therapy.

We prepared C-CPE–fused protein synthesis inhibitory factor (PSIF) derived from *Pseudomonas aeruginosa* exotoxin A and found that C-CPE–PSIF is a CL-4–targeting molecule.[30] Intratumoral administration of C-CPE–PSIF suppressed tumor growth in murine breast cancer cell line 4T1.[31] Injection of C-CPE–PSIF also inhibited tumor growth and spontaneous lung metastasis of murine breast cancer without any apparent side effects.[32] CL-3 and/or CL-4 is expressed in various tissues, such as lung, intestine, liver, and kidney. Most CLs in normal cells are contained within the TJ complexes, whereas localization of CLs is deregulated in some cancers.[6,26] The human colon carcinoma cell line Caco-2 forms a polarized cell monolayer with

well-developed TJs when cells reach confluence. This cell line is often used as a model of polarized normal epithelial cell sheets.[33] The polarized cells are more resistant to C-CPE–PSIF treatment than are nonpolarized cells (Fig. 2). C-CPE–PSIF was more cytotoxic to membrane-seeded polarized cells that were treated from the apical side than the basal side.[31] Therefore, CL-targeting molecules may recognize deregulation of cellular polarity, which could lead to fewer side effects.

Proof of concept for CL-targeted mucosal vaccination

Approximately 20 million people die every year from infectious diseases. Most pathogens invade the body through the mucosal epithelium. A standard strategy for overcoming infectious diseases is the prevention of entry of pathogens through the epithelium and the elimination of infected cells. Vaccination strategies are classified as either parenteral or

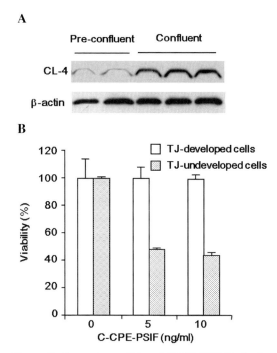

Figure 2. Cytotoxic specificity of C-CPE–PSIF. To develop tight TJs, Caco-2 monolayer cells were grown at confluency for three days. To maintain loose TJs, Caco-2 cells were seeded at subconfluency. The cell lysates were subjected to SDS-PAGE followed by Western blotting with CL-4 antibody (A). The cells were treated with the indicated concentrations of C-CPE–PSIF for 48 h, and then cell viability was measured by WST-8 assay (B). Data are means ± SD ($n = 3$).

mucosal. Parenteral vaccinations activate systemic immune responses, leading to elimination of infected cells. Mucosal vaccinations activate both the systemic and mucosal immune responses, leading to both prevention of the entry of pathogens through the mucosal epithelium and elimination of infected cells.[34–36] Mucosal administration is generally considered the least painful route of administration. Mucosal vaccination is an ideal vaccination strategy; however, mucosal administration of antigen proteins alone often does not activate immune responses.

There are mucosa-associated lymphoid tissues (MALTs) in the intestine, nose, and lungs, as follows: gut-associated lymphoid tissues (GALTs), nasopharynx-associated lymphoid tissues (NALTs), and bronchus-associated lymphoid tissues (BALTs), respectively.[37,38] Antigen-presenting cells, B cells, and T cells, underlie MALTs. Therefore, efficient delivery of antigens into MALTs is one strategy for mucosal vaccination. CL-4 is highly expressed in epithelial cells covering GALTs and NALTs.[7,39] Nasal administration of C-CPE–fused ovalbumin (OVA) induced production of OVA-specific serum IgG and nasal, vaginal, and fecal IgA (Fig. 3, data not shown). A mixture of C-CPE and OVA did not potentiate these immune responses (data not shown). Deletion of the CL-binding domain of OVA–C-CPE attenuated activation of the immune responses. Nasal immunization with OVA–C-CPE showed antitumor activity in mice inoculated with OVA-expressing thymoma cells.[39] Therefore, CL-4–targeting may be a potential strategy for developing mucosal vaccination. C-CPE is a fragment of enterotoxin that is stable in the gastrointestinal tract. C-CPE is a promising lead binder for the development of orally administered vaccines.

Perspective

C-CPE is the most frequently used CL-binding molecule and was 400-fold more potent at enhancing absorption than was an alternative absorption enhancer used clinically. The CL-targeting toxin showed *in vitro* and *in vivo* antitumor activity without any apparent side effects. Nasal immunization with the CL-targeting antigen activated both systemic and mucosal immune responses. Thus, we have demonstrated proof of concept for CL-targeted mucosal absorption, cancer-targeting, and mucosal vaccination by using C-CPE as a CL ligand.

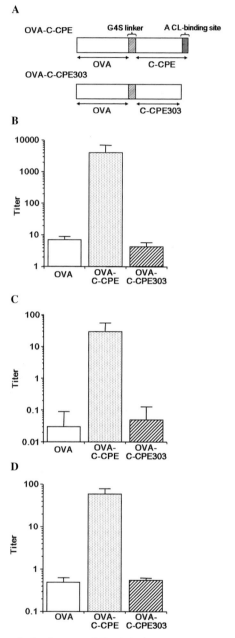

Figure 3. Involvement of CL-4 in immune responses to OVA–C-CPE. (A) Schematic illustration of OVA–C-CPE and its mutant. The C-terminal 16 amino acid-deleted C-CPE mutant (C-CPE303)[25] did not bind to CL-4. To clarify the involvement of CL-4 in the immune response initiated by OVA–C-CPE, OVA was fused with C-CPE303, resulting in OVA–C-CPE303. (B–D) immune responses by OVA–C-CPE or OVA–C-CPE303. Mice were nasally immunized with OVA, OVA–C-CPE, or OVA–C-CPE303 (5 μg OVA) once a week for three weeks. Seven days after the last immunization, the levels of OVA-specific serum IgG (B), nasal IgA (C), and vaginal IgA (D) were measured by ELISA. Data are means ± SD ($n = 4$).

C-CPE is a fragment of enterotoxin. Repeat administration of C-CPE showed production of C-CPE–specific serum IgG and mucosal IgA.[40] Development of biocompatible CL binders, such as chemicals, peptides, and antibodies is needed for future clinical application. CL has very short extracellular loops: the first extracellular loop is approximately 50 amino acids; the second loop is approximately 10 amino acids. CLs are hydrophobic proteins, and preparation of a recombinant protein is currently possible for CL-4 only, and synthetic peptides corresponding to the extracellular loop domains do not fully work as a screening source for CL binders (our unpublished data). Therefore, the development of CL binders, including antibodies, has been very slow. Recently, immunization of autoimmune-prone mice with CL-expressing cells and of rats with CL-coding DNA led to the successful preparation of anti–CL-4 and anti–CL-1 monoclonal antibodies, respectively.[41,42] Importantly, anti–CL-4 antibody showed therapeutic antitumor activity *in vitro* and *in vivo*. Anti–CL-1 antibody inhibited infection of hepatic cells by hepatitis C virus.[41] A dual-targeting monoclonal antibody against CL-3 and CL-4 has also been prepared.[43] CL-displaying BV was developed as a screening reagent for CL binders.[23] A broadly specific CL binder was developed by using a C-CPE library and CL-displaying BV. Thus, the development of CL binders has begun. Proof of concept for CL-targeted drug development is yet to be fully established, but we believe that in the future, step by step, CL-modulating and CL-targeting technology will lead to the clinical application of CL-targeted pharmaceutical therapies.

Acknowledgments

This work was supported in part by a Grant-in-Aid for Scientific Research from the Ministry of Education, Culture, Sports, Science, and Technology of Japan (21689006) and by Health and Labor Sciences Research Grants from the Ministry of Health, Labor, and Welfare of Japan. H.S. and A.T. were supported by Research Fellowships from the Japan Society for the Promotion of Science for Young Scientists.

Conflicts of interest

We declare no conflict interest significant enough to influence the results or interpretation of the submitted paper.

References

1. Powell, D.W. 1981. Barrier function of epithelia. *Am. J. Physiol.* **241:** G275–G288.
2. Wodarz, A. & I. Nathke. 2007. Cell polarity in development and cancer. *Nat. Cell Biol.* **9:** 1016–1024.
3. Schneeberger, E.E. & R.D. Lynch. 2004. The tight junction: a multifunctional complex. *Am. J. Physiol.* **286:** C1213–C1228.
4. Kondoh, M., A. Masuyama, A. Takahashi, *et al.* 2005. A novel strategy for the enhancement of drug absorption using a claudin modulator. *Mol. Pharmacol.* **67:** 749–756.
5. Kominsky, S.L., B. Tyler, J. Sosnowski, *et al.* 2007. *Clostridium perfringens* enterotoxin as a novel-targeted therapeutic for brain metastasis. *Cancer Res.* **67:** 7977–7982.
6. Morin, P.J. 2005. Claudin proteins in human cancer: promising new targets for diagnosis and therapy. *Cancer Res.* **65:** 9603–9606.
7. Tamagawa, H., I. Takahashi, M. Furuse, *et al.* 2003. Characteristics of claudin expression in follicle-associated epithelium of Peyer's patches: preferential localization of claudin-4 at the apex of the dome region. *Lab. Invest.* **83:** 1045–1053.
8. Morita, K., M. Furuse, K. Fujimoto, *et al.* 1999. Claudin multigene family encoding four-transmembrane domain protein components of tight junction strands. *Proc. Natl. Acad. Sci. USA* **96:** 511–516.
9. Sonoda, N., M. Furuse, H. Sasaki, *et al.* 1999. *Clostridium perfringens* enterotoxin fragment removes specific claudins from tight junction strands: evidence for direct involvement of claudins in tight junction barrier. *J. Cell Biol.* **147:** 195–204.
10. McClane, B.A. 1992. *Clostridium perfringens* enterotoxin: structure, action and detection. *J. Food Saf.* **12:** 237–252.
11. Windsor, E. & G.E. Cronheim. 1961. Gastro-intestinal absorption of heparin and synthetic heparinoids. *Nature.* **190:** 263–264.
12. Aungst, B.J. 2000. Intestinal permeation enhancers. *J. Pharm. Sci.* **89:** 429–442.
13. Van Itallie, C.M. & J.M. Anderson. 2006. Claudins and epithelial paracellular transport. *Annu. Rev. Physiol.* **68:** 403–429.
14. Furuse, M., M. Hata, K. Furuse, *et al.* 2002. Claudin-based tight junctions are crucial for the mammalian epidermal barrier: a lesson from claudin-1-deficient mice. *J. Cell Biol.* **156:** 1099–1111.
15. Nitta, T., M. Hata, S. Gotoh, *et al.* 2003. Size-selective loosening of the blood-brain barrier in claudin-5-deficient mice. *J. Cell Biol.* **161:** 653–660.
16. Furuse, M., H. Sasaki, K. Fujimoto, *et al.* 1998. A single gene product, claudin-1 or -2, reconstitutes tight junction strands and recruits occludin in fibroblasts. *J. Cell Biol.* **143:** 391–401.
17. Uchida, H., M. Kondoh, T. Hanada, *et al.* 2010. A claudin-4 modulator enhances the mucosal absorption of a biologically active peptide. *Biochem. Pharmacol.* **79:** 1437–1444.
18. Furuse, M. & S. Tsukita. 2006. Claudins in occluding junctions of humans and flies. *Trends Cell Biol.* **16:** 181–188.

19. Loisel, T.P., H. Ansanay, S. St-Onge, *et al.* 1997. Recovery of homogeneous and functional beta 2-adrenergic receptors from extracellular baculovirus particles. *Nat. Biotechnol.* **15:** 1300–1304.

20. Strehlow, D., S. Jodo & S.T. Ju. 2000. Retroviral membrane display of apoptotic effector molecules. *Proc. Natl. Acad. Sci. USA* **97:** 4209–4214.

21. Hayashi, I., Y. Urano, R. Fukuda, *et al.* 2004. Selective reconstitution and recovery of functional gamma-secretase complex on budded baculovirus particles. *J. Biol. Chem.* **279:** 38040–38046.

22. Masuda, K., H. Itoh, T. Sakihama, *et al.* 2003. A combinatorial G protein-coupled receptor reconstitution system on budded baculovirus. Evidence for Galpha and Galphao coupling to a human leukotriene B4 receptor. *J. Biol. Chem.* **278:** 24552–24562.

23. Kakutani, H., A. Takahashi, M. Kondoh, *et al.* 2011. A novel screening system for claudin binder using baculoviral display. *PLoS One.* **6:** e16611.

24. Takahashi, A., E. Komiya, H. Kakutani, *et al.* 2008. Domain mapping of a claudin-4 modulator, the C-terminal region of C-terminal fragment of *Clostridium perfringens* enterotoxin, by site-directed mutagenesis. *Biochem. Pharmacol.* **75:** 1639–1648.

25. Takahashi, A., M. Kondoh, A. Masuyama, *et al.* 2005. Role of C-terminal regions of the C-terminal fragment of *Clostridium perfringens* enterotoxin in its interaction with claudin-4. *J. Control. Release.* **108:** 56–62.

26. Kominsky, S.L. 2006. Claudins: emerging targets for cancer therapy. *Expert Rev. Mol. Med.* **8:** 1–11.

27. Agarwal, R., T. D'Souza & P.J. Morin. 2005. Claudin-3 and claudin-4 expression in ovarian epithelial cells enhances invasion and is associated with increased matrix metalloproteinase-2 activity. *Cancer Res.* **65:** 7378–7385.

28. Dhawan, P., A.B. Singh, N.G. Deane, *et al.* 2005. Claudin-1 regulates cellular transformation and metastatic behavior in colon cancer. *J. Clin. Invest.* **115:** 1765–1776.

29. Michl, P., M. Buchholz, M. Rolke, *et al.* 2001. Claudin-4: a new target for pancreatic cancer treatment using *Clostridium perfringens* enterotoxin. *Gastroenterology* **121:** 678–684.

30. Ebihara, C., M. Kondoh, N. Hasuike, *et al.* 2006. Preparation of a claudin-targeting molecule using a C-terminal fragment of *Clostridium perfringens* enterotoxin. *J. Pharmacol. Exp. Ther.* **316:** 255–260.

31. Saeki, R., M. Kondoh, H. Kakutani, *et al.* 2009. A novel tumor-targeted therapy using a claudin-4-targeting molecule. *Mol. Pharmacol.* **76:** 918–926.

32. Saeki, R., M. Kondoh, H. Kakutani, *et al.* 2010. A claudin-targeting molecule as an inhibitor of tumor metastasis. *J. Pharmacol. Exp. Ther.* **334:** 576–582.

33. Meunier, V., M. Bourrie, Y. Berger, *et al.* 1995. The human intestinal epithelial cell line Caco-2; pharmacological and pharmacokinetic applications. *Cell Biol. Toxicol.* **11:** 187–194.

34. Boyaka, P.N., M. Marinaro, J.L. Vancott, *et al.* 1999. Strategies for mucosal vaccine development. *Am. J. Trop. Med. Hyg.* **60:** 35–45.

35. Cardenas-Freytag, L., E. Cheng & A. Mirza. 1999. New approaches to mucosal immunization. *Adv. Exp. Med. Biol.* **473:** 319–337.

36. Michels, K.B. & H. zur Hausen. 2009. HPV vaccine for all. *Lancet.* **374:** 268–270.

37. Kiyono, H. & S. Fukuyama. 2004. NALT- versus Peyer's-patch-mediated mucosal immunity. *Nat. Rev. Immunol.* **4:** 699–710.

38. Kunisawa, J., S. Fukuyama & H. Kiyono. 2005. Mucosa-associated lymphoid tissues in the aerodigestive tract: their shared and divergent traits and their importance to the orchestration of the mucosal immune system. *Curr. Mol. Med.* **5:** 557–572.

39. Kakutani, H., M. Kondoh, M. Fukasaka, *et al.* 2010. Mucosal vaccination using claudin-4-targeting. *Biomaterials.* **31:** 5463–5471.

40. Suzuki, H., M. Kondoh, X. Li, *et al.* 2011. A toxicological evaluation of a claudin modulator, the C-terminal fragment of *Clostridium perfringens* enterotoxin, in mice. *Pharmazie.* **66:** 543–546.

41. Fofana, I., S.E. Krieger, F. Grunert, *et al.* 2010. Monoclonal anti-claudin 1 antibodies prevent hepatitis C virus infection of primary human hepatocytes. *Gastroenterology.* **139:** 953–964.

42. Suzuki, M., M. Kato-Nakano, S. Kawamoto, *et al.* 2009. Therapeutic antitumor efficacy of monoclonal antibody against Claudin-4 for pancreatic and ovarian cancers. *Cancer Sci.* **100:** 1623–1630.

43. Kato-Nakano, M., M. Suzuki, S. Kawamoto, *et al.* 2010. Characterization and evaluation of the antitumour activity of a dual-targeting monoclonal antibody against claudin-3 and claudin-4. *Anticancer Res.* **30:** 4555–4562.

Ann. N.Y. Acad. Sci. ISSN 0077-8923

ANNALS OF THE NEW YORK ACADEMY OF SCIENCES
Issue: *Barriers and Channels Formed by Tight Junction Proteins*

Loss of enteral nutrition in a mouse model results in intestinal epithelial barrier dysfunction

Yongjia Feng, Matthew W. Ralls, Weidong Xiao, Eiichi Miyasaka, Richard S. Herman, and Daniel H. Teitelbaum

Department of Surgery, Section of Pediatric Surgery, University of Michigan, Ann Arbor, Michigan

Address for correspondence: Daniel H. Teitelbaum, M.D., Department of Surgery, Section of Pediatric Surgery, University of Michigan, Mott Children's Hospital F3970, Ann Arbor, MI 48109-0245. dttlbm@umich.edu

Total parenteral nutrition (TPN) administration in a mouse model leads to a local mucosal inflammatory response, resulting in a loss of epithelial barrier function (EBF). Although, the underlying mechanisms are unknown, a major contributing factor is a loss of growth factors and subsequent critical downstream signaling. An important component of these is the p-Akt pathway. An additional contributing factor to the loss of EBF with TPN is an increase in proinflammatory cytokine abundance within the mucosal epithelium, including TNF-α and IFN-γ. Loss of critical nutrients, including glutamine and glutamate, may affect EBF, contributing to the loss of tight junction proteins. Finding protective modalities for the small intestine during TPN administration may have important clinical applications. Supplemental glutamine and glutamate may be examples of such agents.

Keywords: small intestine; p-Akt; parenteral nutrition; epithelial barrier function

Introduction

Parenteral nutrition (PN) is commonly used as treatment for many patients, ranging from short-term use in patients with gastrointestinal dysfunction,[1] to long-term use in patients with short bowel syndrome.[2,3] Although it is lifesaving for many, PN use is associated with numerous complications, including an increase in systemic infections, including pneumonia as well as infections of the urinary tract and surgical wounds.[4,5] Although the precise etiology of this increased rate of infections has not been fully established, it is known that organisms arising from enteric flora constitute a large percent of these infections. This suggests that a major loss of epithelial barrier function (EBF) could be a contributing factor leading to this increase in infectious complications. Additional contributing factors include a loss of local growth factors and increase in several proinflammatory mucosal cytokines. This review summarizes the work done by our laboratory and others on total parenteral nutrition (TPN)-associated loss of EBF.

TPN administration results in a proinflammatory cytokine state within the intestinal mucosa

The development of these inflammatory changes is complex and leads to atrophy of small bowel villi, an increase in epithelial cell (EC) apoptosis,[6,7] and a decrease in EC proliferation. This results in a marked reduction in the length of the small and large bowel (Fig. 1). Increases in both tumor necrosis factor-alpha (TNF-α) and interferon gamma (INF-γ) contribute to the loss of EBF found with TPN administration. Blockade of TNF signaling with the use of TNF-receptor knockout mice demonstrate that the major mediator of mucosal atrophy is via the TNF-R1 signaling pathway, and without an intact pathway, mucosal atrophy is significantly prevented.[8] Blockade of IFN-γ signaling using INF-γ–knockout mice has also been shown to be effective in preventing the loss of EBF.[7,9]

Microbiome and TPN

Previous studies from our laboratory and others have shown in a mouse TPN model, in which the

doi: 10.1111/j.1749-6632.2012.06572.x

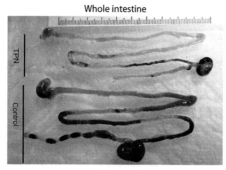

Whole intestine

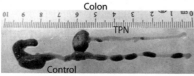

Colon

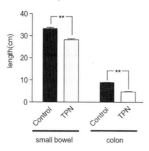

Figure 1. (A) Representative image of harvested intestine from control (chow fed) and TPN mice. Note the significantly reduced length of both small and large intestine. (B) Mean length in centimeters of small and large bowel (colon) from control and TPN mice. $**P < 0.01$.

gut is devoid of all nutrients, a number of significant physical and immunologic changes develop in the intestinal mucosa.[9–13] Data from our laboratory demonstrates that these intestinal changes are intimately associated with alterations in the luminal microbiome.[14] Our laboratory identified a major shift in the microbiome, from a normally benign composition of mostly Firmacutes, to a predominantly Gram-negative, Proteobacteria population. This shift in the microbiome was strongly correlated with a proinflammatory state characterized by an upregulation in Toll-like receptor (TLR) signaling, including TLR2, TLR4, and TLR9.[15] As well, TPN led to a marked downstream increase in NF-κB activation within the lamina propria, with a resultant increase in inflammatory cytokines INF-γ and TNF-α, as well as a decreased population of T regulatory cells.[16] Blockade of this signaling pathway using myeloid differentiation primary response gene

88 (MyD88) knockout mice demonstrated a significant prevention of NF-κB activation, prevention of a proinflammatory state, preservation of T regulatory cell population, and prevention of the loss of EBF.[17] Taken together, this suggests that unique strategies to prevent these major microbiome changes, or block TLR signaling, during TPN administration may prove to be a strategy to prevent the adverse effects of intestinal atrophy and loss of EBF (Fig. 2).

TPN results in a loss of local growth factor production: implications for loss of EBF

The pathophysiology of TPN-driven EC apoptosis, decreased EC proliferation, and loss of EBF is complex. Although conventionally viewed as a mediator of cellular death, TNF-α signaling has also been closely linked with cell survival and regulation of proliferation. In fact, effective epidermal growth factor (EGF) signaling requires both an intact TNF-α[18,19] and ErbB1 receptor pathways. Both EGF and ErbB1 are markedly diminished with TPN administration.[8] The combination of a proinflammatory state, with loss of EGF signaling leads to an imbalance between TNFR1/TNFR2 and EGF signaling. Such a deranged state may well be a major contributing factor to TPN-associated mucosal atrophy and loss of EBF.

Aside from a loss of ErbB ligands, TPN has also been shown to result in a loss of two other key mucosal growth factors, keratinocyte growth factor[20–22] and glucagon-like peptide-2 (GLP-2).[23] Each of these factors are contributors to maintenance of the intestinal barrier function,[24,25] and loss of these factors seem to be major contributors to TPN-associated intestinal atrophy.[23]

TPN-associated loss of EBF[26,12] may well result in increases in bacterial translocation.[27–29] The barrier function of the epithelium is the sum of several physiological processes, including the synthesis and release of mucus from goblet cells, transcytosis of dimeric secretory IgA, luminally directed water movement, and the physical integrity of the epithelial layer itself.[30] Breakdown in EBF and immunologic derangement can lead to the systemic dissemination of pathogens that often lead to the development of multiorgan dysfunction.[31] These findings have been shown in both mouse models and in humans[32] on PN.

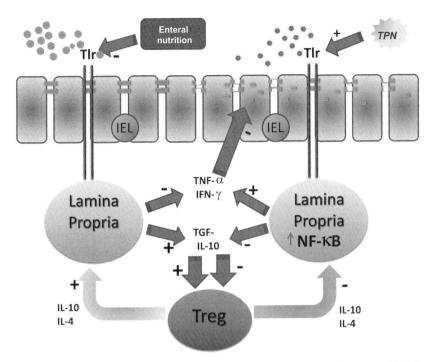

Figure 2. Summary of potential signaling changes with TPN administration. Note loss of nutrients with TPN administration results in a change in the microbiome (represented by the red dots). This leads to enhanced signaling via Toll-like receptors (TLR) activating lamina propria lymphoid tissue leading to NF-κB nuclear transcription and a proinflammatory state within the mucosa, with increases in TNF-α and IFN-γ, and loss of TNF-β and IL-10. These changes lead to a loss of T_{reg} cells and an exacerbation in the proinflammatory state. In total this leads to a loss of epithelial barrier function.

TPN-associated loss of p-Akt abundance and loss of tight junctional proteins

Loss of growth factor signaling may lead to a loss of downstream phosphatidylinositol 3-kinase (PI3K)/p-Akt signaling. This may have significant implications, as a critical role of PI3K/Akt signaling in resistance to apoptosis has been reported in multiple cell types.[33,34] Our previous data has shown that TPN administration led to decreased p-Akt activity in small bowel ECs, contributing to the TPN-associated loss of proliferation and increased apoptosis.[13] T-cell lymphoma-1 (TCL1) is a novel Akt-activating peptide (a head-to-tail dimer of the Akt-binding domain of T cell lymphoma-1).[35] We used this peptide, or an inactive mutant sequence TCL1G, conjugated to a transactivator of transcription peptide sequence (TAT), to promote intracellular uptake. This led to a significant increase in p-Akt, as well as prevention in both the loss of EC proliferation and rise in apoptosis with TPN.[11]

The PI3K/p-Akt signaling may well have implications on preservation of EBF. PI3K/p-Akt signaling has been reported to affect electrogenic intestinal transport.[36] However, this action may be partial in that these authors showed that the application of the PI3 kinase inhibitor, Wortmannin or LY294002, did not significantly alter transepithelial potential difference or resistance. Junctional proteins play a crucial role in the epithelial barrier, and these can be reflected by the measurement of the transepithelial potential difference and resistance. Zonula occludens protein 1 (ZO-1) acts as a functional cross-linker between the E-Cadherin/catenin complex and actin-based cytoskeleton.[37] Other critical factors that modulate EBF include the family of claudins, and these may change significantly with various disease processes, such as Crohn's disease.[38,39] In our TPN mice using Western immunoblots to measure junctional protein expression, we found a significant reduction in the abundance of tight junction proteins ZO-1 and occludin, as well as the adhesion junctional protein E-Cadherin compared to sham (enterally fed) groups. TCL-1 administration (which drives p-Akt abundance) partially prevented the loss of ZO-1 and E-Cadherin expression; however, TCL-1 did not impact the loss of occludin expression. As well, transepithelial resistance was not significantly affected in the TCL-1 TPN group compared to control (non-TCL1) TPN mice. Thus, the partial prevention of the loss of p-Akt, was unable to fully prevent the loss of EBF in this TPN model. Claudins also were differentially expressed with TPN, including a loss of claudins 7 and 15.[12]

As mentioned earlier, our laboratory's data also showed that downregulation of E-Cadherin expression in TPN mice is tightly related to a decrease in p-Akt activity.[13] In addition, we showed an association linking the loss of EC proliferation and increase of EC apoptosis, to a decline in the downstream signaling of the p-Akt pathway via GSK3-β and mTOR pathways.[11] Interestingly, supplementation of TPN mice with intravenous glutamine (GLN) prevented the loss of p-Akt abundance, and resulted in a recovery of EC proliferation as well as barrier function.[40] (see next section on nutrient supplementation). These findings are similar to the McKay group, which recently reported that PI3K and protein kinase C were mediators of IFN-γ–induced increases in epithelial paracellular permeability and the transcytosis of noninvasive, nonpathogenic *Escherichia coli* passage across monolayers of human gut derived ECs. Although Akt is an important mediator of many PI3K-dependent events, this event was Akt independent.[41,42] This may explain our lack of impacting EBF by driving p-Akt abundance.

Recently, it has been reported that TLR2-induced tight junction (TJ) modulation strongly interrelates with promotion of intestinal EC survival through the PI3K/Akt pathway.[43] An *ex vivo* culture model of primary intestinal epithelial cells showed functionally that TLR2 activation promotes cell survival by inhibiting apoptosis. This effect requires the adaptor molecule MyD88, which contains a p85-binding site,[44] as a putative mechanism to recruit PI3K downstream on ligand stimulation. TLR2 stimulation efficiently preserves ZO-1–associated barrier integrity against stress-induced damage, which is controlled by the PI3K/Akt-pathway via MyD88. As well, intestinal EC TLR2 functions via PI3K/Akt to attenuate the MAPK/NF-κB–signaling cascade. Therefore, the complex interactions of an upregulation of TLR2 in our TPN model may well interrelate with the loss of EC proliferation and EBF. Because a major signaling mechanism for TLR4 is via lipopolysaccharide (LPS), much attention has been focused on how to block this signal. Intestinal alkaline phosphatase (IAP), which is released by the brush border of ECs can neutralize LPS and normalize the intestinal microbiota.[45] It will be fascinating to explore whether nutrient deprivation, with TPN, can influence IAP production.

Loss of critical nutrients may be a contributing factor to the development of EBF loss

With the comprehensive clinical application of TPN, it would seem that clinicians have achieved consensus about the optimal TPN formula. However, even with the prerequisite of enough caloric intake, there still are two problems not well handled in the current approach to clinical TPN administration: loss of critical nutrients in the TPN solution and the lost luminal sensing for nutrients in the gut. Each of these may contribute to TPN-associated EBF disruption. These two issues are an emerging research focus in this field.

To address the first issue, we compared the difference between TPN versus TPN plus partial (25%) oral feeding on the intestinal immune system.[46] This small amount of enteral feeding significantly prevented atrophy of the intestinal epithelium and changes in intraepithelial lymphocytes (IEL) phenotypes seen with TPN. As there was a prevention in the upregulation of proinflammatory cytokines (which downregulated normal tight junction function) with partial enteral feeding, the results suggest that the major factor responsible for TPN-associated loss of EBF, bacterial translocation, and IEL-changes, is the lack of enteral feeding and not the administration of the TPN solution itself.

Of great interest is the identification of critical nutrients that could be involved in EBF protective mechanisms, thereby helping us greatly improve the formula of TPN administration. Because GLN has been shown to be an important amino acid in the modulation of EBF under various luminal threats, such as endotoxin lipopolysaccharide, we investigated the benefits of GLN supplementation on TPN mice.[40] Intravenous GLN supplementation significantly prevented the loss of EBF and mucosal atrophy in TPN mice. This was shown with an accompanied decreased intestinal permeability, increased expression of tight junction proteins (including ZO-1, occluding, and junctional adhesion molecule-1), as well as prevention of changes in IEL-derived cytokine profile. It seems that GLN showed this EBF protective effect in TPN mice by regulating the phenotype and function of IEL, potentially via an immunomodulatory effect of GLN.[40]

The GLN studies described earlier were supplemented intravenously. It is also possible that the lack of luminal nutrients with TPN administration not only leads to some certain nutritional deficiencies, but may also result in a loss of gut-nutrient sensing. An increasing literature has shown gut taste receptors (TRs), located throughout the gastrointestinal tract, to be involved in gastrointestinal mucosal defense mechanisms.[47] It is clear that a variety of TRs are present throughout the gastrointestinal tract, and that these are predominately located in enteroendocrine cells and ECs on the luminal side of the mucosa, facing a variety of nutritive and nonnutritive stimuli signals.[48] Interestingly, growing evidence has demonstrated that certain sweet receptors and umami receptors may be under dynamic regulation under different metabolic (glucose/glutamate [GLM]) and luminal stimuli (acid injury/luminal microbiota); and may have particular relevance in disease conditions such as diabetes.[49] This suggests that a compensatory mechanism may exist, which balances nutrient sensing and absorption.[50] Furthermore, through activating TRs, taste stimulants supplementation could enhance the function of the intestinal immune system.[51] Potentially, this stimulation could enhance EBF.

Recently, the umami receptor stimulant, GLM has been shown to enhance duodenal mucosal defense mechanisms through the TRs T1R1/T1R3 and GLM receptors.[52] The Wang group reported that perfusion of L-Glu/IMP (inosine 5′-monophosphate) increased portal venous concentrations of GLP-2 and GLP-1 and increases duodenal bicarbonate secretion.[53] Another recent *in vitro* study demonstrated that similar to GLN, GLM treatment reduced the permeability of IEC monolayers,[54] thereby enhancing EBF. Although there are no publications examining the action of GLM on EBF during TPN ad-

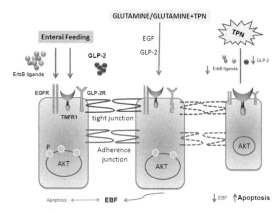

Figure 3. Summary of the potential mechanisms, which may be involved in sustaining epithelial barrier function via the p-Akt signaling pathway. Note loss of local growth factors, including EGF and GLP2, along with respective receptors results in loss of EBF. Restitution of EBF, along with epithelial cell proliferation and prevention of apoptosis, occurs in TPN mice with supplementation of glutamine or glutamate, as well as with exogenous growth factors.

ministration, it is quite possible that stimulation of these TRs may act in this way, and thus, restoring gut nutritional sensing could be a potential strategy to improve EBF in TPN patients. A summary of these findings is given in Figure 3.

Conclusions

Although providing the essential macro- and micronutrients required during a prolonged fasting state, the administration of TPN also leads to small bowel atrophy, increase in EC apoptosis, and increases in proinflammatory mediators. The resultant effect is a significant loss of EBF, predominantly via a loss of critical tight junction proteins. The mechanisms for these actions are complex, but decreases in the p-Akt signaling pathway, and an increase in proinflammatory cytokines, likely play important roles. The additional administration of GLN, the principal energy source for enterocytes, helps prevent these deleterious effects TPN has on the small bowel. Conceivably directed enteral stimulates, such as GLM, may also demonstrate intestinal protection via signaling of intestinal TRs. By continuing to examine the mechanisms of TPN related complications in the small bowel, and potential protective modalities, the harmful effects of this life-prolonging therapy may be eliminated.

Acknowledgment

This work was supported by NIH-R01 AI-44076–11 and -12.

Conflicts of interest

The authors declare no conflicts of interest.

References

1. Braga, M. *et al.* 2009. ESPEN Guidelines on Parenteral Nutrition: surgery. *Clin. Nutr.* **28:** 378–386.
2. Duro, D., D. Kamin & C. Duggan. 2008. Overview of pediatric short bowel syndrome. *J. Pediatr. Gastroenterol Nutr.* **47**(Suppl. 1): S33–S36.
3. Spencer, A. *et al.* 2005. Mortality and outcomes of pediatric short bowel syndrome: redefining predictors of success. *Ann. Surg.* **242:** 1–10.
4. Gogos, C.A. & F. Kalfarentzos. 1995. Total parenteral nutrition and immune system activity: a review. *Nutrition* **11:** 339–344.
5. The Veterans Affairs Total Parenteral Nutrition Cooperative Study Group. 1991. Perioperative total parenteral nutrition in surgical patients. *N. Engl. J. Med.* **325:** 525–532.
6. Wildhaber, B.E. *et al.* 2002. TPN-induced apoptosis in mouse intestinal epithelium: regulation by the BCL-2 protein family. *Ped. Surg. Int.* **18:** 570–575.
7. Yang, H. *et al.* 2002. Interferon-gamma expression by intraepithelial lymphocytes results in a loss of epithelial barrier function in a mouse model of total parenteral nutrition. *Ann. Surg.* **236:** 226–234.
8. Feng, Y. & D.H. Teitelbaum. 2012. Epidermal growth factor/TNF-alpha transactivation modulates epithelial cell proliferation and apoptosis in a mouse model of parenteral nutrition. *Am. J. Physiol. Gastrointest Liver Physiol* **302:** G236–G249.
9. Yang, H. & D.H. Teitelbaum. 2003. Intraepithelial lymphocyte-derived interferon-gamma evokes enterocyte apoptosis with parenteral nutrition in mice. *Am. J. Physiol. Gastrointest Liver Physiol.* **284:** G629–G637.
10. Li, J. *et al.* 1995. Effect of parenteral and enteral nutrition on gut-associated lymphoid tissue. *J. Trauma* **39:** 44–51.
11. Feng, Y., J.E. McDunn & D.H. Teitelbaum. 2010. Decreased phospho-Akt signaling in a mouse model of total parenteral nutrition: a potential mechanism for the development of intestinal mucosal atrophy. *Am. J. Physiol. Gastrointest Liver Physiol.* **298:** G833–G841.
12. Sun, X. *et al.* 2008. Decline in intestinal mucosal IL-10 expression and decreased intestinal barrier function in a mouse model of total parenteral nutrition. *Am. J. Physiol. Gastrointest Liver Physiol.* **294:** G139–G147 PMID: 17991705.
13. Feng, Y. *et al.* 2009. Dissociation of E-Cadherin and beta-catenin in a mouse model of total parenteral nutrition: a mechanism for the loss of epithelial cell proliferation and villus atrophy. *J. Physiol. (London)* **587:** 641–654.
14. Miyasaka, E. *et al.* 2011. Total parenteral nutrition (TPN) in a mouse model leads to major population shifts in the intestinal microbiome. *Gastroenterology* **140:** 1.
15. Miyasaka, E. *et al.* 2010. Removal of enteral nutrition with total parenteral nutrition in mice leads to changes in bacterial flora and an associated increased toll-like receptors in the small intestinal lamina propria. *Gastroenterology* **138:** S-608–S-609.
16. Miyasaka, E. & D. Teitelbaum. 2010. Loss of small-intestine lamina propria T-regulatory cells in a mouse-model of total parenteral nutrition (TPN). *J. Surg. Res.* **158:** 330.
17. Miyasaka, E., Y. Feng & D. Teitelbaum. 2011. Total Parenteral nutrition in a mouse model results in a proinflammatory state in the lamina propria: a Myd88-dependent mechanisms of action [abstract]. *Surgery* In press.
18. McElroy, S.J. *et al.* 2008. Tumor necrosis factor inhibits ligand-stimulated EGF receptor activation through a TNF receptor 1-dependent mechanism. *Am. J. Physiol. Gastrointest. Liver Physiol.* **295:** G285–G293.
19. Yamaoka, T. *et al.* 2008. Transactivation of EGF receptor and ErbB2 protects intestinal epithelial cells from TNF-induced apoptosis. *Proc. Natl. Acad. Sci. USA* **105:** 11772–11777.
20. Yang, H. *et al.* 2004. Intestinal intraepithelial lymphocyte gamma delta-T cell-derived keratinocyte growth factor modulates epithelial growth in the mouse. *J. Immunol.* **172:** 4151–4158.
21. Yang, H., B.E. Wildhaber & D.H. Teitelbaum. 2003. Keratinocyte growth factor improves epithelial function after massive small bowel resection. *J. Parenter Enteral Nutr.* **27:** 198–206.
22. Yang, H. *et al.* 2002. 2002 Harry M. Vars Research Award. Keratinocyte growth factor stimulates the recovery of epithelial structure and function in a mouse model of total parenteral nutrition. *JPEN J. Parenter Enteral Nutr.* **26:** 333–340; discussion 340–331.
23. Feng, Y., J. Holst & D. Teitelbaum. 2011. Total parenteral nutrition (TPN)-associated atrophy is associated with loss of intestinal epithelial cell (EC) migration: modulation of action by epidermal growth factor (EGF) and glucagon-like peptide-2(GLP-2). *Gastroenterology* **140:** S-170–S-171.
24. Tsai, C., M. Hill & K. Drucker. 1997. Biological determinants of intestinotrophic properties of GLP-2 *in vivo. Am. J. Physiol.* **272:** G662–G668.
25. Brubaker, P. *et al.* 1997. Circulating and tissue forms of the intestinal growth factor, glucagon-like peptide-2. *Endocrinology* **138:** 4837–4843.
26. Yang, H., R. Finaly & D.H. Teitelbaum. 2003. Alteration in epithelial permeability and ion transport in a mouse model of total parenteral nutrition. *Crit. Care Med.* **31:** 1118–1125.
27. Kudsk, K.A. *et al.* 1992. Enteral versus parenteral feeding. Effects on septic morbidity after blunt and penetrating abdominal trauma. *Ann. Surg.* **215:** 503–511; discussion 511–503.
28. Kiristioglu, I. *et al.* 2002. Total parenteral nutrition-associated changes in mouse intestinal intraepithelial lymphocytes. *Dig. Dis. Sci.* **47:** 1147–1157.
29. Kiristioglu, I. & D.H. Teitelbaum. 1998. Alteration of the intestinal intraepithelial lymphocytes during total parenteral nutrition. *J. Surg. Res.* **79:** 91–96.
30. Clayburgh, D.R., L. Shen & J.R. Turner. 2004. A porous defense: the leaky epithelial barrier in intestinal disease. *Lab Invest.* **84:** 282–291.

31. Kristof, K. *et al.* 2011. Impact of molecular mimicry on the clinical course and outcome of sepsis syndrome. *Mol. Immunol.* **49:** 512–517.

32. Buchman, A.L. *et al.* 1995. Parenteral nutrition is associated with intestinal morphologic and functional changes in humans. *JPEN J. Parenter Enteral Nutr.* **19:** 453–460.

33. Chang, F. *et al.* 2003. Involvement of PI3K/Akt pathway in cell cycle progression, apoptosis, and neoplastic transformation: a target for cancer chemotherapy. *Leukemia* **17:** 590–603.

34. Bouchard, V. *et al.* 2008. B1 integrin/Fak/Src signaling in intestinal epithelial crypt cell survival: integration of complex regulatory mechanisms. *Apoptosis* **13:** 531–542.

35. McDunn, J. *et al.* 2008. Peptide-mediated activation of Akt and extracellular regulated kinase signaling prevents lymphocyte apoptosis. *FASEB J.* **22:** 561–568.

36. Rexhepaj, R. *et al.* 2007. PI3-kinase-dependent electrogenic intestinal transport of glucose and amino acids. *Pflugers Arch.* **453:** 863–870.

37. Itoh, M. *et al.* 1997. Involvement of ZO-1 in cadherin-based cell adhesion through its direct binding to alpha catenin and actin filaments. *J. Cell Biol.* **138:** 181–192.

38. Zeissig, S. *et al.* 2007. Changes in expression and distribution of claudin 2, 5 and 8 lead to discontinuous tight junctions and barrier dysfunction in active Crohn's disease. *Gut* **56:** 61–72.

39. Markov, A.G. *et al.* 2010. Segmental expression of claudin proteins correlates with tight junction barrier properties in rat intestine. *J. Comp. Physiol. B.* **180:** 591–598.

40. Nose, K. *et al.* 2010. Glutamine prevents total parenteral nutrition-associated changes to intraepithelial lymphocyte phenotype and function: a potential mechanism for the preservation of epithelial barrier function. *J. Interferon Cytokine Res.* **30:** 67–80.

41. Smyth, D. *et al.* 2011. Interferon-gamma-induced increases in intestinal epithelial macromolecular permeability requires the Src kinase Fyn. *Lab Invest.* **91:** 764–777.

42. McKay, D.M. *et al.* 2007. Phosphatidylinositol 3'-kinase is a critical mediator of interferon-gamma-induced increases in enteric epithelial permeability. *J. Pharmacol Exp. Ther.* **320:** 1013–1022.

43. Cario, E., G. Gerken & D. Podolsky. 2007. Toll-like receptor 2 controls mucosal inflammation by regulating epithelial barrier function. *Gastroenterology* **132:** 1359–1374.

44. Arbibe, L. *et al.* 2000. Toll-like receptor 2-mediated NF-kappa B activation requires a Rac1-dependent pathway. *Nat. Immunol.* **1:** 533–540.

45. Malo, M.S. *et al.* 2010. Intestinal alkaline phosphatase preserves the normal homeostasis of gut microbiota. *Gut* **59:** 1476–1484.

46. Wildhaber, B.E. *et al.* 2005. Lack of enteral nutrition–effects on the intestinal immune system. *J. Surg. Res.* **123:** 8–16.

47. Yasumatsu, K. *et al.* 2009. Multiple receptors underlie glutamate taste responses in mice. *Am. J. Clin. Nutr.* **90:** 747S–752S.

48. Wu, S.V. *et al.* 2002. Expression of bitter taste receptors of the T2R family in the gastrointestinal tract and enteroendocrine STC-1 cells. *Proc. Natl. Acad. Sci. USA* **99:** 2392–2397.

49. Fujita, Y. *et al.* 2009. Incretin release from gut is acutely enhanced by sugar but not by sweeteners *in vivo*. *Am. J. Physiol. Endocrinol. Metab.* **296:** E473–E479.

50. Margolskee, R.F. *et al.* 2007. T1R3 and gustducin in gut sense sugars to regulate expression of Na+-glucose cotransporter 1. *Proc. Natl. Acad. Sci. USA* **104:** 15075–15080.

51. Xue, H. & C.J. Field. 2011. New role of glutamate as an immunoregulator via glutamate receptors and transporters. *Front Biosci. (Schol Ed).* **3:** 1007–1020.

52. Akiba, Y. *et al.* 2009. Luminal L-glutamate enhances duodenal mucosal defense mechanisms via multiple glutamate receptors in rats. *Am. J. Physiol. Gastrointest Liver Physiol.* **297:** G781–G791.

53. Wang, J.H. *et al.* 2011. Umami receptor activation increases duodenal bicarbonate secretion via glucagon-like peptide-2 release in rats. *J. Pharmacol Exp. Ther.* **339:** 464–473.

54. Vermeulen, M.A. *et al.* 2011. Glutamate reduces experimental intestinal hyperpermeability and facilitates glutamine support of gut integrity. *World J. Gastroenterol.* **17:** 1569–1573.

Ann. N.Y. Acad. Sci. ISSN 0077-8923

The protein C pathway in intestinal barrier function: challenging the hemostasis paradigm

Silvia D'Alessio, Marco Genua, and Stefania Vetrano

Division of Gastroenterology, IRCCS Istituto Clinico Humanitas, Rozzano, Milan, Italy

Address for correspondence: Stefania Vetrano, Ph.D., Division of Gastroenterology, IBD Center, Laboratory of Immunology and Inflammation, Istituto Clinico Humanitas-IRCCS in Gastroenterology, Viale Manzoni 56, 20089, Rozzano, Milan, Italy. stefania.vetrano@humanitasresearch.it

The protein C (PC) pathway is a well-characterized anticoagulant system. Produced mainly by the liver as a zymogen, PC is activated on the vascular endothelial cell surface by thrombin–thrombomodulin complex. Once activated, PC inactivates two important cofactors of the coagulation cascade, factors Va and VIIIa, which are crucial for thrombin generation. For many years, this pathway has been studied for the clotting process, but only recently great progress has been made in understanding other functions of the PC system. Indeed, much work demonstrates that this pathway exerts several activities not only involved in the coagulative process but also in inflammation, cell proliferation, apoptosis, stabilization of endothelial barrier, and fibrinolysis. In addition, a recent study has shed light on a new role of the PC system in controlling intestinal permeability function by regulating tight junction molecules and promoting mucosal healing. This review highlights these recent insights in the context of the complex pathogenesis of inflammatory bowel disease.

Keywords: intestinal barrier; protein C; inflammatory bowel disease (IBD); coagulation; activated PC

Introduction

Protein C (PC) pathway is one of the major systems that bridges inflammation and coagulation.[1,2] This pathway is composed of thrombomodulin (TM), endothelial cell protein C receptor (EPCR), PC, and protease-activated receptor-1 (PAR-1). It is traditionally acknowledged that TM and EPCR are mainly expressed by vascular endothelial cells, forming a complex on the surface of these cells that converts circulating PC into its active form. Although the function of the PC pathway is classically considered to be anticoagulative, mounting evidence indicates that it also plays a dominant role in inflammation, with each component of the pathway displaying remarkable potent anti-inflammatory activity.[3] TM, EPCR, PAR-1, and activated PC (aPC), individually or together, are emerging as crucial controllers of inflammation in multiple organs, a notion that is leading to the development of novel therapeutic approaches for a wide range of inflammatory disorders.[2,4] In particular, TM di-

rectly inhibits leukocyte adhesion to the activated endothelium, and there is an increased leukocyte influx in mice expressing a nonfunctional TM mutant, in several models of inflammation.[5] Soluble form of EPCR has been reported to bind the integrin Mac-1(CD11b/CD18) on activated neutrophils thus decreasing tight adhesion of neutrophils to activated endothelium.[6] The survival of mice with a severe deficiency of EPCR is reduced following administration of LPS compared with LPS-administered wild-type mice, with a marked increase in neutrophil tissue infiltration and enhanced chemokine production.[7] Finally, aPC inhibits the release of inflammatory cytokines such as TNF-α, IL-6, and IL-8 in experimental models of endotoxin-induced inflammation, as well as limiting leukocyte extravasation into tissues.[8] In patients suffering from inflammatory bowel disease (IBD), aPC acts as a potent anti-inflammatory molecule on human intestinal microvasculature endothelial cells (HIMEC), inhibiting the expression of cell adhesion molecules (CAMs), the production

doi: 10.1111/j.1749-6632.2012.06557.x

of chemokines, and the adhesion of leukocytes.[9] Genetic manipulation of each component of the PC pathway has underscored the vital role of this system in the control of inflammation.[2] In addition to its anti-inflammatory properties, PC has been shown to exert other cytoprotective functions. Indeed, insights collected from recent *in vitro* and *in vivo* studies of the direct cytoprotective effects of aPC include beneficial alterations in gene expression profiles, anti-inflammatory actions, antiapoptotic activities, and stabilization of the endothelial barrier.[2,10] Activated PC achieves its cytoprotective effects by signaling through PAR-1 on endothelial cells, and thereby reinforces endothelial barrier function. The surprising observation that, besides its classical and traditionally well-known expression in the gut mucosal microcirculation, the PC system is expressed by intestinal epithelial cells and controls intestinal barrier function has revolutionized our understanding of this pathway.[11] The unexpected role in controlling intestinal homeostasis and inflammation by regulating barrier function, and in promoting intestinal mucosal healing, paves the way for new therapeutic approaches for the treatment of intestinal bowel disease. This short review will summarize several activities of the PC system and discuss the new emerging role of this pathway in controlling intestinal permeability function.

The PC pathway: anticoagulant effects

PC is the zymogen of the anticoagulant serine protease, activated PC (aPC), whose synthesis occurs mainly in the liver and begins with a single-chain precursor molecule.[12] The proteolytic activation of PC occurs on the surface of endothelial cells and involves two membrane receptors: TM and EPCR.[13] Binding of thrombin to TM on the endothelial surface forms a complex able to cleave PC, which becomes activated. This activation is facilitated by binding of PC to EPCR.[14] Subsequent activation of PC triggers the inactivation of factors Va and VIIIa[15,16] leading to a block in thrombin generation. The inactivation reaction only proceeds efficiently in the presence of Ca^{2+} ions and the cofactor protein S.[16] Indeed, protein S enhances the aPC-mediated cleavage of factor Va by addressing the aPC closer to the membrane, presumably thereby placing it in better proximity to the cleavage site of factor Va.[17] Defects in these aPC cofactors have been linked to increased risk of venous thrombosis.[18] Be-

cause the anticoagulant properties of PC system are widely known and well described in the literature, we will not delve into this aspect.

The PC pathway: cytoprotective effects

The PC system exerts anti-inflammatory effects independently from its activity on the coagulation cascade. Important therapeutic opportunities are emerging based on the manipulation of the PC system.[2,19] In several experimental models in which the conversion of PC into activated PC is impaired, including asthma, sepsis, ischemia-reperfusion injury, and intestinal inflammation, administration of recombinant aPC consistently inhibited cytokine production, CAM expression, and leukocyte influx into tissues.[20–26] Furthermore, systemic injection of aPC prevents the lethal effects of *Escherichia coli* in animal models of sepsis and significantly reduces endotoxin-induced pulmonary vascular injury.[8,27] Intratracheal administration of aPC inhibits bleomycin-induced lung fibrosis in mice, whereas inhalation of recombinant aPC inhibits airway hyperresponsiveness and inflammation in a murine model of asthma.[21,26,28] In addition, administration of aPC displayed a potent anti-inflammatory and neuroprotective effect in a murine model of focal ischemic stroke.[29] Finally, in a model of intestinal ischemia-reperfusion injury, administration of aPC significantly reduced leukocyte rolling and adhesion to the mesenteric vessels, and prevented leukocyte influx in a model of kidney and gastric injury.[20,24,25] The further involvement of aPC in the inflammatory response was studied in transgenic mice expressing 1–18% PC in an endogenous $PC^{-/-}$ background. The onset and severity of PC-deficiency phenotypes in these mice varied significantly but were strongly dependent on their circulating levels of PC.[30] LPS challenge studies in these mice demonstrated that genetic dosing of PC strongly correlated with survival.[31] Mice expressing low levels of endogenous PC presented an enhanced inflammation relative to wild-type (WT) mice, and reconstitution of low-PC mice with aPC improved the disease status and survival outcome.

The anti-inflammatory vascular effects of aPC can be divided into its effects on endothelial cells and on leukocytes. The first includes the inhibition of inflammatory mediator release and downregulation of vascular adhesion molecules, thus reducing leukocyte adhesion and infiltration in the tissues.

In addition, aPC protects and maintains endothelial barrier function and reduces the chemotactic potential of several potent chemotactic agents.[23,27,32–35] Alternatively, the mechanisms by which aPC exerts anti-inflammatory effects on leucocytes are not completely clear; however, evidence points to the important role of EPCR,[21,36–38] which is present on the surface of monocytes, CD56$^+$ natural killer cells, neutrophils but not T and B cells.[33,39–41]

Insights gleaned from *in vitro* and *in vivo* studies have reported direct cytoprotective effects of aPC including beneficial alterations in gene expression profiles, antiapoptotic activities, and stabilization of the endothelial barrier. Each of these activities is distinct and may or may not involve the same intracellular mechanism depending on different factors, such as receptor profile and cell location. Most of aPC known cytoprotective actions seem to be mediated by the EPCR and PAR-1 signaling pathways, but further studies are necessary to clarify these mechanisms. Modulation of gene expression patterns induced *in vitro* by aPC treatment revealed regulation of gene expression for the major pathways of inflammation and apoptosis. aPC suppresses nuclear transcription factor κB (NF-κB)-modulated genes by directly reducing NF-κB expression and functional activity leading to the inhibition of pro-inflammatory cytokine signaling such as TNF-α.[42–45]

With regard to antiapoptotic activity of aPC, this requires the enzymatic active site of aPC and its receptors, EPCR and PAR-1, and reduces many characteristic apoptosis features, including caspase-3 activation, DNA degradation.[46–49] In stressed human brain endothelial cells, aPC reduces the amounts of p53 protein and mRNA and maintains levels of protective Bcl-2 protein, thereby stimulating the intrinsic apoptotic pathway.[46] However, the antiapoptotic effects of aPC are not limited to the intrinsic apoptotic pathway, as aPC counteracts the neurotoxicity of tissue plasminogen activator, which exerts proapoptotic activity via the extrinsic pathway.[48] It has been showed that aPC prolongs the lifespan of *in vivo* circulating monocytes by inhibition of spontaneous monocyte apoptosis,[44] but the molecular mechanisms by which aPC exerts these effects remain still unclear. Reduction of apoptosis is associated with improved survival in sepsis, suggesting a potential important role for aPC-dependent antiapoptotic activity in reducing mortality in sepsis.[50]

Finally, aPC also functions as a cell barrier protective protein, directly through colocalization of aPC/EPCR on lipid rafts, and activation of PAR-1. In turn, aPC/EPCR/PAR-1 cross-activates the sphingosine-1-phosphate receptor 1 (S1P1)[32,51] and/or Tie2 signaling pathways.[52] A direct interaction of EPCR with S1P1 was postulated, but a mechanistic contribution to barrier protective effects remains unclear.[34] It is currently unclear whether the S1P cell survival signals contribute to aPC antiapoptotic activity or other cytoprotective activities. Indirectly, aPC, by downregulating thrombin, inhibits thrombin/PAR-1–mediated vascular permeability,[32,51,53] a process that involves, at least in dendritic cells, cross-activation of S1P3. Many of these observations were made in *in vitro* studies, but recently aPC/EPCR/PAR-1/S1P1 and thrombin/PAR-1/S1P3 pathways were shown to function *in vivo* in endotoxemia models of sepsis.[54] Therefore, aPC and thrombin counterbalance vascular endothelial barrier functions, even though these functions are manifested through the same surface receptor, PAR-1.

The PC pathway in intestinal inflammation

Intestinal barrier function in IBD pathogenesis

The intestinal epithelium plays an active role in the maintenance of mucosal homeostasis.[55,56] The primary function of the epithelial layer is to constitute a tight, highly selective barrier to antigens, solutes, and macromolecules, and regulate the passive movement of solutes through spaces between adjacent epithelial cells. This function is mediated by an apical junctional complex, which is composed of both tight and subadjacent adherens junctions. The tight junctions limit solute flux along the paracellular pathway, functioning as gatekeepers of mucosal permeability.[57,58] Loss of proper barrier function may lead to intestinal inflammation, most likely due to the excessive luminal translocations of fecal antigens leading to the inappropriate activation of the mucosal immune system, as it occurs in IBD patients.[56] The inflammatory bowel diseases represented mainly by ulcerative colitis (UC) and Crohn's disease (CD) are multifactorial chronic inflammatory disorders that include a complex, and not clearly defined, dysfunctional interaction between the microflora of the gut and the mucosal immune system, which leads to progressive destructive damage and defective repair of the gastrointestinal tract.[59] The relationship between barrier dysfunction and IBD has been assessed by several reports. Indeed, a compromise in barrier integrity has been observed in both CD and UC patients, but the mechanisms still remain unclear.[60,61] Patients with Crohn's disease were found to have increased flux of antigen across both the inflamed and uninflamed areas of the mucosa.[62,63] Similarly, UC patients exhibited increased epithelial permeability correlated with the induction of nitric oxide synthase (iNOS) activity.[60,64,65] Furthermore, increased permeability observed in healthy first degree relatives of CD patients, implies that barrier dysfunction could be a very early event in the disease process,[61] whereas in UC patients altered intestinal permeability is most likely the consequence of inflammatory events.[60,64] In addition, in UC patients, the barrier loss is quantitatively massive and different from the defined changes in paracellular permeability that occur in CD patients and their relatives,[64] giving rise to the hypothesis that a different mechanism is responsible for the barrier defects in the two major forms of IBD. TNF-α, IFN-γ, and IL-13 are demonstrated to be the principal pro-inflammatory factors involved in the destruction of the epithelial barrier function.[66,67] TNF-α, which is markedly elevated in IBD patients, seems to cause tight junction barrier dysfunction via a process that requires myosin light chain kinase (MLCK) activation.[68] MLCK activation occurs as a result of increased enzymatic activity and acts on cytoskeleton regulation.[69,70] Both MLCK expression and activity are increased in intestinal epithelial cells of patients with IBD.[68] However, many aspects of these processes still remain unknown.

PC regulation of intestinal inflammation

Although progress has been made in understanding the mechanisms leading to the destruction of barrier function, little is known about the potential pathways involved in enhancing epithelial barrier integrity. Our recent study identified the PC system as a novel pathway involved in the regulation of the intestinal barrier.[11] Surprisingly, besides its classical and traditionally well-known function, PC system is expressed by intestinal epithelial cells and controls epithelial homeostasis by regulating tight junction molecules.[11] Emerging evidence indicates that this system is actively involved in the pathogenesis of several human conditions, including IBD.[2] Alterations in the PC system lead to an imbalance in thrombin generation, which results in abnormal

thrombus formation. Patients suffering from IBD have been associated with an increased risk for extraintestinal thrombosis,[71,72] probably due to an impaired activation of PC. Indeed, the downregulation of TM and EPCR observed in the endothelium of the inflamed mucosa of IBD patients impaired the conversion of PC into aPC, thus enhancing not only extraintestinal thrombosis but also the inflammatory processes in the microvasculature of the inflamed colonic mucosa. Treatment with recombinant aPC reduced the inflammatory response *in vitro* and ameliorated DSS-induced colitis *in vivo*.[9] Although the expression of PC system components on intestinal endothelial cells was predictable, expression on intestinal epithelial cells was unexpected. Although PC is predominately synthesized by the liver, other sources of PC were described, including epididymus, kidney, lung, brain, and male reproductive tissue.[73] Furthermore, positive staining reaction for PC was also found on keratinocytes, where it promoted cell growth and wound healing.[74,75]

There are no studies demonstrating the existence of PC isoforms or comparative analyses between PC isoforms derived from different sources. Therefore, it remains unclear whether PC produced by the liver and PC derived from nonconventional sources are similar in terms of function and structure.

Except for TM, all components of the PC system are expressed by intestinal epithelial cells of noninflamed mucosa and remarkably downregulated in the inflamed mucosa of IBD patients. This confined downregulation of the PC pathway in active inflamed areas suggests that the dysregulation of the PC system components in intestinal epithelial cells is an inflammation-dependent outcome and does not preclude genetic mutations. The presence of EPCR and PAR-1, and the absence of TM expression demonstrate that the intestinal epithelium is able to respond to aPC, but is not capable of converting PC into its active form. The distribution of EPCR and PAR-1 on epithelial cells is still unknown. Epithelial staining of EPCR appears to be distributed at the cell surface in a patchy or punctuate manner.[11] Similar staining pattern of EPCR was also described on endothelial cells, where EPCR is localized on the cell surface in microdomains defined as lipid rafts in association with PAR-1 and TM.[76] Whether PAR-1 and EPCR also associate in lipid rafts on epithelial cells needs to be investigated. Because PC is accumulated in intestinal epithelial cells, it is plausible to think that PC could be converted from other types of TM-expressing cells.[11] However, a TM soluble form, derived from the endothelial membrane is released in plasma, but it is presently unknown whether it is functional.[77] Why the epithelial cells express the inactive form (PC), but are not able to convert it, still needs to be clarified. In the gut, aPC plays an important role in governing intestinal homeostasis. Indeed, an impaired PC activation of intestinal microvasculature endothelial cells induces the release of proinflammatory cytokines and favors increased leukocyte adhesion. Transgenic mice expressing low levels of PC develop spontaneous intestinal inflammation and a more pronounced experimental colitis phenotype after challenge with DSS, thus reinforcing the evidence that low levels of PC in these mice induce a proinflammatory state. Furthermore, besides being associated with high mortality, prothrombotic, and proinflammatory phenotypes, these transgenic mice display an altered intestinal barrier function due to altered expression of ZO-1, JAM-A, and claudin-3,[11] three tight junction proteins important for intestinal barrier function.[78,79] Interestingly, recombinant aPC *in vitro* is able to abrogate TNF-dependent barrier function alteration reverting ZO-1 and JAM-A downregulation. Therefore, aPC acts as a gatekeeper of the intestinal barrier function enhancing the expression of tight junction proteins. Intracellular pathways triggered by aPC-mediated stabilization of the epithelial barrier are still unknown. The protective effects of aPC on endothelial barrier integrity are mediated through its interaction with EPCR and the subsequent cleavage and activation of PAR-1.[80]

So far there is no evidence of the direct involvement of PAR-1 in regulating the intestinal barrier function. PAR-1 is expressed on intestinal epithelial cells and its downregulation upon inflammation correlates with decreased expression of EPCR,[11] but there are no data demonstrating its functional role on intestinal epithelial cells. Further studies are necessary to address this point. The aPC/EPCR/PAR-1 complex on endothelial cells cross-activates sphingosine 1 phosphate (S1P), which is a signaling sphingolipid. It has been reported that S1P is involved in the regulation of proliferation, survival, differentiation, and migration of many cells types, including endothelial cells.[81] Whether S1P signaling mediates aPC effects on epithelial cells needs to be better investigated. S1P, similar to aPC, reverted the TNF-dependent effect on epithelial cells, enhancing transepithelial electric resistance, but these effects were not mediated by S1P receptor 1, which has been found to mediate endothelial barrier stabilization. Therefore, further studies will be necessary to explore the link between aPC and S1P signaling pathway on epithelial cells. Activated PC maintains intestinal homeostasis not only by enhancing tight junction expression, but also by promoting the proliferation and migration of intestinal epithelial cells. As observed in the skin, topical administration of aPC stimulates wound healing, by activating skin re-epithelialization[82] and attenuates cutaneous lupus erythematosus;[83] in the intestine as well, exogenous administration of aPC improves mucosal wound healing. Indeed, intrarectal administration of recombinant aPC reduced a number of ulcers in DSS-induced colitis, thus

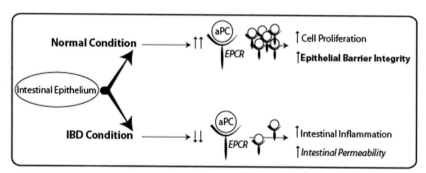

Figure 1. Diagram of PC pathway activity in intestinal epithelial cells. In the normal condition, the PC pathway is highly expressed by intestinal epithelial cells. Activated PC (aPC) bound to endothelial protein C receptor (EPCR) maintains intestinal barrier integrity, enhancing tight junction expression and promoting cell growth. In the IBD condition, activation of PC is impaired and the expression of PC pathway components is downregulated. Consequently, tight junction fence function is reduced and became leaky, leading to increased intestinal permeability and triggering intestinal inflammation.

accelerating the re-epithelialization process without provoking mucosal bleeding[11] (Fig. 1). These data suggest that aPC, by reinforcing tight junction protein expression and acting as a growth factor for intestinal epithelial cells, could be considered as a new therapeutic approach for IBD treatment.

Conclusions

Besides its anticoagulant activity, the PC pathway exerts intestinal cytoprotective effects not only acting on the endothelial compartment but, unexpectedly, by controlling intestinal homeostasis. Indeed, aPC maintains intestinal barrier integrity by enhancing tight junction expression and promoting cell growth. The PC system therefore could be considered as a novel pathway involved in the regulation of intestinal barrier function. Indeed, the alteration of PC expression in the gut epithelium leads to spontaneous altered intestinal permeability and intestinal inflammation, which are well-recognized components of the pathogenesis of IBD. These data, in line with previous findings, shift the axis of interest of PC pathway from its hemostatic role to anti-inflammatory functions, in particular the cytoprotective activity on intestinal epithelial cells. The manipulation of the PC pathway components in the intestinal epithelium may shed light on the development of new ways of treating IBD and mucosal inflammation.

Acknowledgments

The authors thank Carmen Correale for her assistance with the manuscript. This study was supported by grants from IBD Research Foundation (Mini Grant 2010), European Crohn's and Colitis Organization (ECCO) (Grant 2009), Ministero della Salute (GR -2009 convenzione 76) to S.V.

Conflicts of interest

The authors declare no conflicts of interest.

References

1. Esmon, C.T. 2006. Inflammation and the activated protein C anticoagulant pathway. *Semin. Thromb. Hemost.* **32**:(Suppl 1) 49–60.
2. Danese, S. *et al.* 2010. The protein C pathway in tissue inflammation and injury: pathogenic role and therapeutic implications. *Blood* **115**: 1121–1130.
3. Esmon, C.T. 2002. New mechanisms for vascular control of inflammation mediated by natural anticoagulant proteins. *J. Exp. Med.* **196**: 561–564.
4. Okajima, K. 2004. Regulation of inflammatory responses by activated protein C: the molecular mechanism(s) and therapeutic implications. *Clin. Chem. Lab. Med.* **42**: 132–141.
5. Conway, E.M. *et al.* 2002. The lectin-like domain of thrombomodulin confers protection from neutrophil-mediated tissue damage by suppressing adhesion molecule expression via nuclear factor kappaB and mitogen-activated protein kinase pathways. *J. Exp. Med.* **196**: 565–577.
6. Kurosawa, S., C.T. Esmon & D.J. Stearns-Kurosawa. 2000. The soluble endothelial protein C receptor binds to activated neutrophils: involvement of proteinase-3 and CD11b/CD18. *J. Immunol.* **165**: 4697–4703.
7. Iwaki, T. *et al.* 2005. A cardioprotective role for the endothelial protein C receptor in lipopolysaccharide-induced endotoxemia in the mouse. *Blood* **105**: 2364–2371.
8. Murakami, K. *et al.* 1997. Activated protein C prevents LPS-induced pulmonary vascular injury by inhibiting cytokine production. *Am. J. Physiol.* **272**(2 Pt 1): L197–L202.
9. Scaldaferri, F. *et al.* 2007. Crucial role of the protein C pathway in governing microvascular inflammation in inflammatory bowel disease. *J. Clin. Invest.* **117**: 1951–1960.
10. Crawley, J.T., & M. Efthymiou. 2008. Cytoprotective effect of activated protein C: specificity of PAR-1 signaling. *J. Thromb. Haemost.* **6**: 951–953.
11. Vetrano, S. *et al.* 2011. Unexpected role of anticoagulant protein C in controlling epithelial barrier integrity and intestinal inflammation. *Proc. Natl. Acad. Sci. USA* **108**: 19830–19835.
12. Esmon, C.T., J. Stenflo & J.W. Suttie. 1976. A new vitamin K-dependent protein. A phospholipid-binding zymogen of a serine esterase. *J. Biol. Chem.* **251**: 3052–3056.
13. Esmon, C.T. 2003. The protein C pathway. *Chest* **124**(3 Suppl): 26S–32S.
14. Stearns-Kurosawa, D.J. *et al.* 1996. The endothelial cell protein C receptor augments protein C activation by the thrombin-thrombomodulin complex. *Proc. Natl. Acad. Sci. USA* **93**: 10212–10216.
15. Kalafatis, M., M.D. Rand & K.G. Mann. 1994. The mechanism of inactivation of human factor V and human factor Va by activated protein C. *J. Biol. Chem.* **269**: 31869–31880.
16. Nicolaes, G.A. *et al.* 1995. Peptide bond cleavages and loss of functional activity during inactivation of factor Va and factor VaR506Q by activated protein C. *J. Biol. Chem.* **270**: 21158–21166.
17. Yegneswaran, S. *et al.* 1997. Protein S alters the active site location of activated protein C above the membrane surface. A fluorescence resonance energy transfer study of topography. *J. Biol. Chem.* **272**: 25013–25021.
18. Koeleman, B.P., P.H. Reitsma, C.F. Allaart & R.M. Bertina. 1994. Activated protein C resistance as an additional risk factor for thrombosis in protein C-deficient families. *Blood* **84**: 1031–1035.
19. Macias, W.L., S.B. Yan, M.D. Williams, *et al.* 2005. New insights into the protein C pathway: potential implications for the biological activities of drotrecogin alfa (activated). *Crit. Care* **9**(Suppl 4): S38–S45.
20. Mizutani, A., K. Okajima, M. Uchiba & T. Noguchi. 2000. Activated protein C reduces ischemia/reperfusion-induced

renal injury in rats by inhibiting leukocyte activation. *Blood* **95:** 3781–3787.

21. Yuda, H. *et al.* 2004. Activated protein C inhibits bronchial hyperresponsiveness and Th2 cytokine expression in mice. *Blood* **103:** 2196–2204.

22. Grinnell, B.W. & D. Joyce. 2001. Recombinant human activated protein C: a system modulator of vascular function for treatment of severe sepsis. *Crit. Care. Med.* **29**(7 Suppl): S53–60; discussion S60–1.

23. Nick, J.A. *et al.* 2004. Recombinant human activated protein C reduces human endotoxin-induced pulmonary inflammation via inhibition of neutrophil chemotaxis. *Blood* **104:** 3878–3885.

24. Schoots, I.G. *et al.* 2004. Inhibition of coagulation and inflammation by activated protein C or antithrombin reduces intestinal ischemia/reperfusion injury in rats. *Crit. Care. Med.* **32:** 1375–1383.

25. Isobe, H. *et al.* 2004. Activated protein C reduces stress-induced gastric mucosal injury in rats by inhibiting the endothelial cell injury. *J. Thromb. Haemost.* **2:** 313–320.

26. Suzuki, K. *et al.* 2004. Protective role of activated protein C in lung and airway remodeling. *Crit. Care. Med.* **32**(5 Suppl): S262–S265.

27. Murakami, K. *et al.* 1996. Activated protein C attenuates endotoxin-induced pulmonary vascular injury by inhibiting activated leukocytes in rats. *Blood* **87:** 642–647.

28. Yasui, H. *et al.* 2001. Intratracheal administration of activated protein C inhibits bleomycin-induced lung fibrosis in the mouse. *Am. J. Respir. Crit. Care Med.* **163:** 1660–1668.

29. Shibata, M. *et al.* 2001. Anti-inflammatory, antithrombotic, and neuroprotective effects of activated protein C in a murine model of focal ischemic stroke. *Circulation* **103:** 1799–1805.

30. Lay, A.J. *et al.* 2005. Mice with a severe deficiency in protein C display prothrombotic and proinflammatory phenotypes and compromised maternal reproductive capabilities. *J. Clin. Invest.* **115:** 1552–1561.

31. Lay, A.J. *et al.* 2007. Acute inflammation is exacerbated in mice genetically predisposed to a severe protein C deficiency. *Blood* **109:** 1984–1991.

32. Feistritzer, C. & M. Riewald. 2005. Endothelial barrier protection by activated protein C through PAR1-dependent sphingosine 1-phosphate receptor-1 crossactivation. *Blood* **105:** 3178–3184.

33. Feistritzer, C. *et al.* 2003. Endothelial protein C receptor-dependent inhibition of human eosinophil chemotaxis by protein C. *J. Allergy Clin. Immunol.* **112:** 375–381.

34. Finigan, J.H. *et al.* 2005. Activated protein C mediates novel lung endothelial barrier enhancement: role of sphingosine 1-phosphate receptor transactivation. *J. Biol. Chem.* **280:** 17286–17293.

35. Zeng, W. *et al.* 2004. Effect of drotrecogin alfa (activated) on human endothelial cell permeability and Rho kinase signaling. *Crit. Care Med.* **32**(5 Suppl): S302–S308.

36. Sadlack, B. *et al.* 1993. Ulcerative colitis-like disease in mice with a disrupted interleukin-2 gene. *Cell* **75:** 253–261.

37. Shimizu, S. *et al.* 2003. Activated protein C inhibits the expression of platelet-derived growth factor in the lung. *Am. J. Respir. Crit. Care Med.* **167:** 1416–1426.

38. Stephenson, D.A. *et al.* 2006. Modulation of monocyte function by activated protein C, a natural anticoagulant. *J. Immunol.* **177:** 2115–2122.

39. Galligan, L. *et al.* 2001. Characterization of protein C receptor expression in monocytes. *Br. J. Haematol.* **115:** 408–414.

40. Joyce, D.E., D.R. Nelson & B.W. Grinnell. 2004. Leukocyte and endothelial cell interactions in sepsis: relevance of the protein C pathway. *Crit. Care Med.* **32**(5 Suppl): S280–S286.

41. Sturn, D.H. *et al.* 2003. Expression and function of the endothelial protein C receptor in human neutrophils. *Blood* **102:** 1499–1505.

42. Franscini, N. *et al.* 2004. Gene expression profiling of inflamed human endothelial cells and influence of activated protein C. *Circulation* **110:** 2903–2909.

43. Joyce, D.E. *et al.* 2001. Gene expression profile of antithrombotic protein c defines new mechanisms modulating inflammation and apoptosis. *J. Biol. Chem.* **276:** 11199–11203.

44. Joyce, D.E. & B.W. Grinnell. 2002. Recombinant human activated protein C attenuates the inflammatory response in endothelium and monocytes by modulating nuclear factor-kappaB. *Crit. Care Med.* **30**(5 Suppl): S288–S293.

45. Riewald, M. & W. Ruf. 2005. Protease-activated receptor-1 signaling by activated protein C in cytokine-perturbed endothelial cells is distinct from thrombin signaling. *J. Biol. Chem.* **280:** 19808–19814.

46. Cheng, T. *et al.* 2003. Activated protein C blocks p53-mediated apoptosis in ischemic human brain endothelium and is neuroprotective. *Nat. Med.* **9:** 338–342.

47. Joyce, D.E. *et al.* 2001. Gene expression profile of antithrombotic protein c defines new mechanisms modulating inflammation and apoptosis. *J. Biol. Chem.* **276:** 11199–11203.

48. Liu, D. *et al.* 2004. Tissue plasminogen activator neurovascular toxicity is controlled by activated protein C. *Nat. Med.* **10:** 1379–1383.

49. Mosnier, L.O. & J.H. Griffin. 2003. Inhibition of staurosporine-induced apoptosis of endothelial cells by activated protein C requires protease-activated receptor-1 and endothelial cell protein C receptor. *Biochem. J.* **373**(Pt 1): 65–70.

50. Hotchkiss, R.S. *et al.* 2000. Caspase inhibitors improve survival in sepsis: a critical role of the lymphocyte. *Nat. Immunol.* **1:** 496–501.

51. Feistritzer, C. *et al.* 2006. Protective signaling by activated protein C is mechanistically linked to protein C activation on endothelial cells. *J. Biol. Chem.* **281:** 20077–20084.

52. Minhas, N. *et al.* 2009. Activated protein C utilizes the angiopoietin/Tie2 axis to promote endothelial barrier function. *FASEB J* **24:** 873–881.

53. Bae, J.S. *et al.* 2007. Engineering a disulfide bond to stabilize the calcium-binding loop of activated protein C eliminates its anticoagulant but not its protective signaling properties. *J. Biol. Chem.* **282:** 9251–9259.

54. Niessen, F. *et al.* 2008. Dendritic cell PAR1-S1P3 signalling couples coagulation and inflammation. *Nature* **452:** 654–658.

55. Shao, L., D. Serrano & L. Mayer. 2001. The role of epithelial cells in immune regulation in the gut. *Semin. Immunol.* **13:** 163–176.

56. Yu, Y., S. Sitaraman & A.T. Gewirtz. 2004. Intestinal epithelial cell regulation of mucosal inflammation. *Immunol. Res.* **29:** 55–68.

57. Matter, K. & M.S. Balda. 2003. Signalling to and from tight junctions. *Nat. Rev. Mol. Cell Biol.* **4:** 225–236.

58. Schneeberger, E.E. & R.D. Lynch. 2004. The tight junction: a multifunctional complex. *Am. J. Physiol. Cell Physiol.* **286:** C1213–C1228.

59. Danese, S. & C. Fiocchi. 2006. Etiopathogenesis of inflammatory bowel diseases. *World J. Gastroenterol.* **12:** 4807–4812.

60. Schmitz, H. *et al.* 2000. Epithelial barrier and transport function of the colon in ulcerative colitis. *Ann. N.Y. Acad. Sci.* **915:** 312–326.

61. Katz, K.D. *et al.* 1989. Intestinal permeability in patients with Crohn's disease and their healthy relatives. *Gastroenterology* **97:** 927–931.

62. Adenis, A., *et al.* 1992. Increased pulmonary and intestinal permeability in Crohn's disease. *Gut* **33:** 678–682.

63. Hollander, D. *et al.* 1986. Increased intestinal permeability in patients with Crohn's disease and their relatives. A possible etiologic factor. *Ann. Intern. Med.* **105:** 883–885.

64. Gitter, A.H. *et al.* 2001. Epithelial barrier defects in ulcerative colitis: characterization and quantification by electrophysiological imaging. *Gastroenterology* **121:** 1320–1328.

65. Guihot, G. *et al.* 2000. Inducible nitric oxide synthase activity in colon biopsies from inflammatory areas: correlation with inflammation intensity in patients with ulcerative colitis but not with Crohn's disease. *Amino Acids* **18:** 229–237.

66. Ma, T.Y. *et al.* 2004. TNF-alpha-induced increase in intestinal epithelial tight junction permeability requires NF-kappa B activation. *Am. J. Physiol. Gastrointest. Liver Physiol.* **286:** G367–G376.

67. Heller, F. *et al.* 2005. Interleukin-13 is the key effector Th2 cytokine in ulcerative colitis that affects epithelial tight junctions, apoptosis, and cell restitution. *Gastroenterology* **129:** 550–564.

68. Blair, S.A. *et al.* 2006. Epithelial myosin light chain kinase expression and activity are upregulated in inflammatory bowel disease. *Lab Invest* **86:** 191–201.

69. Zolotarevsky, Y. *et al.* 2002. A membrane-permeant peptide that inhibits MLC kinase restores barrier function in in vitro models of intestinal disease. *Gastroenterology* **123:** 163–172.

70. Su, L. *et al.* 2009. Targeted epithelial tight junction dysfunction causes immune activation and contributes to development of experimental colitis. *Gastroenterology* **136:** 551–563.

71. Irving, P.M., K.J. Pasi & D.S. Rampton. 2005. Thrombosis and inflammatory bowel disease. *Clin. Gastroenterol. Hepatol.* **3:** 617–628.

72. van Bodegraven, A.A. *et al.* 2001. Hemostatic imbalance in active and quiescent ulcerative colitis. *Am. J. Gastroenterol.* **96:** 487–493.

73. Yamamoto, K. & D.J. Loskutoff. 1998. Extrahepatic expression and regulation of protein C in the mouse. *Am. J. Pathol.* **153:** 547–555.

74. Xue, M., D. Campbell & C.J. Jackson. 2007. Protein C is an autocrine growth factor for human skin keratinocytes. *J. Biol. Chem.* **282:** 13610–13616.

75. Xue, M. *et al.* 2004. Activated protein C stimulates proliferation, migration and wound closure, inhibits apoptosis and upregulates MMP-2 activity in cultured human keratinocytes. *Exp. Cell Res.* **299:** 119–127.

76. Bae, J.S., L. Yang & A.R. Rezaie. 2007. Receptors of the protein C activation and activated protein C signaling pathways are colocalized in lipid rafts of endothelial cells. *Proc. Natl. Acad. Sci. USA* **104:** 2867–2872.

77. Ishii, H. & P.W. Majerus. 1985. Thrombomodulin is present in human plasma and urine. *J. Clin. Invest.* **76:** 2178–2181.

78. Poritz, L.S. *et al.* 2007. Loss of the tight junction protein ZO-1 in dextran sulfate sodium induced colitis. *J. Surg. Res.* **140:** 12–19.

79. Vetrano, S. *et al.* 2008. Unique role of junctional adhesion molecule-a in maintaining mucosal homeostasis in inflammatory bowel disease. *Gastroenterology* **135:** 173–184.

80. Young, N. & J.R. Van Brocklyn. 2006. Signal transduction of sphingosine-1-phosphate G protein-coupled receptors. *ScientificWorldJournal* **6:** 946–966.

81. Rosen, H. *et al.* 2007. Tipping the gatekeeper: S1P regulation of endothelial barrier function. *Trends Immunol.* **28:** 102–107.

82. Whitmont, K. *et al.* 2008. Treatment of chronic leg ulcers with topical activated protein C. *Arch. Dermatol.* **144:** 1479–1483.

83. Lichtnekert, J. *et al.* 2011. Activated protein C attenuates systemic lupus erythematosus and lupus nephritis in MRL-Fas(lpr) mice. *J. Immunol.* **187:** 3413–3421.

Ann. N.Y. Acad. Sci. ISSN 0077-8923

ANNALS OF THE NEW YORK ACADEMY OF SCIENCES

Issue: *Barriers and Channels Formed by Tight Junction Proteins*

Analysis of absorption enhancers in epithelial cell models

Rita Rosenthal,[1] Miriam S. Heydt,[2] Maren Amasheh,[3] Christoph Stein,[2] Michael Fromm,[1] and Salah Amasheh[1]

[1]Institute of Clinical Physiology, Campus Benjamin Franklin, Charité – Universitätsmedizin Berlin, Freie Universität and Humboldt Universität, Berlin, Germany. [2]Department of Anesthesiology and Critical Care Medicine, Campus Benjamin Franklin, Charité Universitätsmedizin Berlin, Freie Universität and Humboldt Universitt, Berlin, Germany. [3]Department of Gastroenterology, Division of Nutritional Medicine, Campus Benjamin Franklin, Charité Universitätsmedizin Berlin, Freie Universität and Humboldt Universität, Berlin, Germany

Address for correspondence: Salah Amasheh, Institute of Clinical Physiology, Charité, Campus Benjamin Franklin, Hindenburgdamm 30, 12203 Berlin, Germany. salah.amasheh@charite.de

A variety of chemical compounds are currently being discussed as novel drug delivery strategies. One promising strategy is to selectively open the paracellular pathway of epithelia for the passage of macromolecules. A prerequisite for this effect is a rapid and reversible action of these compounds, to allow a marked translocation of a drug, but also to avoid unwanted adverse effects, such as the translocation of noxious agents. Bioactive molecules that elevate paracellular permeability include Ca^{2+} chelators, bacterial toxins, and other compounds, some of which perturb the structural basis of epithelial barrier function—the tight junction. Within the tight junction, organ- and tissue-specific barrier properties are determined mainly by claudins. The majority of members of the claudin protein family seal the paracellular pathway. This paper focuses on recent approaches concerning absorption-enhancing effects, with regard to selectivity and mechanism.

Keywords: tight junctions; claudins; chitosan; HT-29/B6

Tight junctions as targets for drug absorption enhancement

Research of epithelial absorption enhancers dates back half a century, when the Ca^{2+} chelator ethylenediaminetetraacetic acid (EDTA) was shown to increase absorption of heparin in rats and dogs.[1] Presently, a number of absorption enhancers are discussed or already employed in combination with drugs to facilitate absorption, for example, via the nasal route,[2] the skin,[3] and the gastrointestinal tract.[4]

Three parameters determine epithelial barrier properties, namely trans- and paracellular permeability, and local peaks of permeability caused by apoptoses (Fig. 1).

Transcellular transport via receptors or transporters is energy dependent, vectorial, and is strongly dependent on specific substrate recognition and molecular size, which limits the range of desired applications.[5] In contrast, the paracellular route provides passive permeability that varies between different epithelia on functional and molecular levels, and can be perturbed to induce macromolecule permeability.[4] Therefore, the paracellular pathway is a promising target for tissue-specific permeabilization to enhance the absorption of hydrophilic compounds, such as peptide and protein-based medications. These substances cannot permeate across cell membranes and therefore are hindered from achieving certain desired therapeutic responses, compared to lipophilic drugs.

An exception might be specific membrane transporters, as for example, demonstrated by the intestinal oligopeptide transporter PepT1, which recognizes therapeutics such as beta-lactam antibiotics and 5-aminolevulinic as substrates.[5,6] This specific recognition of substrates is due to molecular properties resembling those of di- and tripeptides.[7] However, as a plethora of therapeutic substances do not meet the requirements for selective

doi: 10.1111/j.1749-6632.2012.06562.x

Ann. N.Y. Acad. Sci. 1258 (2012) 86–92 © 2012 New York Academy of Sciences.

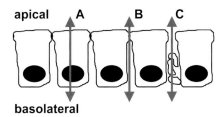

Figure 1. Trans- and paracellular pathways. Epithelial barrier properties are determined by three components, namely transcellular permeability (A), the paracellular pathway (B), and locally limited peaks of permeability caused by apoptoses (C).

substrate recognition of transcellular transporters, and at least two sites of selective transcellular translocation are involved—namely the apical and the basolateral side of epithelia—a promising approach and more versatile strategy is to enhance the absorption of hydrophilic macromolecular drugs by co-administration of absorption-enhancing agents that reversibly open the paracellular barrier.

The structural correlate of the paracellular pathway is the tight junction (TJ), which interconnects the apicolateral membranes of neighboring cells by tetraspan TJ proteins. These proteins, including occludin, tricellulin, and the family of claudins, have been shown to determine paracellular barrier properties.[8–10] Within different organs and tissues, specific predominant expression of single TJ proteins can be regarded responsible for local barrier properties.[11,17]

Members of the claudin family are key determinants for epithelial barrier function in many different epithelia, as demonstrated for intestine,[11] skin,[12] brain capillary endothelium,[13] and nasal epithelium (Fig. 2).[14] Moreover, the discovery of further relevant sites continues, such as mammary epithelium[15] and pleura.[16] In rat intestine, the segment-specific localization of claudins has been demonstrated to reflect different barrier properties along the longitudinal axis in detail, showing differential expression in duodenum, jejunum, ileum, and colon. In these segments, the highest barrier function was measured in colon, followed by duodenum, jejunum, and ileum. Barrier properties were in accordance with expression levels of different claudins in these segments, and the same correlation has been reported for claudin expression along the different segments of the nephron.[17]

Within the tight junction complex interconnecting two neighboring epithelial cells, claudin-based

tight junction strands primarily determine barrier properties.[18] In contrast, occludin, though ubiquitously expressed within the majority of epithelial tight junctions, is not essential for barrier properties as a complete lack does not result in a perturbation of barrier function.[19]

Besides the finding that the majority of claudins contribute to the sealing of the epithelial barrier, single members of the claudin family can functionally contribute to specific permeability. For example, claudin-2 mediates paracellular permeability as a channel for small cations and water,[20,21] and claudin-17 provides an anion channel.[22] For specific permeabilization of the paracellular pathway, tightening claudins are regarded as a promising target, as their perturbation may lead to a temporary opening of this route. However, although an increase of the paracellular channel claudin-2 can be observed in inflamed tissue,[16,23] an augmented expression of claudin-2 alone does not lead to enhanced paracellular permeability for macromolecules.[18] This may at least require further supporting factors, which then aggravate and sustain inflammation. Therefore, effects on other sealing TJ proteins, such as claudin-1, claudin-3, and claudin-5, represent more promising strategies for induction of drug permeability in different epithelia,[24–26] and perturbation of these proteins may markedly induce changes of barrier properties in different tissues, such as intestine,[11] skin,[12] brain capillary endothelium,[13] and nasal epithelium (Fig. 2).[14]

Furthermore, a number of claudins have been reported to be dysregulated in tumors.[27] Apart from detection of claudins as tumor markers, this highlights a crucial role for cell fate, and therefore is discussed as a link for specific targeting and eradication of malignant tissues.[28] Moreover, current drug development strategies focus on virus entry as discussed for hepatitis C virus and HIV, which have been reported to be dependent on the expression of claudin-1 and -5, respectively.[29,30] In addition, inflammatory responses are highly dependent on barrier properties, and perturbation of the tight junction can be antagonized by secondary plant compounds. [31]

Methods for analyses of absorption enhancer effects

For *in vitro* analyses of epithelial barrier properties, the intestinal cell lines HT-29/B6 and Caco-2

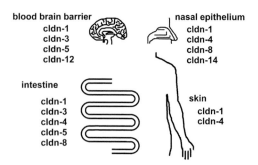

Figure 2. Putative barrier builders as tissue-specific targets. The cartoon represents typical and predominant tissue-specific expression of claudins with a sealing function in BBB, nasal epithelium, gastrointestinal tract, and skin.

have been established as reliable models representing intestinal barrier properties.[4,32] Both cell types are derived from human colorectal carcinoma and grow as confluent monolayers on permeable supports. HT-29/B6 cells are mucin-producing cells with inducible chloride secretion and a broad expression of claudins within the TJ, which also determine the *in vivo* barrier in human epithelia.[32] The use of HT-29/B6 monolayer as intestinal model epithelium has previously been validated by parallel studies in native rat and human colon epithelia. The results of these studies were in accordance with the cell model concerning responses to proinflammatory cytokines perturbing the epithelial barrier, the counter-regulatory preventive effects of secondary plant compounds, and general physiological barrier regulation.[31,32]

Cells grown on permeable supports are suitable for functional barrier analyses, as well as molecular techniques and laser-scanning immunofluorescent microscopy. To examine possible cytotoxic effects on cultured monolayers mediated by absorption enhancers, cell viability tests such as lactate dehydrogenase (LDH) release might be performed.[4] For electrophysiological measurements, confluent monolayers grown on permeable filters can be mounted in Ussing chambers custom-designed for the insertion of supports. Apart from manual measurement with a chopstick electrode, this technique does not only allow exact transepithelial resistance (R^t, TER) measurements, but also voltage clamp (0 mV) to avoid substrate or marker translocation driven by charge in unidirectional flux measurements. Moreover, in modified Ussing chambers two-path impedance spectroscopy mea-

surements can be performed, which discriminate between the trans- and the paracellular resistance (R^{trans} and R^{para}).[18] To evaluate translocation of macromolecular hypothetical therapeutics, dextran flux of different molecular weights can be performed (e.g., FD-4, FD-10, FD-20 with 4, 10, and 20 kDa, respectively).[4] However, to directly analyze selective permeability for one single molecule, further techniques have to be combined.

Mechanisms opening the paracellular pathway

A number of different molecules have been considered as enhancers for drug absorption. These include organic bioactive compounds, such as Ca^{2+} chelators, fatty acids, bacterial toxins, and other molecules (examples listed in Table 1). Mechanisms underlying the perturbation of the epithelial barrier vary: some compounds directly interact with single TJ proteins,[33] whereas others disturb barrier function by a general modification of TJ strands,[34] or may act without detectable change of TJ proteins.[4]

Ca^{2+} chelation

Ca^{2+} chelators, such as ethylenediaminetetraacetic acid (EDTA) and ethylene glycol tetraacetic acid (EGTA), are known as potent absorption enhancers with high specificity for the paracellular pathway. Extracellular chelation of Ca^{2+} triggers condensation of the intracellular perijunctional actomyosin ring, resulting in an internalization of TJ proteins, which are connected with the cytoskeleton via the PDZ-domain binding scaffolding zonula occludens-1 protein (ZO-1).[35,36] Surprisingly, a loss of ZO-1 expression does not markedly affect transepithelial resistance (R^t), indicating that the epithelial cell may maintain its function by recruiting additional scaffolding proteins, for example, ZO-2.[37] The specific effect of Ca^{2+} chelators can be used to identify a selective opening of the paracellular pathway in *in vitro* studies, as demonstrated for two-path impedance spectroscopy.[34]

Clostridium perfringens enterotoxin

Clostridium perfringens enterotoxin (CPE) causes a common food borne illness via binding to claudin-3, -4, -6, -7, -8, and -14, but not claudin-5 and -10.[35] A prominent example of this binding is represented by claudin-3 and claudin-4, formerly known as CPE-R1 and CPER, which stands for Clostridium

Table 1. Paracellular absorption enhancers: compounds, mechanisms, and selectivity

Compound	Mechanism	Selectivity	Reference
CPE	claudin binding	4 kDa FITC dextran	33, 59
chitosan	charge	10 kDa FITC dextran	4
EGTA/EDTA	via Ca^{2+}	10 kDa FITC dextran	34
decanoylcarnitin	via Ca^{2+}	4 kDa FITC dextran	60
sodium caprate	via Ca^{2+}	4 kDa FITC dextran	60
ZOT	via PKC	PEG 4000	35

perfringens receptor and reflects the susceptibility of these membrane proteins for CPE. After binding to the second extracellular loop (ECL2) of claudins via the Clostridium perfringens enterotoxin C-terminal domain (cCPE), the N-terminal domains induce pore formation of the mucosal plasma membrane inducing apoptosis.[38] The noncytotoxic cCPE alone targets single claudin subtypes and thereby increases paracellular permeability.[33,39] Moreover, full-length CPE might be employed to target cells overexpressing claudins, known as tumor markers, as a novel approach in anticancer therapies.[40]

Chitosan

Chitosan is a nontoxic biopolymer, which has been reported as an enhancer of drug delivery in gastrointestinal, dermal, nasal, ocular, and pulmonary epithelia. Moreover, chitosan-based microemulsions were tested in mice for drug delivery through the blood–brain barrier (BBB).[41] The mechanisms underlying the effects on the paracellular pathway across the intestinal epithelial layer have recently been analyzed in detail in HT-29/B6 and Caco-2 cells.[4,42] In Caco-2 cells, a reversible opening of TJs by chitosan resulted in both decreased transepithelial electrical resistance (R^t) and increased paracellular marker fluxes.[43] Further studies reported a reorganization of the actin cytoskeleton in Caco-2 cells induced by chitosan.[44,45] In contrast, others did not observe morphological changes in the actin cytoskeleton.[46,47] In addition, changes in the distribution of occludin and the TJ-associated scaffolding protein ZO-1 were observed in these cells.[45,46] Chitosan appears to activate a protein kinase C (PKC)-dependent signaling pathway that affects the integrity of the TJ in Caco-2 cells, as the chitosan-mediated decrease in R^{trans} and changes in the cellular localization of ZO-1 could be inhibited by

blocking of PKC.[48] Moreover, in Caco-2 cells, the opening of the TJ was associated with an increased expression and altered localization of claudin-1 and -4.[42,49]

Chitosan has a pK_a of 5.5–7.0 and therefore is positively charged at a pH lower than 7.0. Several studies demonstrated that the effect of chitosan on paracellular permeability is dependent on this positive charge.[43] Therefore, a number of chitosan derivatives have been developed to increase chitosan's efficacy as absorption enhancer at neutral and alkaline pH values, such as trimethyl chitosan (TMC) or mono-*N*-carboxymethyl chitosan (MCC), which had comparable effects on the TJs.[50,51] Furthermore, chitosan nanoparticles, and chitosan-coated microspheres were developed for macromolecular drug delivery via endocytosis and transcytosis.[52] Chitosan nanoparticles or chitosan-coated phospholipid vesicles also caused a decrease in R^t, and an increase in permeability of paracellular flux markers, demonstrating an additional effect on paracellular permeability.[53]

In HT-29/B6 cells chitosan induced a rapid and reversible decrease in R^t, and two-path impedance spectroscopy revealed effects on both the paracellular (R^{para}) and the transcellular (R^{trans}) resistance.[4] pH-dependence and inhibition of both effects by a negatively charged macromolecule, heparin, indicated that chitosan acted only in the protonated form. The decrease in R^{trans} was mediated by activation of a chloride-bicarbonate exchanger involved in intracellular pH regulation. This activation was coupled to the decrease in R^{para} which was associated with an increase in permeability for ions and for paracellular flux markers up to 10 kDa. No effects on expression and subcellular distribution of TJ proteins or the actin cytoskeleton were observed.

These results demonstrate that the absorption-enhancing effect of chitosan in HT-29/B6 cells is due to changes in intracellular pH caused by the activation of a chloride-bicarbonate exchanger resulting in an opening of the TJ, facilitating the passage of macromolecules of up to 10 kDa.

Fatty acids

Sodium caprate is a medium chain fatty acid, which is clinically employed as an enhancer for intestinal absorption and as a component of a rectal suppository.[54] Functional effects on intestinal barrier properties have been analyzed in Caco-2 cells.[55] Moreover, it is also discussed for transdermal applications.[56] One important advantage of the compound is its safety profile, as an intact epithelial barrier is essential for physiological homeostasis and defense against extrinsic antigens.[57] However, sodium caprate compares well with other dietary constituents and endogenous secretions and, because of its safety profile, is superior to other agents, as for example, aspirin or alcohol.[58] On a molecular level, a decrease of claudin-1 and occludin within the TJ has been reported, with a dispersion into fragmented patterns on the cell–cell contact region and translocation as granular structures in the cytoplasm.[56] However, it remains to be elucidated whether this mechanism is sufficient to explain the rapid and reversible mechanism of paracellular macromolecule passage in intestine, and whether the observed mechanisms can be transferred to the intestine *in vivo*.

Concluding remarks

The majority of TJ proteins seal the paracellular pathway in a tissue- and organ-specific way. Therefore, they represent a promising target for paracellularly acting absorption enhancers to promote drug delivery in different epithelia, as discussed for the nasal route, the skin, the gastrointestinal tract, and the brain capillary endothelium, which constitute the BBB. Effects of absorption enhancers can differ, with effects on single claudins, the perijunctional actomyosin ring, and further mechanisms. An important prerequisite for the use as an absorption enhancer, however, is a rapid and reversible action to avoid unwanted side effects. Intestinal cell models, such as monolayers of HT-29/B6 cells, represent valuable tools to analyze absorption enhancer effects on epithelial barrier properties in detail.

Conflicts of interest

The authors declare no conflicts of interest.

References

1. Windsor, E. & G.E. Cronheim. 1961. Gastro-intestinal absorption of heparin and synthetic heparinoids. *Nature* **190**: 263–264.
2. Duan, X. & S. Mao. 2010. New strategies to improve the intranasal absorption of insulin. *Drug Discov. Today* **15**: 416–427.
3. Zhou, X., D. Liu, H. Liu, *et al.* 2010. Effect of low molecular weight chitosans on drug permeation through mouse skin: 1. Transdermal delivery of baicalin. *J. Pharm. Sci.* **99**: 2991–2998.
4. Rosenthal, R., D. Günzel, C. Finger, *et al.* 2012. The effect of chitosan on transcellular and paracellular mechanisms in the intestinal epithelial barrier. *Biomaterials* **33**: 2791–2800.
5. Döring, F., J. Walter, J. Will, *et al.* 1998. Delta-aminolevulinic acid transport by intestinal and renal peptide transporters and its physiological and clinical implications. *J. Clin. Invest.* **101**: 2761–2767.
6. Amasheh, S., U. Wenzel, M. Boll, *et al.* 1997. Transport of charged dipeptides by the intestinal H$^+$/peptide symporter PepT1 expressed in Xenopus laevis oocytes. *J. Membr. Biol.* **155**: 247–256.
7. Döring, F., J. Will, S. Amasheh, *et al.* 1998. Minimal molecular determinants of substrates for recognition by the intestinal peptide transporter. *J. Biol. Chem.* **273**: 23211–23218.
8. Furuse, M., T. Hirase, M. Itoh, *et al.* 1993. Occludin: a novel integral membrane protein localizing at tight junctions. *J. Cell Biol.* **123**: 1777–1788.
9. Ikenouchi, J., M. Furuse, K. Furuse, *et al.* 2005. Tricellulin constitutes a novel barrier at tricellular contacts of epithelial cells. *J. Cell Biol.* **171**: 939–945.
10. Furuse, M., K. Fujita, T. Hiiragi, *et al.* 1998. Claudin-1 and -2: novel integral membrane proteins localizing at tight junctions with no sequence similarity to occludin. *J. Cell Biol.* **141**: 1539–1550.
11. Markov, A.G., A. Veshnyakova, M. Fromm, *et al.* 2010. Segmental expression of claudin proteins correlates with tight junction barrier properties in rat intestine. *J. Comp. Physiol. B* **180**: 591–598.
12. Yuki, T., A. Hachiya, A. Kusaka, *et al.* 2011. Characterization of tight junctions and their disruption by UVB in human epidermis and cultured keratinocytes. *J. Invest. Dermatol.* **131**: 744–752.
13. Nitta, T., M. Hata, S. Gotoh, *et al.* 2003. Size-selective loosening of the blood-brain barrier in claudin-5-deficient mice. *J. Cell Biol.* **161**: 653–660.
14. Takano, K., T. Kojima, M. Go, *et al.* 2005. HLA-DR- and CD11c-positive dendritic cells penetrate beyond well-developed epithelial tight junctions in human nasal mucosa of allergic rhinitis. *J. Histochem. Cytochem.* **53**: 611–619.
15. Markov, A.G., N.M. Kruglova, Y.A. Fomina, *et al.* 2011. Altered expression of tight junction proteins in mammary epithelium after discontinued suckling in mice. *Pflügers Arch.* **463**: 391–398.

16. Markov, A.G., M.A. Voronkova, G.N. Volgin, *et al.* 2011. Tight junction proteins contribute to barrier properties in human pleura. *Respir. Physiol. Neurobiol.* **175:** 331–335.

17. Amasheh, S., M. Fromm & D. Günzel. 2011. Claudins of intestine and nephron—a correlation of molecular tight junction structure and barrier function. *Acta Physiol. (Oxf)* **201:** 133–140.

18. Furuse, M., K. Furuse, H. Sasaki & S. Tsukita. 2001. Conversion of zonulae occludentes from tight to leaky strand type by introducing claudin-2 into Madin-Darby canine kidney I cells. *J. Cell Biol.* **153:** 263–272.

19. Schulzke, J.D., A.H. Gitter, J. Mankertz, *et al.* 2005. Epithelial transport and barrier function in occludin-deficient mice. *Biochim. Biophys. Acta* **1669:** 34–42.

20. Amasheh, S., N. Meiri, A.H. Gitter, *et al.* 2002. Claudin-2 expression induces cation-selective channels in tight junctions of epithelial cells. *J. Cell Sci.* **115:** 4969–4976.

21. Rosenthal, R., S. Milatz, S.M. Krug, *et al.* 2010. Claudin-2, a component of the tight junction, forms a paracellular water channel. *J. Cell Sci.* **123:** 1913–1921.

22. Krug, S.M., D. Günzel, M.P. Conrad, *et al.* 2012. Claudin-17 forms tight junction channels with distinct anion selectivity. *Cell. Mol. Life Sci.* DOI: 10.1007/s00018-012-0949-x

23. Amasheh, S., S. Dullat, M. Fromm, *et al.* 2009. Inflamed pouch mucosa possesses altered tight junctions indicating recurrence of inflammatory bowel disease. *Int. J. Colorectal Dis.* **24:** 1149–1156.

24. Furuse, M., M. Hata, K. Furuse, *et al.* 2002. Claudin-based tight junctions are crucial for the mammalian epidermal barrier: a lesson from claudin-1-deficient mice. *J. Cell Biol.* **156:** 1099–1111.

25. Milatz, S., S.M. Krug, R. Rosenthal, *et al.* 2010. Claudin-3 acts as a sealing component of the tight junction for ions of either charge and uncharged solutes. *Biochim. Biophys. Acta Biomembr.* **1798:** 2048–2057.

26. Amasheh, S., T. Schmidt, M. Mahn, *et al.* 2005. Contribution of claudin-5 to barrier properties in tight junctions of epithelial cells. *Cell Tiss. Res.* **321:** 89–89.

27. Tsukita, S., Y. Yamazaki, T. Katsuno, *et al.* 2008. Tight junction-based epithelial microenvironment and cell proliferation. *Oncogene* **27:** 6930–6938.

28. Kondoh, M., T. Yoshida, H. Kakutani & K. Yagi. 2008. Targeting tight junction proteins-significance for drug development. *Drug. Discov. Today* **13:** 180–186.

29. Evans, M.J., T. von Hahn, D.M. Tscherne, *et al.* 2007. Claudin-1 is a hepatitis C virus co-receptor required for a late step in entry. *Nature* **446:** 801–805.

30. Lu, T.S., H.K. Avraham, S. Seng, *et al.* 2008. Cannabinoids inhibit HIV-1 Gp120-mediated insults in brain microvascular endothelial cells. *J. Immunol.* **181:** 6406–6416.

31. Amasheh, M., A. Fromm, S.M. Krug, *et al.* 2010. TNFalpha-induced and berberine-antagonized tight junction barrier impairment via tyrosine kinase, Akt and NFkappaB signaling. *J. Cell Sci.* **123:** 4145–4155.

32. Amasheh, S., S. Milatz, S.M. Krug, *et al.* 2009. Na$^+$ absorption defends from paracellular back-leakage by claudin-8 upregulation. *Biochem. Biophys. Res. Commun.* **378:** 45–50.

33. Fujita, K., J. Katahira, Y. Horiguchi, *et al.* 2000. Clostridium perfringens enterotoxin binds to the second extracellular loop of claudin-3, a tight junction integral membrane protein. *FEBS Lett.* **476:** 258–261.

34. Krug, S.M., M. Fromm & D. Günzel. 2009. Two-path impedance spectroscopy for measuring paracellular and transcellular epithelial resistance. *Biophys. J.* **97:** 2202–2211.

35. Fasano, A., S. Uzzau, C. Fiore & K. Mararetten. 1997. The enterotoxic effect of Zonula Occludens toxin on rabbit small intestine involves the paracellular pathway. *Gastroentrology* **112:** 839–846.

36. Rothen-Rutishauser, B., F.K. Riesen, A. Braun, *et al.* 2002. Dynamics of tight and adherens junctions under EGTA treatment. *J. Membr. Biol.* **188:** 151–162.

37. Yu, D., A.M. Marchiando, C.R. Weber, *et al.* 2010. MLCK-dependent exchange and actin binding region-dependent anchoring of ZO-1 regulate tight junction barrier function. *Proc. Natl. Acad. Sci. USA* **107:** 8237–8241.

38. Sonoda, N., M. Furuse, H. Sasaki, *et al.* 1999. Clostridium perfringens enterotoxin fragment removes specific claudins from tight junction strands: evidence for direct involvement of claudins in tight junction barrier. *J. Cell Biol.* **147:** 195–204.

39. Takahashi, A., Y. Saito, M. Kondoh, *et al.* 2012. Creation and biochemical analysis of a broad-specific claudin binder. *Biomaterials* **33:** 3464–3474.

40. Kominsky, S.L., M. Vali, D. Korz, *et al.* 2004. Clostridium perfringens enterotoxin elicits rapid and specific cytolysis of breast carcinoma cells mediated through tight junction proteins claudin 3 and 4. *Am. J. Pathol.* **164:** 1627–1633.

41. Yao, J., J.P. Zhou, Q.N. Ping, *et al.* 2008. Distribution of nobiletin chitosan-based microemulsions in brain following i.v. injection in mice. *Int. J. Pharm.* **352:** 256–262.

42. Yeh, T.H., L.W. Hsu, M.T. Tseng, *et al.* 2011. Mechanism and consequence of chitosan-mediated reversible epithelial tight junction opening. *Biomaterials* **32:** 6164–6173.

43. Kotze, A.F., H.L. Luessen, A.G. de Boer, *et al.* 1999. Chitosan for enhanced intestinal permeability: prospects for derivatives soluble in neutral and basic environments. *Eur J. Pharm. Sci.* **7:** 145–151.

44. Schipper, N.G., S. Olsson, J.A. Hoogstraate, *et al.* 1997. Chitosans as absorption enhancers for poorly absorbable drugs 2: mechanism of absorption enhancement. *Pharm. Res.* **14:** 923–929.

45. Ranaldi, G., I. Marigliano, I. Vespignani, *et al.* 2002. The effect of chitosan and other polycations on tight junction permeability in the human intestinal Caco-2 cell line. *J. Nutr. Biochem.* **13:** 157–167.

46. Smith, J., E. Wood & M. Dornish. 2004. Effect of chitosan on epithelial cell tight junctions. *Pharm. Res.* **21:** 43–49.

47. Dodane, V. & B. Kachar. 1996. Identification of isoforms of G proteins and PKC that colocalize with tight junctions. *J. Membr. Biol.* **149:** 199–209.

48. Smith, J.M., M. Dornish & E.J. Wood. 2005. Involvement of protein kinase C in chitosan glutamate-mediated tight junction disruption. *Biomaterials* **26:** 3269–3276.

49. Dorkoosh, F.A., C.A. Broekhuizen, G. Borchard, *et al.* 2004. Transport of octreotide and evaluation of mechanism of opening the paracellular tight junctions using superporous hydrogel polymers in Caco-2 cell monolayers. *J. Pharm. Sci.* **93:** 743–752.

50. Thanou, M., J.C. Verhoef & H.E. Junginger. 2001. Chitosan and its derivatives as intestinal absorption enhancers. *Adv. Drug Deliv. Rev.* **50:** S91–101.

51. Thanou, M., M.T. Nihot, M. Jansen, *et al.* 2001. Mono-N-carboxymethyl chitosan (MCC), a polyampholytic chitosan derivative, enhances the intestinal absorption of low molecular weight heparin across intestinal epithelia *in vitro* and in vivo. *J. Pharm. Sci.* **90:** 38–46.

52. Sandri, G., M.C. Bonferoni, S. Rossi, *et al.* 2010. Insulin-loaded nanoparticles based on N-trimethyl chitosan: *in vitro* (Caco-2 model) and ex vivo (excised rat jejunum, duodenum, and ileum) evaluation of penetration enhancement properties. *AAPS PharmSciTech.* **11:** 362–371.

53. Kudsiova, L. & M.J. Lawrence. 2008. A comparison of the effect of chitosan and chitosan-coated vesicles on monolayer integrity and permeability across Caco-2 and 16HBE14o-cells. *J. Pharm. Sci.* **97:** 3998–4010.

54. Lindmark, T., J.D. Soderholm, G. Olaison, *et al.* 1997. Mechanism of absorption enhancement in humans after rec-

tal administration of ampicillin in suppositories containing sodium caprate. *Pharm. Res.* **14:** 930–935.

55. Lindmark, T., Y. Kimura & P. Artursson. 1998. Absorption enhancement through intracellular regulation of tight junction permeability by medium chain fatty acids in Caco-2 cells. *J Pharmacol. Exp. Ther.* **284:** 362–369.

56. Kurasawa, M., S. Kuroda, N. Kida, *et al.* 2009. Regulation of tight junction permeability by sodium caprate in human keratinocytes and reconstructed epidermis. *Biochem. Biophys. Res. Commun.* **381:** 171–175.

57. Bjarnason, I. 1994. Intestinal permeability. *Gut* **35:** S18–S22.

58. Wolfe, M.M., D.R. Lichtenstein & G. Singh. 1999. Gastrointestinal toxicity of nonsteroidal antiinflammatory drugs. *N. Engl. J. Med.* **340:** 1888–1899.

59. Kondoh, M., A. Masuyama, A. Takahashi, *et al.* 2005. A novel strategy for the enhancement of drug absorption using a claudin modulator. *Mol. Pharmacol.* **67:** 749–756.

60. Tomita, M., M. Hayashi & S. Awazu. 1996. Absorption-enhancing mechanism of EDTA, caprate, and decanoylcarnitine in Caco-2 cells. *J. Pharm. Sci.* **85:** 608–611.

Ann. N.Y. Acad. Sci. ISSN 0077-8923

ANNALS OF THE NEW YORK ACADEMY OF SCIENCES
Issue: *Barriers and Channels Formed by Tight Junction Proteins*

Calcium regulation of tight junction permeability

Markus Bleich, Qixian Shan, and Nina Himmerkus

Physiologisches Institut der Christian-Albrechts-Universität zu Kiel, Kiel, Germany

Address for correspondence: Markus Bleich, Physiologisches Institut, Olshausenstraße 40, D-24098 Kiel, Germany.
m.bleich@physiologie.uni-kiel.de

Calcium transport in the kidney is a key element in Ca^{2+} homeostasis. Ca^{2+} concentration, or more precisely the activity of freely dissociated Ca^{2+} ions, is a prerequisite for the appropriate function of virtually every cell. Along the renal tubule, about 85% of the filtered Ca^{2+} is transported across tight junctions at the paracellular route of reabsorption. Therefore, claudins, which form the conductive and selective part of the tight junctions, have moved into the focus of interest with respect to regulatory events in the control of Ca^{2+} transport. This control is of particular interest for the kidney since it has to defend itself against nephrocalcinosis and kidney stones. Tight junction proteins provide pathways, driving forces, and regulatory targets for Ca^{2+} transport. Direct regulation of tight junctions by changing Ca^{2+} concentrations allows fast and efficient feedback loops to adapt Ca^{2+} transport to the requirements of kidney function and plasma Ca^{2+} concentration.

Keywords: isolated perfused tubule; kidney; ion transport; claudin; calcium-sensing receptor

Plasma Ca^{2+} concentration is maintained within a narrow range to secure a variety of Ca^{2+}-dependent cellular functions. Ca^{2+} homeostasis thereby becomes a crucial factor for overall body operation, growth, and survival. In concert with intestinal absorption and bone metabolism, renal epithelial Ca^{2+}, HCO_3^-, and phosphate transport represent the control units in the regulation of plasma Ca^{2+} concentration.[1] Major transport pathways affecting Ca^{2+} homeostasis depend both on parathyroid hormone (PTH) and calcitonin, which are both released from the parathyroid gland under the reciprocal control by the extracellular calcium-sensing receptor (CaSR),[2,3] and on calcitriol. CaSR mediates the suppression of PTH release at rising extracellular Ca^{2+} concentrations via cAMP decrease and intracellular PLC/Ca^{2+} signaling, and it engages a variety of further intracellular signaling systems.[2] The importance of this regulatory pathway becomes evident in mice defective for CaSR. These knockout animals die early after birth, but they survive if the production of PTH is suppressed in the same animals by the combined knockout of PTH.[4] This indicates a primary role of CaSR in the regulation of PTH release, but does not exclude an additional function in renal transport regulation.

By glomerular filtration in the kidney, nonprotein-bound Ca^{2+} is delivered to the tubular fluid. Ca^{2+} reabsorption back from the tubule lumen via the basolateral interstitium into the plasma occurs at three major sites:[5] 60% of filtered Ca^{2+} is reabsorbed by the proximal tubule (PT), 25% by the thick ascending limb of the loop of Henle (TAL), and final excretion is tuned by transport in the distal convoluted- and connecting tubule (DCT and CNT).[6] While Ca^{2+} transport in DCT and CNT occurs across the cells (transcellular) via luminal TRPV5 cation channels,[7] bulk reabsorption in PT and TAL takes the paracellular route. The conductive part of this paracellular pathway is made by claudins. They form a charge- and size-selective paracellular pore meshwork, and every nephron segment has its specific signature of expressed claudins that confer the permeability properties to its tight junctions.[8,9] Thereby, claudins determine if tight junctions of a nephron segment operate as a barrier or as a conduit for Ca^{2+}. Since renal Ca^{2+} transport is a function of plasma as well as tubular Ca^{2+} concentration, paracellular Ca^{2+} transport mechanisms in PT and TAL would be the regulatory targets. And as Ca^{2+} homeostasis is so important for organ function, it is self evident that, in addition to PTH control, the kidney

doi: 10.1111/j.1749-6632.2012.06539.x

tubular system itself must have means of regulation and feedback mechanisms at different levels. For example, in cases of high Ca^{2+} intake, the mere reduction of PTH has been insufficient to prevent an increase in plasma Ca^{2+} concentration.[10]

How can Ca^{2+} transport be affected at the tight junction level? There are at least three ways to control Ca^{2+} absorption in PT and TAL nephron segments: (1) by changes in the protein composition of tight junctions; (2) by controlling the driving forces for Ca^{2+} absorption across the tight junctions; and (3) by direct biophysical interference with either claudins or transcellular transport proteins at the molecular level.

Control of Ca^{2+} absorption by changes in tight junction protein composition

The synthesis of data on claudin expression along the nephron, and from functional data obtained in epithelial cell lines expressing the respective claudins,[11–13] resulted in the view that the protein composition of tight junctions strongly influences the permeability properties of the paracellular pathway.[8] According to these cell culture experiments, claudin-2 and claudin-10 are the major determinants of high ion permeability with cation selectivity in the PT. Fortunately, there are also data available from isolated perfused PT from claudin-2 knockout mice.[14] The knockout of claudin-2 nicely demonstrates how strongly a change in cation selectivity affects PT function. These mice lose cation permeability in PT, and cation selectivity (P_{Na}/P_{Cl}) changes in favor of Cl^-. To understand the consequence of these changes on PT paracellular transport, we have to take a look at the electrophysiology along the PT.

Based on preferred HCO_3^- absorption in PT, Cl^- concentration along the tubule increases with respect to the peritubular side, and the late PT develops a small and lumen-positive transepithelial Cl^- diffusion voltage. This diffusion voltage depends on P_{Na}/P_{Cl} and is therefore higher in claudin-2 knockout mice. Hence, claudin-2 knockout mice are facing two oppositely directed changes with consequences for Ca^{2+} absorption in PT. On the one hand, the diffusion voltage gets more positive and increases the driving force for Ca^{2+} absorption; on the other hand, the pathway for Ca^{2+} becomes more resistive. In these mice, the latter effect is predominant and leads to a threefold loss of Ca^{2+} in claudin-2 knockout mice (Fig. 1).[14] A regulatory decrease in

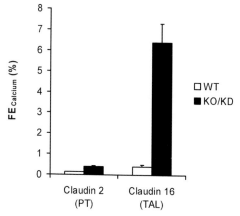

Figure 1. Fractional excretion (FE) of Ca^{2+} (% of the glomerular filtrate) in wild-type mice (WT) and in the respective littermates defective for claudin-2 (knockout, KO) or claudin-16 (RNA-interference-knockdown, KD). Data are redrawn from Muto *et al.*[14] and Himmerkus *et al.*[18] Both KO and KD have a significant reduction in Ca^{2+} absorption, which is only partially compensated by downstream Ca^{2+} transport. Claudin-2 is predominantly expressed in PT, claudin 16 in TAL.

claudin-2 expression would therefore be a tool to impede Ca^{2+} transport at its major site of tubular absorption.

In the thick ascending limb, functional studies from knockdown and knockout mice have shown that the tight junction proteins claudin-16 and claudin-19[15,16] interfere in two ways with paracellular Ca^{2+} transport.[17–20] On the one hand, their expression is necessary to provide cation selectivity to the pathway that secures the driving force for Ca^{2+} absorption by the generation of a lumen-positive diffusion voltage; on the other hand, the knockout of claudin-16 is correlated with the reduction of Ca^{2+} permeability.[20] Both reduced lumen-positive voltage and decreased permeability synergistically result in a limitation of Ca^{2+} absorption and a corresponding hypercalciuria (Fig. 1). It is important to note that the defect in TAL cation permeability has a much higher impact on Ca^{2+} handling compared to the PT, pointing to the critical role of this segment. Corresponding to claudin-2 in the PT, regulatory decrease of claudin-16 and -19 expression in TAL would therefore be a powerful tool to impede Ca^{2+} transport at this site of tubular absorption.

Combined studies on function and expression of claudins in PT and TAL under the influence of chronic changes in whole-body Ca^{2+} metabolism are needed to clarify the role of paracellular protein composition for Ca^{2+} metabolism.

Control of Ca^{2+} absorption by changes in driving forces across the tight junctions

As evident from studies in the TAL, it is always the combination of driving force and paracellular permeability for Ca^{2+} that finally determines the rate of its reabsorption. In the TAL, the transepithelial voltage of the lumen with respect to the interstitial side is in the range of +10 mV and is generated by secondary active transcellular NaCl transport.[21] It finally increases to about +30 mV if the tubular fluid becomes dilute by continuous NaCl absorption. In this case, the voltage source is a diffusion potential across the cation-selective paracellular junction.[17,22] A regulation of Ca^{2+} absorption via driving forces could therefore act directly on the mechanisms of transcellular NaCl transport. In the TAL, the respective transport proteins are ROMK, NKCC2, ClC-K, Na^+, and K^+-ATPase, with their respective regulatory proteins and pathways. An inhibition of NKCC2, for example, would result in a breakdown of lumen-positive transepithelial voltage, since Cl^- transport current is inhibited and tubular fluid would not be diluted anymore. Hence, there would be not driving forces for Ca^{2+} absorption. A regulation of Ca^{2+} transport via driving forces could also act on the composition and properties of claudins, which generate the diffusion voltage by their cation selectivity (P_{Na}/P_{Cl}) at a given dilution gradient. The TAL tight junctions are composed of at least 7–9 claudins, among which claudin-16 and -19 have been in focus as prime candidates for the regulation of Ca^{2+} transport.[23] Others are likely to follow within a short time (unpublished observations, and see Ref. 24).

In the PT, as mentioned above, the situation seems to be more complicated. The transepithelial voltage is very small, and it changes from lumen negative (which drives Ca^{2+} secretion) to lumen positive in the later PT (which drives Ca^{2+} absorption).[25] In addition, solutes, including Ca^{2+}, are leaving the lumen, dragged by a continuous water flux across the tight junctions. This water flux in turn is driven by an isoosmotic mechanism of solute absorption[26] that lives on transmembrane Na^+ transport in PT. Thereby, transcellular absorption of Na^+, together with organic substrates or HCO_3^-, becomes the main driver of PT Ca^{2+} absorption

and an attractive target for its regulation. The positive effect of thiazide diuretics on Ca^{2+} absorption in the kidney gives a nice example in this context. By their induction of volume contraction, these diuretics stimulate PT Na^+ absorption which is again paralleled by the respective increase in Ca^{2+} transport.[27,28] A very recent study strengthens this view by showing a direct correlation between PT Na^+ transport via Na^+/H^+ exchanger type 3 (NHE3)—which facilitates the uptake of Na^+ in exchange for H^+ in PT luminal brushborder membrane to mediate transcellular Na^+ and HCO_3^- reabsorption and Ca^{2+} absorption.[29] From the perspective of driving forces, targets in PT to modulate Ca^{2+} transport would therefore be claudins involved in paracellular selectivity and cation permeability—e.g., -2, -10, and the proteins mediating bulk transcellular Na^+ transport like NHE3, the basolateral Na^+,HCO_3^- cotransporter and Na^+, K^+-ATPase.

Control of Ca^{2+} absorption by direct Ca^{2+} interference with either claudins or transcellular transport proteins at the molecular level

Direct interference of Ca^{2+} with those proteins involved in its transport would be an attractive regulatory option. In fact, early studies in gallbladder epithelium showed that Ca^{2+} affects paracellular properties by its putative interference with charged residues at tight junction structures,[30,31] rendering the tight junction itself to be a Ca^{2+} sensor. The critical question is whether this occurs at physiologically relevant Ca^{2+} concentrations. The CaSR fulfills this criterion and has been reviewed extensively as the (exclusive) key sensor of Ca^{2+} concentration in the kidney.[32–35] The CaSR is described to take over the critical step of communicating either luminal (PT) or basolateral (TAL) Ca^{2+} activity to the respective transport proteins via its intracellular second messenger pathways.[33] However, there are few data from renal tubules, and there is increasing evidence that CaSR is not the only Ca^{2+} sensor influencing tubular Ca^{2+} absorption, salt/water transport, and thereby Ca^{2+} homeostasis. We and others have studied isolated perfused TAL of mouse and rabbit for the influence of Ca^{2+} on NaCl transport and tight junction properties (Fig. 2). The experiments show that basolateral Ca^{2+} increases paracellular resistance and decreases paracellular cation selectivity. As mentioned above, this feeds into

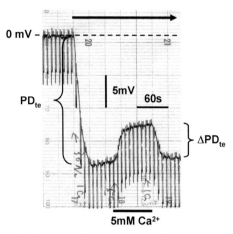

Figure 2. Transepithelial measurement of voltage versus time in an isolated perfused segment of TAL. In the continuous presence of furosemide (50 μM), transcellular transport is inhibited, and the experimentally applied transepithelial NaCl gradient (arrow) from the lumen (145 mM) to the bath (30 mM) generates a transepithelial diffusion voltage PD_{te}. Increase of basolateral Ca^{2+} concentration from 1.3 to 5 mM reduces the absolute value of PD_{te}. This change (ΔPD_{te}) reflects the reduction of paracellular cation selectivity. The effect is fast, reversible, and independent of intracellular signaling.

reduced Ca^{2+} absorption via reduced driving force and reduced pathway conductance. It is important to note that these effects were acute and independent of CaSR signaling. Neither interference with cAMP concentrations[36] nor pharmacological modulation of CaSR signaling (unpublished) had any influence on the effect of Ca^{2+} on tight junction properties. Along these lines, Yu *et al.* performed a molecular analysis of tight junction Ca^{2+} regulation in dog renal epithelial cells, expressing claudin-2 and thereby resembling tight junction properties of PT.[37] Similar to TAL, in these experiments, Ca^{2+} increases paracellular resistance and decreases paracellular cation selectivity, the effect being entirely dependent on a molecular signature of the tight junction protein, namely a charged amino acid in the pore region. These findings give the first molecular example of claudins as Ca^{2+} sensors affecting driving force and permeability for Ca^{2+} transport, independent of membrane receptors and respective intracellular signaling pathways. In this context, interpretations of experiments using anionic modulators of CaSR (like neomycin) appear in a new light, since these substances are at the same time candidates to bind at the "Ca^{2+} sensor" of tight junction proteins. Yu *et al.*[37] also offer an alternative expla-

nation for the relation between renal Ca^{2+} load and a consecutive increase in diuresis. So far, this phenomenon has been increasingly explained by the effects of CaSR signaling on transcellular tubular Na^+ absorption. Now, we have learned that Ca^{2+} is able to directly increase paracellular resistance and to directly decrease paracellular Na^+ permeability by interaction with claudins, thereby triggering diuresis by a reduction in paracellular tubular Na^+ transport. The access of solutes to the tight junction protein network is limited by the narrow basolateral cleft. Within this unstirred compartment, this may result in different solute concentrations at the Ca^{2+} binding site. As shown by Yu *et al.*,[37] Na^+ and Ca^{2+} compete for binding at the pore of claudin 2. In addition, NaCl concentration in the basolateral space at the tight junction directly determines Ca^{2+} activity via the ionic strength (c.f., technical note), and it is tempting to also speculate that the electrical field along the paracellular path might have a significant influence on the Ca^{2+} activity at the claudin Ca^{2+} receptor (Fig. 3).

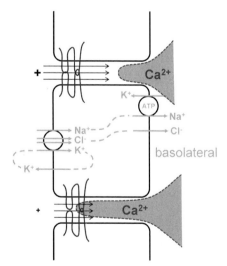

Figure 3. Paracellular pathways under the influence of Ca^{2+}. Cell model for the TAL. To affect the tight junction, Ca^{2+} has to reach its site of action at the respective tight junction proteins. The access of basolateral Ca^{2+} is determined by its activity (gray color of Ca^{2+} "cloud" reaching into the paracellular cleft) and by the electrical field along the basolateral cleft. This electrical field is indicated by the arrows and the repulsion increases with lumen-positive voltage. Thereby, Ca^{2+} activity at the binding site within the claudin network at the tight junction depends on ionic strength (determining Ca^{2+} activity) and on the transepithelial electrical field.

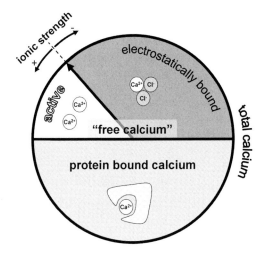

Figure 4. Calcium species in clinical and experimental settings. Total calcium is divided into different fractions: protein bound, electrostatically bound, and active Ca^{2+}. The term "Ca^{2+} concentration" is mostly used for "free calcium," which equals the molar concentration of calcium in protein-free experimental solutions. Only a part of free calcium is active Ca^{2+}. Protein concentration and the ionic strength of the solution determine the proportion between the fractions and thereby the concentration of biologically active Ca^{2+}.

Of course, this does not weaken the role of CaSR as one important element among others in the renal contribution to Ca^{2+} homeostasis. Combined PTH and CaSR knockout mice have been used to further investigate the renal role of CaSR, thereby finding that its major task in renal transport was the limitation of a plasma Ca^{2+} excess at increased ingestion by the suppression of Ca^{2+} transport.[10] In this study, plasma Ca^{2+} in CaSR knockouts had to increase much higher to obtain the same effect on renal Ca^{2+} excretion—i.e., the kidney is able to respond to Ca^{2+} in the absence of CaSR, but CaSR shifts the renal Ca^{2+} response to a more narrow and fine tuned range.

There is another reason for a strict and efficient control of renal tubular Ca^{2+} absorption. Renal concentrating ability bears the thread of Ca^{2+} exceeding the concentration limit of its dissolution product and forming precipitates in the tubular lumen. This requires a feedback control from tubular Ca^{2+} concentration to Ca^{2+} reabsorption or a feedback leading to a limitation of urine concentration. On the other hand, the local enrichment of Ca^{2+} in the peritubular fluid would also challenge kidney structure in phases of Ca^{2+} hyper-absorption by precipitations at the interstitial site. In the discussion of

renal self-defense against nephrocalcinosis and kidney stones, tubular transport, tubular flow rate, peritubular perfusion, renal blood flow, and glomerular filtration rate have to be taken into account. Together with the above-mentioned properties of the paracellular pathways, the composition of tubular and peritubular fluid is the integrated result of all these factors. Further studies in renal claudin and CaSR physiology in native tubular epithelia might help to give more insight into the pathogenesis of nephrocalcinosis, in particular in diuretic salt wasting such as antenatal Bartter syndrome or neonatal furosemide treatment.[38,39]

In conclusion, the renal contribution to Ca^{2+} homeostasis is intrinsically tied to the function and regulation of tight junction proteins in PT and TAL. These proteins provide pathways, driving forces, and regulatory targets for Ca^{2+} transport. Direct Ca^{2+} regulation of tight junctions allows fast and efficient feedback loops to adapt Ca^{2+} transport to the requirements of kidney function and plasma Ca^{2+} concentration.

Practical remark

As in this manuscript, many articles on calcium homeostasis and metabolism use the term "Ca^{2+} concentration" to indicate the physicochemical fraction of calcium that is most relevant for the respective study. However, this might be confusing if absolute values for this parameter are given or if different techniques for calcium measurement and calibration procedures have been applied.

The problem has been nicely addressed by Siggaard-Andersen *et al.*[40] Figure 4 clarifies the definitions. Total calcium concentration is the concentration of all calcium atoms as measured, for example, by atomic absorption spectroscopy or flame photometry. In human plasma, total calcium concentration is in the range of 2.5 mmol/L.[41] About 50% of plasma calcium is bound to proteins, and therefore only 1.25 mmol/L is the concentration of calcium available, for example, after glomerular filtration. This nonprotein-bound fraction of total calcium is frequently named "free calcium concentration." It corresponds to the total calcium concentration of many experimental bath solutions and also to commercially available calibration solutions. Although this fraction of calcium is "free" of protein binding, the majority is still electrostatically bound to small anions. This binding depends on the ionic strength of the respective electrolyte solution. The nonbound part is the active Ca^{2+} concentration. Since tight junction physiology depends on the interaction of active Ca^{2+} ions with the respective binding sites and on concentration gradients of active Ca^{2+}, only Ca^{2+} activity is the concentration that is biologically relevant in this case. Ca^{2+} activity in plasma and many experimental solutions is in the range of 0.4 mmol/L. This calcium activity is also the parameter that is measured by calcium-selective electrodes in clinical chemistry and daily lab routine. Why does the clinical lab deliver Ca^{2+} values of 1.25 mmol/L instead, although the measurements have been done with calcium-selective electrodes? The simple explanation is that calibration of the electrodes has been done with solutions of known free Ca^{2+} concentrations. As long as there is a constant relation between free Ca^{2+} and Ca^{2+} activity, there is nothing to object. However, this is not the case if calibration solutions and samples differ in composition. This pitfall has to be kept in mind dealing with Ca^{2+} sensitive processes and measurements. The activity coefficient for Ca^{2+} depends on the ionic strength of the respective experimental solution under investigation. This means that a simple reduction of the NaCl concentration in the same sample (e.g., in a patient with hyponatremia or in the TAL after dilution of the luminal fluid) results in a substantial increase in Ca^{2+} activity. A dilution of NaCl concentration from 145 to 30 mmol/L causes after all an increase in Ca^{2+} activity from 0.4 to almost 0.7 mmol/L (for calculation of ionic strength and activity coefficient, please refer to the respective textbooks of physical chemistry). Since measurements of calcium concentrations are frequently done with ion-selective electrodes, this means in turn that the calibration solution must have the same ionic strength as the sample of interest. If, for example, experimental samples of different ionic strength are measured on an automated analyzer that is calibrated for standard plasma ionic strength (0.16 mol/kg), calcium values for samples of lower strength will be overestimated, since the readout in clinical chemistry is always free calcium as derived from the calibration solution composition.

Acknowledgments

We thank Dr. Paul Steels (BIOMED, Physiology, University of Hasselt, Belgium) for continuous, fruitful, and inspiring discussions, and Allein Plain for critical reading of the manuscript.

Conflicts of interest

The authors declare no conflicts of interest.

References

1. Lang, F. 1996. Ca^{2+}, Mg^{2+}, and Phosphate Metabolism. In *Comprehensive Human Physiology – From Cellular Mechanisms To Integration*. Greger, R. & U. Windhorst, Eds.: 1595–1609. Springer Verlag. Berlin Heidelberg New York.
2. Riccardi, D. & E.M. Brown. 2010. Physiology and pathophysiology of the calcium-sensing receptor in the kidney. *Am. J. Physiol. Renal. Physiol.* **298:** F485–F499.
3. Fudge, N.J. & C.S. Kovacs. 2004. Physiological studies in heterozygous calcium sensing receptor (CaSR) gene-ablated mice confirm that the CaSR regulates calcitonin release in vivo. *BMC. Physiol.* **4:** 5.
4. Kos, C.H., A.C. Karaplis, J.B. Peng, *et al.* 2003. The calcium-sensing receptor is required for normal calcium homeostasis independent of parathyroid hormone. *J. Clin. Invest.* **111:** 1021–1028.
5. Greger, R. 1996. Renal handling of the individual solutes of glomerular filtrate. In *Comprehensive Human Physiology – From Cellular Mechanisms To Integration*. Greger, R. & U. Windhorst, Eds.: 1517–1544. Springer Verlag. Berlin Heidelberg New York.
6. Hoenderop, J.G. & R.J. Bindels. 2005. Epithelial Ca2+ and Mg2+ channels in health and disease. *J. Am. Soc. Nephrol.* **16:** 15–26.
7. Boros, S., R.J. Bindels & J.G. Hoenderop. 2009. Active Ca(2+) reabsorption in the connecting tubule. *Pflugers. Arch.* **458:** 99–109.
8. Muto, S., M. Furuse & E. Kusano. 2011. Claudins and renal salt transport. *Clin. Exp. Nephrol.* doi: 10.1007/s10157-011-0491-4.
9. Günzel, D. & A.S. Yu. 2009. Function and regulation of claudins in the thick ascending limb of Henle. *Pflugers. Arch.* **458:** 77–88.
10. Kantham, L., S.J. Quinn, O.I. Egbuna, *et al.* 2009. The calcium-sensing receptor (CaSR) defends against hypercalcemia independently of its regulation of parathyroid hormone secretion. *Am. J. Physiol. Endocrinol. Metab.* **297:** E915–E923.
11. Angelow, S., R. Ahlstrom & A.S. Yu. 2008. Biology of Claudins. *Am. J. Physiol. Renal. Physiol.* **295:** F867–F876.
12. Rosenthal, R., S. Milatz, S.M. Krug, *et al.* 2010. Claudin-2, a component of the tight junction, forms a paracellular water channel. *J. Cell Sci.* **123:** 1913–1921.
13. Furuse, M., K. Furuse, H. Sasaki & S. Tsukita. 2001. Conversion of zonulae occludentes from tight to leaky strand type by introducing claudin-2 into Madin-Darby canine kidney I cells. *J. Cell Biol.* **153:** 263–272.
14. Muto, S., M. Hata, J. Taniguchi, *et al.* 2010. Claudin-2-deficient mice are defective in the leaky and cation-selective paracellular permeability properties of renal proximal tubules. *Proc. Natl. Acad. Sci. USA* **107:** 8011–8016.

15. Hou, J., A. Renigunta, M. Konrad, *et al.* 2008. Claudin-16 and claudin-19 interact and form a cation-selective tight junction complex. *J. Clin. Invest.* **118:** 619–628.

16. Hou, J. & D.A. Goodenough. 2010. Claudin-16 and claudin-19 function in the thick ascending limb. *Curr. Opin. Nephrol. Hypertens.* **19:** 483–488.

17. Shan, Q., N. Himmerkus, J. Hou, *et al.* 2009. Insights into driving forces and paracellular permeability from claudin-16 knockdown mouse. *Ann. N. Y. Acad. Sci.* **1165:** 148–151.

18. Himmerkus, N., Q. Shan, B. Goerke, *et al.* 2008. Salt- and acid/base metabolism in claudin-16 knockdown mice – impact for the pathophysiology of FHHNC patients. *Am. J. Physiol. Renal. Physiol.* **295:** F1641–F1647.

19. Hou, J., Q. Shan, T. Wang, *et al.* 2007. Transgenic RNAi depletion of claudin-16 and the renal handling of magnesium. *J. Biol. Chem.* **282:** 17114–17122.

20. Will, C., T. Breiderhoff, J. Thumfart, *et al.* 2010. Targeted deletion of murine Cldn16 identifies extra- and intrarenal compensatory mechanisms of Ca2+ and Mg2+ wasting. *Am. J. Physiol. Renal. Physiol.* **298:** F1152–F1161.

21. Greger, R. 1985. Ion transport mechanisms in thick ascending limb of Henle's loop of mammalian nephron. *Physiol. Rev.* **65:** 760–797.

22. Greger, R. 1981. Cation selectivity of the isolated perfused cortical thick ascending limb of Henle's loop of rabbit kidney. *Pflügers. Arch.* **390:** 30–37.

23. Günzel, D., L. Haisch, S. Pfaffenbach, *et al.* 2009. Claudin Function in the Thick Ascending Limb of Henle's Loop. *Ann. N. Y. Acad. Sci.* **1165:** 152–162.

24. Gong, Y., V. Renigunta, N. Himmerkus, *et al.* 2012. Claudin-14 regulates renal Ca(++) transport in response to CaSR signalling via a novel microRNA pathway. *EMBO J.* doi: 10.1038/emboj.2012.49.

25. Fromter, E. & K. Gessner. 1974. Free-flow potential profile along rat kidney proximal tubule. *Pflugers. Arch.* **351:** 69–83.

26. Larsen, E.H. & N. Mobjerg. 2006. Na+ recirculation and isosmotic transport. *J. Membr. Biol.* **212:** 1–15.

27. Dimke, H., J.G. Hoenderop & R.J. Bindels. 2010. Hereditary tubular transport disorders: implications for renal handling of Ca2+ and Mg2+. *Clin. Sci. (Lond.)* **118:** 1–18.

28. Suki, W.N., R.S. Schwettmann, F.C. Rector, Jr. & D.W. Seldin. 1968. Effect of chronic mineralocorticoid administration on calcium excretion in the rat. *Am. J. Physiol.* **215:** 71–74.

29. Pan, W., J. Borovac, Z. Spicer, *et al.* 2011. The Epithelial Sodium-Proton Exchanger, NHE3, Is Necessary For Renal And Intestinal Calcium (Re)Absorption. *Am. J. Physiol. Renal. Physiol.* doi:10.1152/ajprenal.00504.2010.

30. Wright, E.M. & J.M. Diamond. 1968. Effects of pH and polyvalent cations on the selective permeability of gall-bladder epithelium to monovalent ions. *Biochim. Biophys. Acta.* **163:** 57–74.

31. Wright, E.M., P.H. Barry & J.M. Diamond. 1971. The Mechanism of Cation Permeation in Rabbit Gallbladder. *J. Membr. Biol.* **4:** 331–357.

32. Hebert, S.C. 2006. Therapeutic use of calcimimetics. *Annu. Rev. Med.* **57:** 349–364.

33. Ward, D.T. & D. Riccardi. 2002. Renal physiology of the extracellular calcium-sensing receptor. *Pflugers. Arch.* **445:** 169–176.

34. Brown, E.M. & R.J. MacLeod. 2001. Extracellular calcium sensing and extracellular calcium signaling. *Physiol. Rev.* **81:** 239–297.

35. Ward, D.T. & D. Riccardi. 2012. New concepts in calcium-sensing receptor pharmacology and signalling. *Br. J. Pharmacol.* **165:** 35–48.

36. Di Stefano, A., M. Wittner, B. Gebler & R. Greger. 1988. Increased Ca++ or mg++ concentration reduces relative tight-junction permeability to Na+ in the cortical thick ascending limb of Henle's loop of rabbit kidney. *Ren. Physiol. Biochem.* **11:** 70–79.

37. Yu, A.S., M.H. Cheng & R.D. Coalson. 2010. Calcium inhibits paracellular sodium conductance through claudin-2 by competitive binding. *J. Biol. Chem.* **285:** 37060–37069.

38. Sayer, J.A., G. Carr & N.L. Simmons. 2004. Nephrocalcinosis: molecular insights into calcium precipitation within the kidney. *Clin. Sci. (Lond.)* **106:** 549–561.

39. Pattaragarn, A., J. Fox & U.S. Alon. 2004. Effect of the calcimimetic NPS R-467 on furosemide-induced nephrocalcinosis in the young rat. *Kidney Int.* **65:** 1684–1689.

40. Siggaard-Andersen, O., J. Thode & N. Fogh-Andersen. 1983. What is "ionized calcium"? *Scand. J. Clin. Lab. Invest. Suppl.* **165:** 11–16.

41. Greger, R. & M. Bleich. 1996. Normal values for physiological parameters. In *Comprehensive Human Physiology – From Cellular Mechanism to Integration.* Greger, R. & U. Windhorst, Eds.: 2427–2450. Springer Verlag. Berlin Heidelberg New York.

Ann. N.Y. Acad. Sci. ISSN 0077-8923

ANNALS OF THE NEW YORK ACADEMY OF SCIENCES
Issue: *Barriers and Channels Formed by Tight Junction Proteins*

Effects of quercetin studied in colonic HT-29/B6 cells and rat intestine *in vitro*

Maren Amasheh,[1] Julia Luettig,[1] Salah Amasheh,[2] Martin Zeitz,[1] Michael Fromm,[2] and Jörg-Dieter Schulzke[1]

[1]Department of Gastroenterology, Infectious Diseases, and Rheumatology; Division of Nutritional Medicine, Charité—Universitätsmedizin Berlin, Berlin, Germany. [2]Institute of Clinical Physiology, Charité—Universitätsmedizin Berlin, Berlin, Germany

Address for correspondence: Prof. Dr. Jörg-Dieter Schulzke, Department of Gastroenterology, Infectious Diseases, and Rheumatology, Division of Nutritional Medicine, Charité, Campus Benjamin Franklin, Hindenburgdamm 30, 12203 Berlin, Germany. joerg.schulzke@charite.de

The aim of this study was to analyze the influence of quercetin on intestinal barrier function using the human colonic epithelial cell line HT-29/B6 and rat small and large intestine *in vitro*. Rat native ileum and late distal colon were incubated in Ussing chambers, and the total resistance (R_T) was measured, and expression of tight junction proteins was characterized in immunoblots. By simulating inflammatory conditions with TNF-α, we examined the barrier-preventive effects of quercetin. Incubation with TNF-α led to a decrease of R_T in HT-29/B6 cell monolayers, which could be partially inhibited by quercetin. In accordance with cell culture experiments, quercetin increased mucosal resistance of rat ileum and late distal colon. Thus, barrier disturbance in late distal colon specimens induced by TNF-α and IFN-γ could be partially prevented by coincubation with quercetin. These findings demonstrate that quercetin enhances barrier function in rat small and large intestine and possesses protective effects on cytokine-induced barrier damage.

Keywords: flavonoid; barrier function; tight junction; TNF-α; claudins

Introduction

Previously, our group[1] and Suzuki *et al.*[2] have published that quercetin enhances barrier properties in Caco-2 cell monolayers. Although this cell line originates from a colonic neoplasia, it features small intestinal properties and has widely been used as an *in vitro* model of the human small intestinal mucosa. In the Caco-2 cell model quercetin, induces a transcriptional upregulation of the sealing tight junction protein claudin-4 and its distribution to the tight junction[1] and mediates an additional junctional assembly of ZO-2, occludin, and claudin-1 by PKCδ inhibition.[2] Different absorptive mechanisms for quercetin have been controversely discussed, but in synopsis, a relatively low overall absorption rate in the small intestine has been described,[3,4] resulting in a flow of unabsorbed quercetin into the large intestine. Therefore, we aimed to analyze effects of

quercetin on a typical colonic cell model, namely HT-29/B6 cells.

High dietary intake of fruits and vegetables is closely associated with benefits in human health due to antioxidative and anticancer actions of secondary plant compounds. A main group of bioactive plant chemicals are flavonoids, which consist of 4,000 phenolic compounds. Flavonoids are ubiquitously present in plants as coloring or taste ingredients, and they are synthesized by the plants as a self-defense mechanism against herbivores. Our conclusion, based on differences in molecular structure, is that flavonoids are divided into six different classes, namely flavonols, flavanones, flavones, isoflavones, flavonols, and anthocyanidins. Quercetin belongs to the subgroup of flavonols and is the most common flavonoid in nature, notably present in onions, kale, and apples.[5] Human daily intake of quercetin averages hundreds

doi: 10.1111/j.1749-6632.2012.06609.x

of milligrams,[6] and beneficial effects in different human pathologies include hypertension, cancer, and inflammatory conditions.[7]

Chronic intestinal inflammation, for example, in inflammatory bowel disease (IBD) results in a dysregulation of intestinal barrier function.[8–10] Direct cytokine effects are triggered via cellular receptors and proinflammatory intracellular signaling cascades resulting in an increased apoptotic rate and in tight junctional disturbance.[11–14] Indirect effects were mediated by subepithelial tissue components. Several immunologic[15] and neuronal regulatory processes[16] can modulate the epithelial cells, and further influences come from the subepithelial connective tissue, such as myofibroblasts.[17] Therefore, it is of great interest to elucidate mechanisms of cytokine effects beyond cell culture experiments. With respect to time course and intensity of cytokine effects, we previously established an IBD-model for studying cytokine effects on intestinal barrier function of rat late distal colon in Ussing chamber setups, using Th1 cytokines TNF-α and IFN-γ.[18] Thus, we aimed to define the effects of quercetin on barrier function of HT-29/B6 colon cells and native rat intestine with or without cytokine coincubation with TNF-α and IFN-γ *in vitro*.

Materials and methods

Cell culture techniques and solutions

Confluent monolayers of the human colon carcinoma cell line HT-29/B6 were grown in 25-cm^2 culture flasks containing RPMI1640 medium with stable L-glutamine (Gibco/Invitrogen, Karlsruhe, Germany), 10% fetal calf serum (FCS), and 1% penicillin/streptomycin.[17] Cells were cultured at 37 °C in a humidified 5% CO$_2$ atmosphere. Monolayers were grown on porous polycarbonate culture plate inserts (effective area: 0.6 cm^2; Millicell HA or PCF, Millipore, Eschborn, Germany), and experiments were performed for seven days. 200 μM quercetin was added to the medium on both sides. 1,000 U/mL human recombinant TNF-α was added basolaterally for 24 hours.

Transepithelial resistance (R_T, $\Omega \cdot$ cm^2) of cell monolayers was determined with an ohmmeter (D. Sorgenfrei, Institute of Clinical Physiology, Berlin, Germany) as well as in a four-electrode Ussing chambers, custom-designed for the insertion of Millicell filter supports. Water-jacketed gas lifts were filled with 10 mL circulating fluid on each side. Bathing Ringer's solution contained 113.6 mM NaCl, 2.4 mM Na$_2$HPO$_4$, 0.6 mM NaH$_2$PO$_4$, 21 mM NaHCO$_3$, 5.4 mM KCl, 1.2 mM CaCl$_2$, 1.2 mM MgCl$_2$, 10 mM D(+)-glucose, 10 mM D(+)-mannose, 2.5 mM L-glutamine, and 0.5 mM β-hydroxybutyric acid. Ringer solution was gassed with 95% O$_2$ and 5% CO$_2$, to ensure a pH value of 7.4 at 37 °C. Before each single experiment, the resistance of the bathing solution and the filter support was measured.

Animal studies

Male Wistar rats (250–300 g) were anesthetized and sacrificed by inhalation of a saturated atmosphere of CO$_2$. Then, the ileum and colon were removed, rinsed with Ringer solution, and stripped of serosa and muscular layers. The tissue was then mounted into standard Ussing-type chambers. The bathing solution for the Ussing experiments contained (in mmol/l): Na$^+$, 140; Cl$^-$, 123.8; K$^+$, 5.4; Ca^{+2}, 1.2; Mg^{+2}, 1.2; HPO$_4^{-2}$ 2.4; H2PO$_4^-$, 0.6; HCO$_3^-$, 21; D(+)-glucose, 10; b-OH-butyrate, 0.5; glutamine, 2.5 and D(+)-mannose 10. The solution was gassed with 95% O2 and 5% CO$_2$, and temperature was maintained at 37 °C using water-jacketed reservoirs. The pH was 7.4 in all experiments. Antibiotics (50 mg/L piperacillin and 50 mg/L imipenem) served to prevent bacterial growth and had no effect on short-circuit current and total resistance in the concentrations used. Bathing solution was supplemented with heat-inactivated 10% FCS.

Recombinant rat TNF-α (Pepro Tech, Rocky Hill, NJ) and recombinant rat IFN-γ (Biotrend, Destin, FL) were added 30 min after starting the experiment. Cytokines were given (TNF-α 10^4 U/mL, IFN-γ 100, or 1,000 U/mL) to the basolateral side of the tissue. Serosal bathing solution was supplemented with heat-inactivated 10% FCS. Experiments were stopped after 20 hours.

Western blot

Western blotting was performed as reported previously.[20] Cells were scraped from the permeable supports and lysed in buffer containing 20 mM TRIS, 5 mM MgCl2, 1 mM EDTA, 0.3 mM EGTA, and protease inhibitors (Complete; Boehringer, Mannheim, Germany). Tissue samples were removed from the Ussing chambers and protein content was determined with a BCA protein assay (Pierce, Rockford, IL), using a plate reader (Tecan, Crailsheim,

Germany). Aliquots of 10 μg protein were mixed with SDS buffer (Laemmli), loaded onto 8.5% or 12.5% SDS polyacrylamide gels, electrophoretically separated, and blotted onto plyvinylidene fluoride (PVDF) membranes. Proteins were detected by immunoblotting with antibodies raised against human occludin, claudin-1, -2, -3, -4, and -7. Chemiluminescence was induced with a Lumi-LightPLUS Western blotting kit (Roche, Mannheim, Germany), detected with an imaging system (LAS-1000, Fuji, Tokyo, Japan), and analyzed by quantification software (AIDA, Raytest, Straubenhardt, Germany). Confocal laser scanning microscopy of rat late distal colon specimens was performed as described previously.[18]

Promoter analysis

For reporter gene assays, a 264-bp DNA fragment, including the transcription start point of claudin-2, was amplified from human genomic DNA (Genome walking kit, Clontech, Mountain View, CA). The DNA fragment was cloned into the pGL4.10 reporter gene vector (Promega, Madison, WI). Plasmids were screened by ampicillin selection, plasmid isolation, and agarose gel electrophoresis. Insert sequence and orientation were confirmed by DNA sequencing. HT-29/B6 cells were seeded into 6 well plates (5×10^5 cells per well) 24 h before transfection. Transient transfection of reporter gene constructs along with pGL4.70 coreporter-plasmid (Promega) was analyzed in the absence or presence of 25 μM quercetin. Measurement of luciferase activity was performed with the dual-luciferase reporter assay system (Promega) as previously described.[31] Promoter activities were expressed as relative light units normalized for the activity of renilla luciferase in each setup. The data were calculated as the means of three identical setups.

Chemicals

All chemicals, unless otherwise noted, were purchased from Sigma-Aldrich (St. Louis, MO). Antibodies against claudin-1, -2, -3, -4, -7, and occludin were purchased from Zymed Laboratories (Zymed/Invitrogen, San Francisco, CA).

Statistical analysis

Data are expressed as means ± standard error of the mean (SEM). Statistical analysis was performed by using Student's *t*-test. $P < 0.05$ was considered significant.

Results

Effects of quercetin on transepithelial resistance of HT-29/B6 cells

Mucosal and serosal incubation with 200 μM quercetin did not increase resistance but slightly lowered R_T of HT-29/B6 monolayers after 24 h (quercetin 89% ± 2%, $n = 7$; Fig. 1A). In contrast, TNF-α (1,000 U/mL) induced a marked decrease of R_T to 44% ± 6% after 24 h ($n = 7$, $P < 0.001$). This effect TNF-α was partially inhibited by 200 μM quercetin (R_T 61% ± 3% of initial resistance; $n = 8$, $P < 0.05$; Fig. 1).

Effects of quercetin on the expression of TJ proteins in HT-29/B6 monolayers

Subsequent to electrophysiological experiments, expression of TJ proteins was analyzed by immunoblotting. Densitometric analysis of Western blots revealed a decrease of claudin-2 (65% ± 10% of control, $n = 6$; $P < 0.05$) and claudin-3 expression (47% ± 13% of controls, $n = 6$, $P < 0.01$) with the addition of quercetin. Expression of occludin, claudin-1, -4, and -7 remained unaltered (Fig. 1B).

Effect of quercetin on claudin-2 promoter activity

Luciferase reporter gene assays were performed to study effects of quercetin on the promoter of claudin-2. The claudin-2 promoter was cloned into a pGL4.10 vector bearing a luciferase gene for quantitative measurements of promoter activity. A quercetin concentration of 25 μmol/L decreased the promoter activity to 66% ± 1% ($P < 0.05$; $n = 3$) expressed as the percentage of controls. TNF-α showed a tendency to increase claudin-2 promoter activity, but did not reach significance (107% ± 3% of control, $n = 3$), whereas quercetin together with TNF-α reduced promoter activity to 70% ± 2% (Fig. 1C). Thus, quercetin is able to inactivate the claudin-2 promoter, which can explain the decrease in claudin-2 protein level subsequently. This part of the study was performed on subconfluent HT-29/B6 cells, because an experimental design that includes transfection of HT-29/B6 cells is only efficient on single (still dividing) cells.

Apoptosis in cell culture experiments

Apoptotic rates were not altered significantly under experimental conditions, as analyzed by detection of TUNEL staining. Quercetin did not alter the apoptotic rate of HT-29/B6 cells after 24 h of exposure

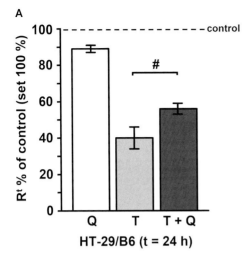

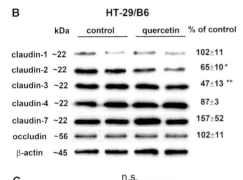

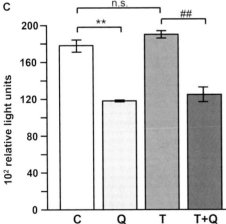

Figure 1. Effects of quercetin on R_T of HT-29/B6 cells. Quercetin (Q; 200 μM) had no marked effects on R_T of HT-29/B6 monolayers. (A) TNF-α (1,000 U/mL) induced a decrease of R_T to 44% ± 6% after 24 h (Q; $n = 7$, $P < 0.001$). These effects of TNF-α were partially inhibited by 200 μM quercetin (Q + T; 61% ± 3% of initial resistance; $n = 8$, $P < 0.05$). (B) Analysis of tight junction protein expression revealed a decrease of claudin-2 at basal levels and inhibits induction of claudin-2 upregulation by TNF-α. (C) Effects on claudin-2 are mediated by a direct genomic downregulation of claudin-2 promoter.

(TUNEL-assay: control: 0.30% ± 0.07%, $n = 4$ vs. quercetin 0.21% ± 0.05%, $n = 3$, n.s.). TNF-α, a cytokine with pro-apoptotic activity, did not upregulate apoptosis after 24 h of exposure (TNF-α 0.25% ± 0.04%, $n = 3$, n.s.). Coincubation of quercetin and TNF-α caused an apoptotic rate not different from controls or after TNF-α alone (quercetin + TNF-α: 0.36% ± 0.07%, $n = 3$, n.s.).

Electrophysiological measurements of rat intestine

Effect of quercetin on rat ileum was analyzed in Ussing chamber experiments. R_T measurements were performed for four hours. Three concentrations of quercetin were tested, namely 50, 100, and 200 μM quercetin. Quercetin was given to the mucosal and serosal bathing solution. Application of 50 μM quercetin increased R_T to 148% ± 5% ($n = 5$, $P < 0.05$), and 100 μM quercetin increased R_T to 145% ± 5% of initial resistance ($n = 5$, $P < 0.01$) versus control condition 121% ± 5% ($n = 7$). 200 μM quercetin increased tissue resistance similar to lower quercetin concentrations, but did not reach significance because of higher value variability (144% ± 27%, $n = 4$, Fig. 2A). In addition, specimens of rat colon were analyzed. Quercetin, given in a concentration of 200 μM on both sides of the bathing solution, increased R_T to 129% ± 4% of initial resistance values after 20 h incubation time (control: 100% ± 8% of initial resistance, $n = 5$, $P < 0.05$).

Application of TNF-α and IFN-γ led to a marked decrease in R_T (52% ± 5% of initial resistance vs. 112% ± 9% under the control condition, $n = 5$, $P < 0.01$). This effect started at about 8–12 h after addition of the cytokines. Quercetin could partially inhibit the cytokine-induced R_T decrease (88% ± 17% of initial resistance, $n = 5$, $P < 0.05$, Fig. 2B).

Tight junction protein expression of rat intestine

Original blots are shown in Fig. 2C. To investigate whether altered tight junction structure contributes to the cytokine-induced decrease of resistance, immunoblot analyses were performed for occludin, claudin-1, -2, and -4. Because these proteins are integral membrane proteins forming the tight junction strand network, crude membrane fractions were used for analysis. Claudin-2 and -4 showed a tendency toward downregulation by quercetin in rat colon epithelia, which is in accordance to cell culture experiments in HT-29/B6 cells. As previously

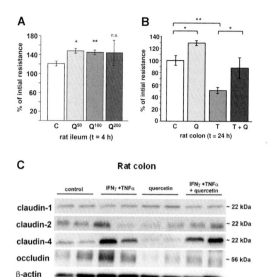

Figure 2. Effects of quercetin on R_T of rat intestine. (A) Quercetin 50 μM (148% ± 5%, $n = 5$, $P < 0.05$) and 100 μM (145% ± 5%, $n = 5$, $P < 0.01$) induced an increase in epithelial resistance of rat small intestine. (B) In rat, late distal colon quercetin was used in a concentration of 200 μM, leading to an increase of resistance to 129% ± 4% ($n = 5$, $P < 0.01$). Application of TNF-α (10^4 U/mL) and IFN-γ (10^3 U/mL) reduced R_T (52% ± 5% of initial resistance, $n = 5$, $P < 0.01$), which could be partially inhibited by quercetin (88% ± 17% of initial resistance, $n = 5$, $P < 0.05$). (C) Analysis of tight junction proteins of rat late distal colon showed claudin-2 to be downregulated by quercetin in coincubation with cytokines (TNF-α+ IFN-γ 171% ± 24% of control, quercetin + TNF-α+ IFN-γ 93% ± 17% of control, $P < 0.05$, $n = 4$); claudin-4 expression was downregulated by quercetin versus control (73% ± 9% of control, $P < 0.05$, $n = 4$); other tight junction proteins remained unchanged.

published by our group, TNF-α increases claudin-2 expression,[18] which in part could explain the decrease of epithelial resistance. This cytokine-induced increase of claudin-2 could be blocked by coincubation with quercetin (TNF-α + IFN-γ 171% ± 24% of control, quercetin + TNF-α + IFN-γ 93% ± 17% of control, $P < 0.05$, $n = 4$). Quercetin alone reduced claudin-4 expression compared to control (73% ± 9% of control, $P < 0.05$, $n = 4$). Other tight junction proteins remained unchanged.

Laser scanning microscopy of E-cadherin and DAPI showed an altered distribution of E-cadherin after TNF-α exposure with a loss of intercellular staining in the basolateral membrane of the enterocytes and a concomitant increase in the amount of intracellular aggregates. Also, crypt architecture and the lining of enterocyte nuclei was disturbed

after cytokine treatment. In quercetin and TNF-α– incubated colon specimens, a large number of apoptotic cells with condensed nuclei was observed that left the epithelial formation (Fig. 3).

Discussion

Several flavonoids have been identified to protect intestinal barrier function and modulate TJ regulation. These effects have been described for green tea catechin epigallocatechingallate (EGCG),[21] quercetin from onions and apples,[1,2] myricetin from grapes, herbs and vegetables,[2] kaempferol from red grapes,[22] and genistein from soybeans.[23] In general, these substances suppress inflammatory pathways and increase the barrier function of epithelia utilizing a variety of molecular mechanisms.

Previously, we described that quercetin enhances barrier function in Caco-2 cell monolayers via transcriptional regulation of the TJ protein claudin-4 and its distribution to the tight junction.[1] So far, the effects of quercetin have not been investigated in native intestine, neither under physiological nor under defined inflamed conditions.

As the central result of this study, quercetin enhanced barrier properties of rat ileum and late distal colon and prevents cytokine-induced barrier loss in the HT-29/B6 model epithelium and rat late distal colon.

Quercetin effects on barrier function in different cell lines

In contrast to our previously published results in Caco-2 cell monolayers, no direct induction of a sealing tight junction proteins by quercetin could be identified, neither in HT-29/B6 cell monolayers nor in rat intestine. It may be hypothesized that a different type of cell differentiation explains the discrepancy of the results in these two cell lines. Although Caco-2 cells are of colonic origin, these cells become differentiated and polarized morphologically and functionally like small intestinal enterocytes with expression of microvilli and brush-border enzymes as well as specific small intestinal transport proteins, such as SGLT1. In contrast, the highly differentiated cell line HT-29/B6 is a subclone of the human colon cancer cell line HT-29 with typical properties of the large intestine like chloride secretion or mucus production and an epithelial barrier similar to colon crypt cells.[19] Second, we hypothesize that the mucus barrier of HT-29/B6 cells could

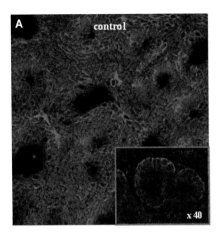

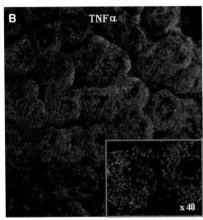

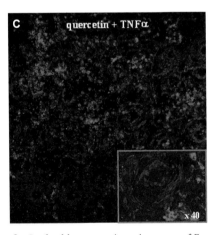

Figure 3. Confocal laser scanning microscopy of E-cadherin in rat intestine. E-cadherin (red) and nuclei (DAPI blue) were stained in specimens of late distal colon of rat with or without TNF-α (10^4 U/mL) and IFN-γ (10^3 U/mL), and quercetin 200 μM. TNF-α and IFN-γ induced an increase in the number of cells with intracellularly condensed E-cadherin and condensed nuclei that had left the epithelial layer (induction of apoptosis). Quercetin treatment did not reverse this but even seemed to intensify it.

block effects driven from the mucosal membrane of the enterocytes. As a third aspect, these different cell populations, although derived from the same tissue, could significantly differ in their response to external stimuli and, indeed, several fundamental differences, for example, exist between NF-κB pathways in HT-29 and Caco-2 cells.[24] In HT-29/B6 cells, NF-κB signaling is involved in barrier modulation by regulating assembly and disassembly of claudin-1 and claudin-2.[11] In addition, NF-κB can induce transcription and translation of certain target genes and proteasome activity, which subsequently could modulate TJ expression and protein quantity[25] and could explain differences in TJ modulation in Caco-2 and HT-29/B6 cells.

Relevance of epithelial apoptosis

Barrier function of epithelia directly correlates with the continuity of the monolayer and intact TJs. In general, after incubation with TNF-α a decrease of TJ strand numbers and an increase of apoptoses are induced in intestinal epithelial cells.[14,26] Previous analysis of local conductances had revealed up to 50% of the TNF-α–induced permeability increase to be associated with apoptoses, whereas the other 50% depended on modulations of TJs in nonapoptotic areas.[14] Quercetin has been shown in various studies to possess pro-apoptotic properties in intestinal cell lines,[27,28] namely via upregulation of apoptosis-related proteins.

In this study, neither TNF-α nor quercetin at the dose and time scheme applied induced a detectable increase in apoptotic rate in TUNEL stainings of HT-29/B6 cells. However, it may be hypothesized that the slight decrease of R_T in HT-29/B6 cells after quercetin coincubation could be due to early stages of epithelial apoptoses, which cannot be completely ruled out using TUNEL stainings because nuclear fragmentation occasionally could occur at late stages of apoptotic events.[29]

Furthermore, Bojarski *et al.*[30] have shown that induction of epithelial apoptosis differently altered claudin-1 and claudin-2 in HT-29/B6 monolayers. Claudin-2 protein was decreased but claudin-1 was unaltered or even increased, an observation corresponding to changes in the TJ pattern in quercetin experiments.

In rat colon, staining of E-cadherin and DAPI showed an increased number of cells with condensed nuclei that have left the epithelium after combined quercetin and cytokine incubation, indicating an increased number of epithelial apoptoses. We hypothesize that quercetin, on the one hand, strengthens epithelial barrier function via downregulation of claudin-2 and on the other hand, according to the literature, induces cell death in predisposed epithelial cells. Therefore, that the effect of quercetin was only a partial and not a complete recovery could be due to a synergistic pro-apoptotic effect together with TNF-α.

Quercetin effects on barrier function of rat intestine

Previously, beneficial effects of two quercetin glycosides, rutin, and quercitrin, have been shown in experimental models of rat colitis,[30,31] whereas no clear effects were evident for the aglycone form.[32] It has been hypothesized that the inability of quercetin to ameliorate colon inflammation in animal models could be due to a high absorption rate in the small intestine, preventing effective doses to reach the colonic mucosa. The setup of our *in vitro* experiments ensure sufficient mucosal concentrations of quercetin, which ameliorate barrier properties in rat small and large intestine, indicated by a 20–30% increase of epithelial resistance. In contrast to our data, in Caco-2 cells with a quercetin-induced increase in claudin-4 expression, analysis of tight junction proteins in rat late distal colon did not show an upregulation of claudin-4 expression but a tendency of increase in claudin-1. According to our experiments in HT-29/B6 cells, the pore-forming claudin-2 is downregulated by quercetin. The latter could explain the quercetin-induced increased in epithelial resistance of rat colon specimens.

Barrier-preventive effects of quercetin

TNF-α, together with IFN-γ, impaired the epithelial barrier function of rat distal colon. Variable effects of both Th1 cytokines, TNF-α or IFN-γ, have been reported on ZO-1 and TJ-structure.[18,25,33] Our conclusion, supporting the hypothesis above, that sufficient amounts of quercetin are needed to achieve an anti-inflammatory effect in inflamed colon, is that quercetin exhibits, in part, barrier preventive effects in our *in vitro* model of rat late distal colon incubated with TNF-α. These effects result in inhibiting cytokine-dependent breakdown of epithelial resistance.

A major factor in inflamed epithelia is the regulation of claudin-2. Claudin-2 has been shown to be upregulated in active Crohn's disease and in the inflamed mucosa of patients with ulcerative colitis.[8,34] Because claudin-2 forms cation-selective channels,[18] its upregulation in IBD has been proposed to contribute to altered barrier function and aggravate diarrhea by a leak flux mechanism caused by an impaired epithelial barrier and an increased back-leak of ions and water into the intestinal lumen.

Previously, we could demonstrate that TNF-α time- and dose-dependently increased claudin-2, attributed to transcriptional activation.[35] Furthermore, we could show, that quercetin downregulates basal claudin-2 expression and inhibits cytokine-derived upregulation of claudin-2. This is mediated by a direct genomic downregulation by quercetin, affecting the claudin-2 promoter.

In this study, we analyzed the preventive effect of quercetin on epithelial barrier function of colonic HT-29/B6 cell monolayers, native rat small and large intestine, and in Th1 cytokine-derived barrier disturbance. These findings reflect new protective and preventive mechanisms and antidiarrheal effects of the flavonoid quercetin, which can be of interest in future designs for the therapy of inflammatory bowel disease.

Acknowledgments

We thank A. Fromm, I.M. Lee, and D. Sorgenfrei for their excellent technical assistance.

Conflicts of interest

The authors declare no conflicts of interest.

References

1. Amasheh, M., S. Schlichter, S. Amasheh, *et al.* 2008. Quercetin enhances epithelial barrier function and increases claudin-4 expression in Caco-2 cells. *J. Nutr.* **138:** 1067–1073.
2. Suzuki, T. & H. Hara. 2009. Quercetin enhances intestinal barrier function through the assembly of zonula occludens-2, occludin, and claudin-1 and the expression of claudin-4 in Caco-2 cells. *J. Nutr.* **139:** 965–974.
3. Murota, K. & J. Terao. 2003. Antioxidative flavonoid quercetin: implication of its intestinal absorption and metabolism. *Arch. Biochem. Biophys.* **417:** 12–17.
4. Shimoi, K., T. Yoshizumi, T. Kido, *et al.* 2003. Absorption and urinary excretion of quercetin, rutin, and alphaG-rutin, a water soluble flavonoid, in rats. *J. Agric. Food Chem.* **51:** 2785–2789.
5. Hertog, M.G.L., P.C.H. Hollman & M.B. Katan. 1992. Content of potentially anticarcinogenic flavonoids of

28 vegetables and 9 fruits commonly consumed in the Netherlands. *J. Agric. Food Chem.* **40:** 2379–2383.

6. Hollman, P.C.A. & I.C.W. Arts. 2000. Flavonoids, flavones and flavanols—nature, occurrence and dietary burden. *J. Sci. Food Agric.* **80:** 1081–1093.

7. Middleton, E., C. Kandaswami & T. Theoharides. 2000. The effects of plant flavonoids on mammalian cells: implications for inflammation, heart disease and cancer. *Pharmacol. Rev.* **52:** 673–751.

8. Zeissig, S., N. Buergel, D. Guenzel, *et al.* 2007. Changes in expression and distribution of claudin 2, 5 and 8 lead to discontinuous tight junctions and barrier dysfunction in active Crohn's disease. *Gut* **56:** 61–72.

9. Gitter, A.H., F. Wullstein, M. Fromm & J.D. Schulzke. 2001. Epithelial barrier defects in ulcerative colitis: characterization and quantification by electrophysiological imaging. *Gastroenterology* **121:** 1320–1328.

10. Suenaert, P., V. Bulteel, L. Lemmens, *et al.* 2002. Anti-tumor necrosis factor treatment restores the gut barrier in Crohn's disease. *Am. J. Gastroenterol.* **97:** 2000–2004.

11. Amasheh, M., A. Fromm, S.M. Krug, *et al.* 2010. TNFalpha-induced and berberine-antagonized tight junction barrier impairment via tyrosine kinase, Akt and NFkappaB signaling. *J. Cell Sci.* **123:** 4145–4155.

12. Bruewer, M., A. Luegering, T. Kucharzik, *et al.* 2003. Proinflammatory cytokines disrupt epithelial barrier function by apoptosis-independent mechanisms. *J. Immunol.* **171:** 6164–6172.

13. Prasad, S., R. Mingrino, K. Kaukinen, *et al.* 2005. Inflammatory processes have differential effects on claudins 2, 3 and 4 in colonic epithelial cells. *Lab. Invest.* **85:** 1139–1162.

14. Gitter, A.H., K. Bendfeldt, J.D. Schulzke & M. Fromm. 2000. Leaks in the epithelial barrier caused by spontaneous and TNF-alpha-induced single-cell apoptosis. *FASEB J.* **14:** 1749–1753.

15. Brandtzaeg, P. 2011. The gut as communicator between environment and host: immunological consequences. *Eur. J. Pharmacol.* **668**(Suppl. 1): S16–S32.

16. Tixier, E., J.P. Galmiche & M. Neunlist. 2006. Intestinal neuro-epithelial interactions modulate neuronal chemokines production. *Biochem. Biophys. Res. Commun.* **344:** 554–561.

17. Hausmann, M. & G. Rogler. 2008. Immune—non immune networks in intestinal inflammation. *Curr. Drug Targets* **9:** 388–394.

18. Amasheh, M., I. Grotjohann, S. Amasheh, *et al.* 2009. Regulation of mucosal structure and barrier function in rat colon exposed to tumor necrosis factor alpha and interferon gamma in vitro: a novel model for studying the pathomechanisms of inflammatory bowel disease cytokines. *Scand. J. Gastroenterol.* **44:** 1226–1235.

19. Kreusel, K.M., M. Fromm, J.D. Schulzke & U. Hegel. 1991. Cl⁻ secretion in epithelial monolayers of mucus-forming human colon cells (HT-29/B6). *Am. J. Physiol.* **261:** C574–C582.

20. Amasheh, S., N. Meiri, A.H. Gitter, *et al.* 2002. Claudin-2 expression induces cation-selective channels in tight junctions of epithelial cells. *J. Cell Sci.* **115:** 4969–4976.

21. Watson, J.L., S. Ansari, H. Cameron, *et al.* 2004. Green tea polyphenol (−)-epigallocatechin gallate blocks epithelial barrier dysfunction provoked by IFN-gamma but not by IL-4. *Am. J. Physiol. Gastrointest. Liver Physiol.* **287:** G954–G961.

22. Suzuki, T., S. Tanabe & H. Hara. 2011. Kaempferol enhances intestinal barrier function through the cytoskeletal association and expression of tight junction proteins in Caco-2 cells. *J. Nutr.* **141:** 87–94.

23. Atkinson, K.J. & R.K. Rao. 2001. Role of protein tyrosine phosphorylation in acetaldehyde-induced disruption of epithelial tight junctions. *Am. J. Physiol. Gastrointest. Liver Physiol.* **280:** G1280–G1288.

24. Jobin, C., S. Haskill, L. Mayer, *et al.* 1997. Evidence for altered regulation of IκBα degradation in human colonic epithelial cells. *J. Immunol.* **158:** 226–234.

25. Ma, T.Y., G.K. Iwamoto, N.T. Hoa, *et al.* 2004. TNF-alpha-induced increase in intestinal epithelial tight junction permeability requires NF-kappa B activation. *Am. J. Physiol. Gastrointest. Liver Physiol.* **286:** G367–G376.

26. Schmitz, H., M. Fromm, C.J. Bentzel, *et al.* 1999. Tumor necrosis factor-alpha (TNFalpha) regulates the epithelial barrier in the human intestinal cell line HT-29/6B. *J. Cell Sci.* **112:** 137–146.

27. Wenzel, U., A. Herzog, S. Kuntz & H. Daniel. 2004. Protein expression profiling identifies molecular targets of quercetin as a major dietary flavonoid in human colon cancer cells. *Proteomics* **4:** 2160–2174.

28. Chen, Y.C., S.C. Shen, J.M. Chow, *et al.* 2004. Flavone inhibition of tumor growth via apoptosis in vitro and in vivo. *Int. J. Oncol.* **25:** 661–70.

29. Wenzel, U. & H. Daniel. 2004. Early and late apoptosis events in human transformed and non-transformed colonocytes are independent on intracellular acidification. *Cell Physiol. Biochem.* **14:** 65–76.

30. Bojarski, C., A.H. Gitter, K. Bendfeldt, *et al.* 2001. Permeability of human HT-29/B6 colonic epithelium as a function of apoptosis. *J. Physiol.* **535:** 541–52.

31. Camuesco, D., M. Comalada, M.E. Rodriguez-Cabezas, *et al.* 2004. The intestinal anti-inflammatory effect of quercitrin is associated with an inhibition in iNOS expression. *Br. J. Pharmacol.* **143:** 908–918.

32. Comalada, M., D. Camuesco, S. Sierra, *et al.* 2005. In vivo quercitrin anti-inflammatory effect involves release of quercetin, which inhibits inflammation through downregulation of the NF-κB pathway. *Eur. J. Immunol.* **35:** 584–592.

33. Nusrat, A. 2000. Molecular physiology and pathophysiology of tight junctions. IV. Regulation of tight junctions by extracellular stimuli: nutrients, cytokines, and immune cells. *Am. J. Physiol. Gastrointest. Liver Physiol.* **279:** G851–G857.

34. Heller, F., P. Florian, C. Bojarski, *et al.* 2005. Interleukin-13 is the key effector Th2 cytokine in ulcerative colitis that affects epithelial tight junctions, apoptosis, and cell restitution. *Gastroenterology* **129:** 550–564.

35. Mankertz, J., S. Tavalali, H. Schmitz, *et al.* 2000. Expression from the human occludin promoter is affected by tumor necrosis factor alpha and interferon gamma. *J. Cell Sci.* **113:** 2085–2090.

Ann. N.Y. Acad. Sci. ISSN 0077-8923

ANNALS OF THE NEW YORK ACADEMY OF SCIENCES

Issue: *Barriers and Channels Formed by Tight Junction Proteins*

Claudin-based paracellular proton barrier in the stomach

Atsushi Tamura,[1] Yuji Yamazaki,[1] Daisuke Hayashi,[1,2] Koya Suzuki,[1] Kazuhiro Sentani,[3] Wataru Yasui,[3] and Sachiko Tsukita[1]

[1]Laboratory of Biological Science, Graduate School of Frontier Biosciences and Graduate School of Medicine, Osaka University, Osaka, Japan. [2]Department of Geriatric Medicine and Nephrology, Graduate School of Medicine, Osaka University, Osaka, Japan. [3]Department of Molecular Pathology, Graduate School of Biomedical Sciences, Hiroshima University, Hiroshima, Japan

Address for correspondence: Sachiko Tsukita, Laboratory of Biological Science, Graduate School of Frontier Biosciences, Osaka University, 2-2 Yamadaoka, Suita, Osaka 565-0871, Japan. atsukita@biosci.med.osaka-u.ac.jp

The claudins comprise a multigene family that consists of at least 27 members. Claudins are responsible for establishing the paracellular barrier—which has permselectivity—at the tight junctions in epithelial cells, and the specific patterns of claudin expression in the epithelial cell sheets that cover the internal and external surfaces of organs contribute to the formation of microenvironments and organs' biological functions. Data on the detailed characterization of individual claudins and their roles in different microenvironments are accumulating. A study on the stomach-specific *claudin-18*–knockout mouse, which has gastritis, recently revealed that the stomach-type claudin-18 specifically forms the proton barrier in the stomach, consistent with previously reported circumstantial evidence. Combined with previous studies on the specific ionic homeostasis by different types of claudins, our findings support the idea that claudins may regulate ion-specific homeostasis *in vivo*.

Keywords: claudin; proton barrier; gastritis; knockout mouse; metaplasia

Biological significance of the large claudin family

Since the first claudin was identified as a tight-junction protein, the claudin family has been shown to include a large number of members, at least 27, in humans and mice (Fig. 1A).[1–3] Why the claudin family is so large is an important question. Given that claudins form the paracellular barrier in various types of epithelial cell sheets in different organs *in vivo*, the many species of claudin are probably important to confer different epithelial cell–sheet barrier properties, including barrier-dependent permselectivity, thereby establishing microenvironments in specific regions of individual organs.[4–7] Together with the specific properties of claudins determined by cell-level analyses, critical roles for claudins *in vivo* have recently been revealed by studying knockout mice and human claudin genes that are mutated in various diseases.[8–10]

Studies involving the transfection or knockdown of individual claudins or their combinations in cul-tured epithelial cells led us and others to speculate that claudin-2, -10b, and -15 contribute to cation channels or pores, whereas claudin-4, -7, and -10a contribute to anion channels or pores or to cation barriers.[2,10–14] The positive or negative electrical charges of the first loops of a claudin molecule appear to function in paracellular epithelial permeation by creating an ion-channel like pore.[10,14] On the other hand, the dynamic properties of tight junctions (TJs) may determine the paracellular permeability for solutes, the mechanism of which is not well understood.[12,15–17]

Mutational analyses of human claudin genes have suggested that neonatal sclerosing cholangi-tis and icthyosis are caused by a *claudin-1* muta-tion.[15,18] Nonsyndromic recessive deafness is at-tributable to a *claudin-14* mutation, which causes dysfunction of the paracellular permselectivity of the inner ear epithelia.[19] Familiar hypomagnesemia is caused by a *claudin-16* mutation, which re-sults in dysfunction of the cation permselectivity of the renal epithelia and familial hypomagnesemia,

doi: 10.1111/j.1749-6632.2012.06570.x

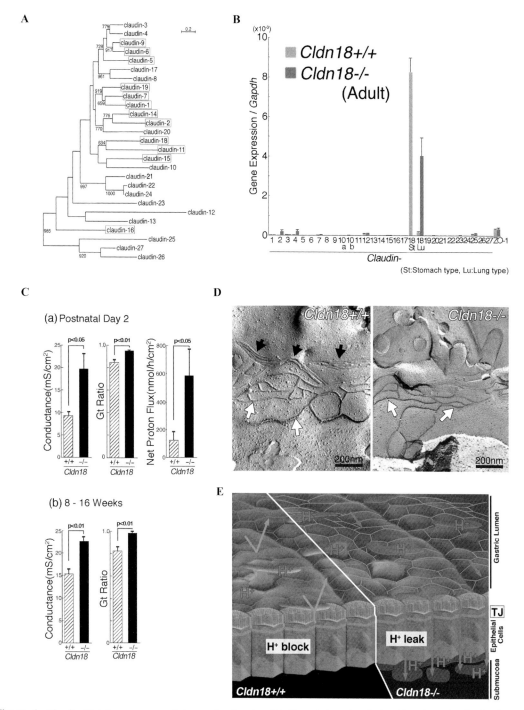

Figure 1. Claudin-18st forms a paracellular proton barrier. (A) The phylogenic relationship of 27 claudin-family members. The boxed claudins are those that have been analyzed in knockout mice for their biological significance *in vivo*. (B) The specific loss of *claudin-18st* in the *Cldn18−/−* stomach. Quantitative RT-PCR. (C) Physiological characterization of the transepithelial conductance and paracellular proton barrier in the *Cldn18+/+* stomach and its loss in the *Cldn18−/−* stomach. The Gt ratio was obtained by dividing the conductance after acid-loading by the conductance before acid-loading. (D) Freeze-fracture electron microscopic images of TJ strands in the *Cldn18+/+* and *Cldn18−/−* stomach. Densely packed TJ strands (black arrows) in the apical region of the *Cldn18+/+* gastric epithelial cells were missing in the *Cldn18−/−* cells, which showed only loosely anastomosing TJ strands (white arrows). (E) Schematic illustrating the loss of the paracellular proton barrier by claudin-18st deficiency.

hypercalciuria and nephrocalcinosis (FHHNC).[20] FHHNC is reportedly caused by a *claudin-19* mutation.[21,22] These findings suggest that claudins are important determinants of specific homeostatic properties.[7–9] Knockout mice can be a powerful tool in clarifying the relationship between the function of claudins in TJs and biological homeostasis *in vivo*.[23,24]

The role of claudins as regulators of homeostasis as revealed by knockout mouse studies

The phenotypes of claudin knockout mice have revealed important roles for specific claudins in regulating the barrier integrity of certain epithelial tissues. For example, *claudin-1* knockout mice die shortly after birth due to disruption of epidermal barrier function and resultant dehydration. In *claudin-5* knockout mice, neonatal lethality is attributable to the disruption of endothelial barrier integrity in the central nervous system and the resultant loss of blood–brain barrier function.[25,26] *Claudin-11* knockout mice are viable, but they have locomotion defects and male sterility due, respectively, to a lack of TJs in the myelin sheaths of the central nervous system, which affects nerve conduction, and between Sertoli cells of the testes, which affects the blood–testis barrier.[27] The knockout of ion-leaky *claudin-15* causes defects in nutrition absorption by perturbing the Na^+ homeostasis that is required for the proper functioning of nutrition absorption transporters.[24] Thus, the claudin-type–specific properties of a permselective paracellular barrier help regulate the homeostasis of different biological systems. Detailed studies of these systems should reveal the mechanisms by which claudins regulate the homeostasis of each microenvironment.

The ion permselectivity of paracellular barriers, and particularly of the permeability to Na^+, K^+, $Mg2^+$, and Cl^-, have been intensively analyzed.[2,24,28,29] Findings in studies of cultured cells suggest that claudins are involved in the proton (H^+) barrier. In one report, exogenously expressed claudin-8 conferred a proton barrier property on MDCK-2 cells; in another, exogenously expressed claudin-18 conferred the ability for cells to form a proton barrier between them.[29,30]

Furthermore, the expression of stomach type of *claudin-18* (*claudin-18st*) is upregulated in human Barrett's esophagus, which might contribute to the high proton barrier seen in this disorder.

The proton-barrier property of claudin-18st, a dominant claudin in the stomach, is thought to be especially important *in vivo* to prevent H^+ leakage from acidic gastric juice into the tissues. To uncover the role of this claudin *in vivo*, we recently generated and analyzed *claudin-18st* knockout mice.[31]

Claudin-18 is a proton paracellular barrier protein in the stomach

We first examined which claudins are expressed in the mouse stomach by qRT-PCR. As shown in Figure 1B, many claudins are expressed in the stomach, but the expression level of *claudin-18* is exclusively high. *Claudin-18* has two alternative splicing forms, the stomach (type 2 splicing form, designated as *claudin-18st* in this manuscript) and lung (type 1 splicing form) types, which use different first exons and the same exons 2–4. The two isoforms are regulated by different tissue-specific promotors.[32] The stomach-type *claudin-18* is thought to regulate the H^+-leakage resistance of the stomach's paracellular barrier. To examine this possibility, we recently generated and analyzed knockout mice of stomach-type claudin-18 (*Cldn18*$^{-/-}$ mice).

The epithelial paracellular barrier function against H^+ was examined in the stomach of *Cldn18*$^{+/+}$ and *Cldn18*$^{-/-}$ mice. Electrophysiological measurements showed that the total ion permeability (conductance) for the buffer containing 150 mM NaCl was higher in the *Cldn18*$^{-/-}$ stomach than the *Cldn18*$^{+/+}$ one, suggesting that the claudin-18 deficiency compromised the total paracellular ionic barrier. While the paracellular barrier against acidity was effective in the *Cldn18*$^{+/+}$ stomach, this was not the case for the *Cldn18*$^{-/-}$ stomach, suggesting that the H^+ leakage was much lower in the *Cldn18*$^{+/+}$ stomach compared to *Cldn18*$^{-/-}$ one. These findings are consistent with the idea that claudin-18st plays a specific role in the paracellular barrier of the stomach to block H^+ leakage (Fig. 1C).

The mode of claudin-18st polymerization creates the gastric proton barrier

The formation of TJ strands, which can be revealed by freeze-fracture electron microscopy, is required for TJs to exert their paracellular barrier function.

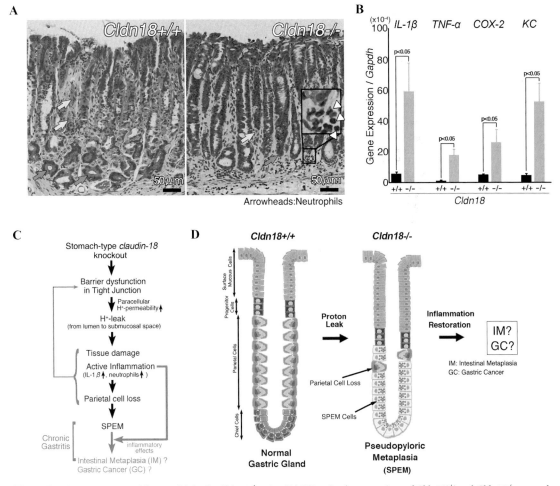

Figure 2. Characterization of the gastritis in the *Cldn18$^{-/-}$* mice. (A) HE-stained preparations of *Cldn18$^{+/+}$* and *Cldn18$^{-/-}$* stomach tissue. Gastritis was recognized in the *Cldn18$^{-/-}$* stomach, in which parietal and chief cells (yellow and pink arrows, respectively) were decreased. Note the presence of neutrophils that had infiltrated the submucosa in the *Cldn18$^{-/-}$* stomach (white arrows in enlarged (×3.3) photographs in the inset). (B) Expressions of inflammatory cytokines IL-1β, TNF-α, COX-2, and KC in the *Cldn18$^{+/+}$* and *Cldn18$^{-/-}$* stomach. The cytokine expressions were upregulated in the *Cldn18$^{-/-}$* stomach compared to the *Cldn18$^{+/+}$* one. (C and D) Schematics showing the process of gastritis onset in the *Cldn18$^{-/-}$* mouse stomach. Proton leakage due to dysfunctional tight junctions may trigger pseudopyloric gastritis in the *Cldn18$^{-/-}$* stomach, with IL-1β–related inflammation.

As it is generally thought that the TJ-strand morphology at least partly reflects TJ function, we next investigated the TJ strands in the *Cldn18$^{-/-}$* stomach using freeze-fracture electron microscopy (Fig. 1D). Tightly packed parallel arrays of TJ strands exist in the adult *Cldn18$^{+/+}$* stomach, and these structures were lost in the *Cldn18$^{-/-}$* stomach, suggesting that these arrays reflect the TJ paracellular proton barrier (Fig. 1E). In contrast, the stomach of the *Cldn18$^{-/-}$* mouse showed much more loosely anastomosing TJ strands, which probably contained claudin species other than claudin-18st. An important future re-

search goal is to determine the unique mode of claudin-18st polymerization that creates the proton paracellular barrier.

Biological significance of the claudin-18st paracellular proton barrier in the stomach

Although *Cldn18$^{-/-}$* mice grow up without gross abnormalities compared to *Cldn18$^{+/+}$* mice, tissue-level examination revealed a critical biological effect of claudin-18st *in vivo*: adult *Cldn18$^{-/-}$* mice have chronic gastritis. Compared to *Cldn18$^{+/+}$*,

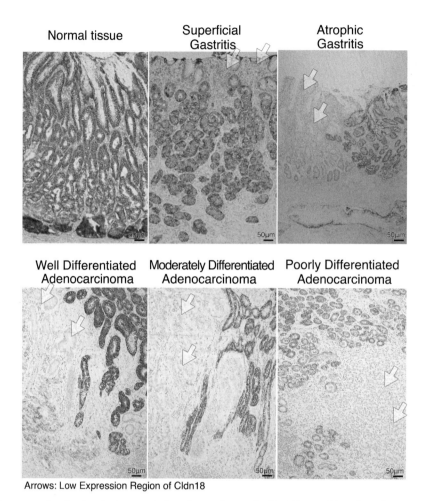

Figure 3. Human samples showing the downregulation of stomach-type claudin-18 in gastritis and gastric cancer. Immunohistochemically claudin-18–stained micrographs of paraffin sections of a normal stomach, stomachs with gastritis, and gastric cancer. Claudin-18 was downregulated in the gastritis and gastric cancer samples.

HE-stained *Cldn18*$^{-/-}$ stomach tissue revealed fewer parietal and chief cells, which had largely been replaced by metaplastic cells (Fig. 2A). Furthermore, inflammatory cells had infiltrated the *Cldn18*$^{-/-}$ submucosal space. Thus, the claudin-18st-based paracellular proton barrier protects the stomach against atrophic inflammation triggered by the undesired leakage of protons into the gastric submucosa.

We next examined the characteristics of the gastritis in the *Cldn18*$^{-/-}$ mice in detail. We found that the levels of proinflammatory cytokines including IL-1β (but not IL-6) and the neutrophil chemoattractant KC were significantly higher in the *Cldn18*$^{-/-}$ stomach than the *Cldn18*$^{+/+}$ stomach, by qRT-PCR (Fig. 2B). The serum protein level of IL-1β

was also significantly upregulated in the *Cldn18*$^{-/-}$ mice. Analysis of the immune cell types by FACS revealed that neutrophils, which are positive for Gr-1, predominated in the gastritic *Cldn18*$^{-/-}$ tissue. This is in agreement with our HE staining and immunofluorescence images showing that neutrophils were significantly increased in the stomach of younger adult *Cldn18*$^{-/-}$ mice examined (< 20 weeks old).

Claudin-18st and gastric metaplasia in mice

In the *Cldn18*$^{-/-}$ mouse stomach, the gastric epithelium was largely occupied by proliferating mucous-like cells that were positive for trefoil factor family 2 (TFF2). Some of these cells were also positive for intrinsic factor, indicating that they were spasmolytic polypeptide-expressing metaplastic (SPEM) cells.[33] Although several reports have shown that SPEM cells are associated with >90% of gastric cancers, neither dysplasia nor gastric cancer was found in the stomach of younger *Cldn18*[-/-] mice (<20 weeks old).[33,34] On the other hand, other chronic inflammation markers such as IL-6 were sporadically upregulated in *Cldn18*[-/-] mice (Suzuki *et al.*, unpublished data), hinting at a relationship between the claudin-18st–based SPEM and dysplasia/cancer (Fig. 2C and D).

Claudin-18 and gastritis and gastric cancer in humans

Since previous studies reported that claudin-18 expression is downregulated in human gastric cancer, we also investigated the expression levels of claudin-18 in human chronic and autoimmune gastritis by immunohistochemical methods. Claudin-18 was found to be downregulated in human gastritis specimens, at foci that showed atrophy and metaplasia. In superficial gastritis, the claudin-18 level was also decreased. Thus, it would appear that claudin-18 is downregulated pathologically in human gastritis as it is in gastric cancer (Fig. 3). These findings point to the possibility that claudin-18 loss induces gastritis, which creates a setting for dysplasia and/or cancer. This finding leads to important questions on the role of claudins in cancer that will be addressed in future research.

Summary

Given the conceptual consensus that the large multi-gene claudin family plays a critical role in creating various microenvironments in organ systems, particular attention is being paid to effects of claudin-based regulation of paracellular barriers and the barrier-based ion permeability between epithelial cells on homeostasis *in vivo*. We recently showed that claudin-18 forms the paracellular proton barrier that prevents gastritis in the stomach. Further studies will be focused on revealing the function of claudins in metaplasia, dysplasia, and cancer. As other claudins also regulate biological functions, further analyses at the cellular level and on the *in vivo* functions of claudin-family proteins should elucidate the physiological significance of claudin-family members in various biological systems.

Conflicts of interest

The authors declare no conflicts of interest.

References

1. Tsukita, S., M. Furuse & M. Itoh. 2001. Multifunctional strands in tight junctions. *Nat. Rev. Mol. Cell Biol.* **2:** 285–293.
2. Van Itallie, C.M. & J.M. Anderson. 2006. Claudins and epithelial paracellular transport. *Annu. Rev. Physiol.* **68:** 403–429.
3. Mineta, K. *et al.* 2011. Predicted expansion of the claudin multigene family. *FEBS Lett.* **585:** 606–612.
4. Angelow, S., R. Ahlstrom & A.S. Yu. 2008. Biology of claudins. *Am. J. Physiol. Renal. Physiol.* **295:** F867–F876.
5. Fujita, H. *et al.* 2006. Differential expression and subcellular localization of claudin-7, -8, -12, -13, and -15 along the mouse intestine. *J. Histochem. Cytochem.* **54:** 933–944.
6. Holmes, J.L. *et al.* 2006. Claudin profiling in the mouse during postnatal intestinal development and along the gastrointestinal tract reveals complex expression patterns. *Gene Expr. Patterns* **6:** 581–588.
7. Soini, Y. 2011. Claudins in lung diseases. *Respir. Res.* **12:** 70.
8. Marchiando, A.M., W.V. Graham & J.R. Turner. 2010. Epithelial barriers in homeostasis and disease. *Annu. Rev. Pathol.* **5:** 119–144.
9. Gupta, I.R. & A.K. Ryan. 2010. Claudins: unlocking the code to tight junction function during embryogenesis and in disease. *Clin. Genet.* **77:** 314–325.
10. Van Itallie, C.M., A.S. Fanning & J.M. Anderson. 2003. Reversal of charge selectivity in cation or anion-selective epithelial lines by expression of different claudins. *Am. J. Physiol. Renal. Physiol.* **285:** F1078–F1084.
11. Amasheh, S. *et al.* 2002, Claudin-2 expression induces cation-selective channels in tight junctions of epithelial cells. *J. Cell Sci.* **115:** 4969–4976.
12. Furuse, M., K. Furuse, H. Sasaki & S. Tsukita. 2001. Conversion of zonulae occludentes from tight to leaky strand type by introducing claudin-2 into Madin-Darby canine kidney I cells. *J. Cell Biol.* **153:** 263–272.
13. Van Itallie C.M. *et al.* 2006. Two splice variants of claudin-10 in the kidney create paracellular pores with different ion selectivities. *Am. J. Physiol. Renal. Physiol.* **291:** F1288–F1299.
14. Yu, A.S. *et al.* 2009. Molecular basis for cation selectivity in claudin-2-based paracellular pores: identification of an electrostatic interaction site. *J. Gen. Physiol.* **133:** 111–127.
15. Sasaki, H. *et al.* 2003. Dynamic behavior of paired claudin strands within apposing plasma membranes. *Proc. Natl. Acad. Sci. USA* **100:** 3971–3976.
16. Madara, J.L. 1998. Regulation of the movement of solutes across tight junctions. *Annu. Rev. Physiol.* **60:** 143–159.
17. Shen, L. *et al.* 2011. Tight junction pore and leak pathways: a dynamic duo. *Annu. Rev. Physiol.* **73:** 283–309.
18. Hadj-Rabia, S. *et al.* 2004. Claudin-1 gene mutations in neonatal sclerosing cholangitis associated with ichthyosis: a tight junction disease. *Gastroenterology* **127:** 1386–1390.
19. Wilcox, E.R. *et al.* 2001. Mutations in the gene encoding tight junction claudin-14 cause autosomal recessive deafness DFNB29. *Cell* **104:** 165–172.

20. Weber, S. *et al.* 2000. Familial hypomagnesaemia with hypercalciuria and nephrocalcinosis maps to chromosome 3q27 and is associated with mutations in the PCLN-1 gene. *Eur. J. Hum. Genet.* **8:** 414–422.

21. Simon, D.B. *et al.* 1999. Paracellin-1, a renal tight junction protein required for paracellular Mg^{2+} resorption. *Science* **285:** 103–106.

22. Naeem, M., S. Hussain & N. Akhtar. 2011. Mutation in the tight-junction gene claudin 19 (CLDN19) and familial hypomagnesemia, hypercalciuria, nephrocalcinosis (FHHNC) and severe ocular disease. *Am. J. Nephrol.* **34:** 241–248.

23. Anderson, J.M. & C.M. Van Itallie. 2009. Physiology and function of the tight junction. *Cold. Spring. Harb. Perspect. Biol.* **1:** a002584.

24. Tamura, A. *et al.* 2011. Loss of claudin-15, but not claudin-2, causes Na^+ deficiency and glucose malabsorption in mouse small intestine. *Gastroenterology* **140:** 913–923.

25. Furuse, M. *et al.* 2002. Claudin-based tight junctions are crucial for the mammalian epidermal barrier: a lesson from claudin-1-deficient mice. *J. Cell Biol.* **156:** 1099–1111.

26. Nitta, T. *et al.* 2003. Size-selective loosening of the blood-brain barrier in claudin-5-deficient mice. *J. Cell Biol.* **161:** 653–660.

27. Gow, A. *et al.* 1999. CNS myelin and sertoli cell tight junction strands are absent in Osp/claudin-11 null mice. *Cell* **99:** 649–659.

28. Hou, J. *et al.* 2006. Study of claudin function by RNA interference. *J. Biol. Chem.* **281:** 36117–36123.

29. Angelow, S., K.J. Kim & A.S. Yu. 2006. Claudin-8 modulates paracellular permeability to acidic and basic ions in MDCK II cells. *J. Physiol.* **571:** 15–26.

30. Jovov, B. *et al.* 2007. Claudin-18: a dominant tight junction protein in Barrett's esophagus and likely contributor to its acid resistance. *Am. J. Physiol. Gastrointest. Liver Physiol.* **293:** G1106–G1113.

31. Hayashi, D. *et al.* 2012. Deficiency of claudin-18 causes paracellular $H^{(+)}$ leakage, up-regulation of interleukin-1β, and atrophic gastritis in mice. *Gastroenterology* **142:** 292–304.

32. Niimi, T. *et al.* 2001. Claudin-18, a novel downstream target gene for the T/EBP/NKX2.1 homeodomain transcription factor, encodes lung-and stomach-specific isoforms through alternative splicing. *Mol. Cell Biol.* **21:** 7380–7390.

33. Weis, V.G. & J.R. Goldenring. 2009. Current understanding of SPEM and its standing in the preneoplastic process. *Gastric Cancer* **12:** 189–197.

34. Oshima, H. & M. Oshima. 2010. Mouse models of gastric tumors: Wnt activation and PGE2 induction. *Pathol. Int.* **60:** 599–607.

Ann. N.Y. Acad. Sci. ISSN 0077-8923

ANNALS OF THE NEW YORK ACADEMY OF SCIENCES

Issue: *Barriers and Channels Formed by Tight Junction Proteins*

Regulation of epithelial proliferation by tight junction proteins

Attila E. Farkas, Christopher T. Capaldo, and Asma Nusrat

Epithelial Pathobiology and Mucosal Inflammation Research Unit, Department of Pathology and Laboratory Medicine, Emory University, Atlanta, Georgia

Address for correspondence: Asma Nusrat, Epithelial Pathobiology and Mucosal Inflammation Research Unit, Dept. of Pathology and Laboratory Medicine, Whitehead Bld Rm 105m, Emory University, Atlanta, GA, 30322. anusrat@emory.edu

The epithelial tight junction (TJ) is the apical-most intercellular junction and serves as a gatekeeper for the paracellular pathway by permitting regulated passage of fluid and ions while restricting movement of large molecules. In addition to these vital barrier functions, TJ proteins are emerging as major signaling molecules that mediate crosstalk between the extracellular environment, the cell surface, and the nucleus. Biochemical studies have recently determined that epithelial TJs contain over a hundred proteins that encompass transmembrane proteins, scaffolding molecules, cytoskeletal components, regulatory elements, and signaling molecules. Indeed, many of these proteins have defined roles in regulating epithelial polarity, differentiation, and proliferation. This review will focus on recent findings that highlight a role for TJ proteins in controlling cell proliferation during epithelial homeostasis, wound healing, and carcinogenesis.

Keywords: tight junctions; proliferation; signaling; epithelial

Epithelial tight junction structure and function

Epithelial sheets are located at boundaries of biological compartments where they form strictly regulated sealing barriers that control the trafficking of solutes and immune cells while maintaining overall tissue homeostasis. As a case in point, a single layer of columnar intestinal epithelial cells (IECs) protects the underlying tissue compartments from luminal contents while at the same time allowing absorption and secretion of fluid and ions. Such polarized epithelial cells are linked together by a series of intercellular junctions that include the tight junction (TJ), adherens junction (AJ), and desmosomes.[1] Given the close physical proximity and functional crosstalk between the TJ and AJ, they are collectively referred to as the *apical junctional complex*. The TJ is the apical-most intercellular junction and serves as a gatekeeper for the paracellular pathway, where it permits regulated passage of fluid and ions while restricting movement of large molecules.[2] Subjacent to the TJ, the AJ controls cell–cell adhesion.[3]

While barrier properties of TJ proteins have been well appreciated, it is now becoming evident that these proteins regulate numerous key cellular processes responsible for epithelial polarity and tissue homeostasis; encompassing cell proliferation, differentiation, and migration.

By electron microscopy, the TJ is visualized as an apical region of close membrane apposition between adjoining polarized epithelial cells. Structurally, the TJ is composed of anastomosing strands of transmembrane proteins, which span the extracellular space to interact with TJs of adjacent cells (Fig. 1A). Over the past two decades the number of proteins identified in the TJ has grown immensely. In fact, a recent biochemical study determined that epithelial TJs contain hundreds of proteins including transmembrane and scaffolding proteins, cytoskeletal components, regulatory elements, and signaling molecules.[4] Within the junction, transmembrane proteins associate with specific scaffold proteins, which both link transmembrane proteins to intracellular signaling molecules and stabilize transmembrane proteins

doi: 10.1111/j.1749-6632.2012.06556.x

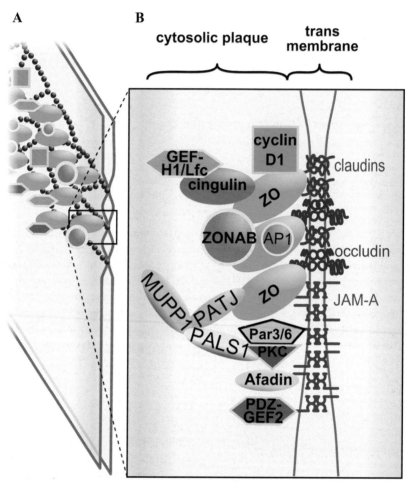

Figure 1. Structure of the TJ. (A) TJ transmembrane proteins form strands that associate with cytoplasmic plaque proteins. (B) The TJ protein complex consists of several transmembrane and associated plaque proteins, including structural and adaptor molecules, signaling molecules, and transcription factors.

within the junction.[5] Key TJ transmembrane proteins include the claudin family of proteins, occludin, junctional adhesion molecule-A (JAM-A), and other CTX family members such as the coxsackie and adenovirus receptor (CAR) (Fig. 1). Scaffolding components, which include zonula occludens (ZO) proteins, associate with transmembrane proteins, bind to actin, and sequester regulatory proteins. Each class of TJ proteins has the potential to regulate other TJ constituents, either directly or indirectly through scaffold protein mediators. Furthermore, it is becoming evident that there is a more global crosstalk between TJ proteins and the nucleus, so as to coordinate overall epithelial/endothelial cell adhesion and tissue homeostasis.[6–9] Extensive reviews of TJ protein constituents and their interactions under physiological and pathological conditions have been provided by several research groups, including ours.[8,10–16] This short review will highlight the emerging role of TJ proteins in regulation of epithelial proliferation.

TJ transmembrane protein regulation of epithelial proliferation

The last decades saw the discovery of several single pass TJ transmembrane proteins that have important signaling functions to control epithelial homeostasis. These proteins fall into two broad categories: immunoglobulin superfamily (IgSF), including JAMs, CAR, coxsackie- and adenovirus receptor-like membrane protein (CLMP), or endothelial cell-selective adhesion molecule (ESAM);

and non-IgSF proteins like crumbs homolog 3 (CRB3).[17–21]

The most extensively studied IgSF TJ protein is JAM-A. JAM-A is expressed in a wide variety of cells, and in polarized epithelial cells is localized to TJs, where it regulates barrier function.[22,23] In fact, loss of JAM-A influences the composition of TJ proteins with increased expression of "leaky" claudins -10 and -15.[22,23] JAM-A has two extracellular immunoglobulin loops, of which the membrane distal loop (D1) dimerizes to regulate epithelial function, which includes maintenance of paracellular permeability, cell migration, and proliferation.[24,25] Recent studies have also revealed that JAM-A regulates epithelial cell migration by modulating β_1 integrin expression. This effect is mediated by a small GTPase Rap-1.[26] The importance of JAM-A in the outside-in signaling of epithelial cells is further highlighted by its physical interaction with afadin and PDZ-GEF2.[27]

In a recent study using complementary *in vitro* and *in vivo* experimental approaches, we demonstrated that JAM-A regulates IEC proliferation.[25] Increased proliferation of intestinal crypt epithelial cells was observed in JAM-A knockout mice where the proliferative zone is expanded along the crypt-luminal axis. This study demonstrated a central role of Akt/β-catenin signaling in mediating JAM-A control of IEC proliferation. Loss of JAM-A in model intestinal epithelial cell lines in culture and in JAM-A knockout mice, results in phosphorylation and activation of Akt and its downstream target protein β-catenin.[25] Activation of this signaling pathway promotes phospho-β-catenin Ser552 nuclear translocation and β-catenin/TCF/LEF transcriptional activity.[28] The increased cell proliferation seen in JAM-A knockout animals can be partially rescued by pharmacological inhibition of Akt. This study has identified Akt as a key signaling element that mediates crosstalk between JAM-A and β-catenin nuclear activity.[25] Lastly, in keeping with the outside-in signaling of JAM-A, dimerization of its membrane distal loop plays a pivotal role in controlling IEC proliferation. Taken together, JAM-A dimerization in TJs serves to suppress IEC proliferation. Consistent with these results we observe a gradient of JAM-A expression along the intestinal crypt-luminal axis with increased JAM-A expression in nonproliferating luminal epithelial cells.

The tetraspan integral membrane claudin proteins play a vital role in regulating paracellular permeability in epithelial and endothelial cells. Claudins form TJ-like strands when expressed in cells otherwise devoid of TJs, and knockout mice lacking claudin family members demonstrate barrier defects.[29] Extracellular loops of claudin proteins are responsible for homo- and heterotypic intercellular interactions and form charge selective pores between cells.[30,31] In fact, the first extracellular loop of claudins controls such charge and fluid selectivity of the paracellular pathway, and the second extracellular loop is involved in claudin–claudin interaction and TJ formation.[32–35] Claudin family proteins exhibit organ and tissue-specific expression. For example, specific subsets of claudins exhibiting differential permselectivity are responsible for the specific barrier properties observed in the small versus large intestine.[36]

While claudins regulate endothelial and epithelial barrier function, their contribution to controlling cellular homeostasis is now becoming evident. Such properties are not mutually exclusive, as barrier function influences the epithelial microenvironment and therefore overall homeostasis.[37] This notion is supported by a report that claudin-15 knockout mice have an enlarged upper small intestinal phenotype or megaintestine.[38] This phenotype was not observed at birth but becomes evident at 9 weeks of age. Several morphological changes were discernible in the upper small intestine. The length and width of villi doubled, the number of epithelial cells per villus was increased while the size of epithelial cells remained unchanged. The crypt epithelial proliferative zone was enlarged twofold compared to normal mice while apoptosis was unaffected. Surprisingly the expression and subcellular localization of other claudins was unaffected while development of TJ strands was decreased. Given this complex phenotype, the authors suggested that the increased cellular proliferation is secondary to an altered intestinal epithelial microenvironment, such as translocation of luminal contents (e.g., electrolyte and growth factors) into the underlying mucosa. However, a direct effect of claudin-15 knockdown on IEC proliferation was not experimentally excluded. The notion of mucosal microenvironment in regulating IEC proliferation is also in keeping with our recent study that demonstrated the distinct influence of mucosal inflammation on IEC proliferation.[39]

While direct mechanisms linking claudin proteins to cell proliferation are lacking, altered regulation frequently correlates with states of increased cell growth. Indeed, overexpression of claudin-11 induces proliferation in oligodendrocytes,[40] while claudin-3 overexpression increases tubulogenesis and BrdU incorporation in mIMCD-3 cells (mouse inner medullary collecting duct cells).[41] Additional reports have revealed that overexpression of claudins 6, 7, or 9 enhances invasiveness and proliferation of an adenocarcinoma cell line and, importantly, claudin misregulation is commonly found in human carcinomas.[42–45]

The occludin family of TJ transmembrane proteins includes occludin, tricellulin, and MARVELD3.[46–48] These proteins share a MARVEL (MAL and related proteins for vesicle trafficking and membrane link) domain that consists of four transmembrane helixes.[49] The first transmembrane TJ protein identified, occludin, forms short TJ-like strands by itself.[50] Occludin knockout mice do not have epithelial barrier defects. However, occludin has been reported to function as a signaling protein and regulate epithelial barrier function in *in vitro* studies. These findings suggest that compensatory mechanisms likely come into play in the occludin knockout mice.[51] A role for occludin in cancer pathogenesis has been suggested.[52,53] In model salivary gland epithelial cells, we have observed a central role for occludin in suppressing epithelial mesenchymal transition (EMT) induced by increased Raf activity.[54] Raf, a downstream effector of Ras oncogene, suppresses occludin expression by upregulating the transcription factor Slug, which directly binds to and represses the occludin promoter activity.[55] The suppression of Raf activity was mediated by the second extracellular loop of occludin.[56] Recently, further evidence has been found regarding the role of occludin in the regulation of proliferation: occludin has been found to localize in centrosomes during interphase, and phosphorylation at Ser 490 increases mitotic entry, while a Ser 490Ala mutation impedes centrosome separation, delays mitotic entry and inhibits proliferation.[57] Another MARVEL domain-containing protein, MARVELD1, has also been found to play an important role in cell-cycle progression.[58] Blood vessel epicardial substance (Bves), a TJ protein containing three transmembrane domains, was originally described in the developing heart.[59] It has been identified in epithelia, skeletal, and smooth muscle.[60] Bves contributes to the establishment and mainte-

nance of epithelial integrity,[61] it regulates RhoA and ZONAB/DbpA activity in corneal epithelial cells,[62] and also prevents EMT. Thus Bves downregulation may be an important step in the pathogenesis of colon cancer.[63]

Recent evidence also indicates a role for transmembrane TJ proteins in the regulation of proliferation through Eph/Ephrin receptor signaling.[64,65] Ephrin receptors modulate intercellular signaling related to cell growth through PI3K and MAPK signaling, which may involve a complex reciprocal regulation with claudins 1 and 4.[66,64] Furthermore, it is interesting to note that the receptor serine/threonine kinase TGF-βR (transforming growth factor receptor) type I and II colocalize with ZO-1 in the TJ during TGF-β–dependent EMT, and the direct phosphorylation of Par6 by TGF-βRII is required for this process.[67] This finding places a well-studied modulator of epithelial proliferation among the transmembrane proteins of the TJ.[68]

In summary, recent studies have implicated a role for TJ transmembrane proteins in controlling cell growth either by directly regulating signaling pathways or secondary to controlling the paracellular barrier and local microenvironment of cells. Another key mechanism by which TJ transmembrane proteins impact cellular processes is through sequestration of intercellular plaque proteins and signaling protein mediators. The latter mechanisms are detailed below.

TJ plaque protein regulation of cell proliferation

Transmembrane TJ proteins interact with a variety of cytoplasmic scaffolds, largely mediated by PDZ (PSD-95/DLG/ZO-1) protein-binding domains. The first TJ protein to be discovered was the membrane-associated guanylate kinase homolog (MAGUK) family protein zonula occludens 1 (ZO-1),[69] an interacting partner of claudins 1–8,[70] occludin,[71] JAM-A,[72] the actin cytoskeleton,[73] as well as other plaque proteins, such as ZO-2 and ZO-3,[74] AF-6,[75] and cingulin.[76] In addition, plaque scaffold proteins like the ZOs serve to localize a variety of regulatory proteins such as kinases, GTP exchange factors, and transcription factors at TJs[6] (Fig. 1).

Several MAGUK family TJ plaque proteins have been implicated in the regulation of cell proliferation. The first was ZO-1, which has been shown to sequester the transcription factor ZONAB at the

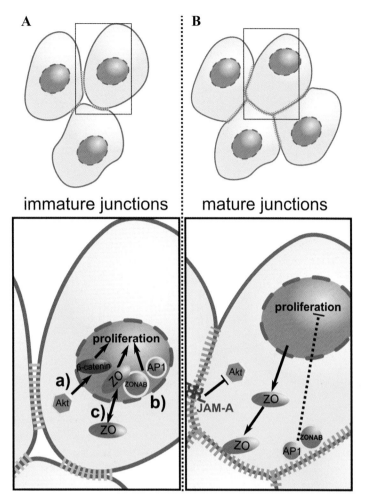

Figure 2. Regulation of epithelial proliferation by tight junction proteins. TJ proteins are emerging as regulators of outside-in signaling and have been observed to play important roles in epithelial homeostasis that encompasses proliferation, migration, and differentiation. (A) In cells with immature junctions, pro-mitogenic signaling molecules, (a) transcription factors, (b) and TJ proteins (c) translocate to the nucleus and influence cell proliferation. (B) These proteins are sequestered by mature TJs in differentiated cells. TJ proteins actively regulate cell proliferation during physiological tissue renewal and wound restitution.

TJs in an epithelial confluence–dependent manner.[77] Furthermore, ZO family proteins have been observed to translocate to the nucleus during biological states of increased cell proliferation. For example, ZOs have been observed in the nucleus of subconfluent epithelial cells, while exclusively at TJs in high confluence cultures (see Fig. 2, our unpublished data).[78,79] While the role of ZO-1 in the nucleus is unclear, ZO-2 translocates to the nucleus where it inhibits transcription of the G_1 cyclin, cyclin D1.[80] Importantly, correct spatial and temporal regulation of cyclin D1 is not only vital for its proper function as a mediator of cell cycle advancement, aberrant cyclin D1 localization is thought to con-tribute to carcinogenesis.[81] Cyclin D1 promotes cell cycle progression, along with its associated kinase Cdk4, by signaling to downstream targets in the nucleus in order to promote S-phase entry.[82] Indeed, several recent reports have found that ZO family proteins are regulators of cyclin D1 gene expression and protein stability.[68,83–85] Additionally, we have recently shown that ZO-3 stabilizes cyclin D1 at TJs during mitosis, thereby promoting iterative cell proliferation within IECs that have intact junctions.[85] This occurs through direct, PDZ domain-mediated interactions between cyclin D1 and ZO-3. How does this mechanism promote cell growth? During conditions of rapid cell proliferation, such as in subconfluent cell cultures or during wound healing events, cyclin D1 begins to accumulate in G_2 in order to promote rapid G_1-to-S-phase transition.[86] Enhanced protein stability of cyclin D1 due to ZO-3 protein binding would promote G_1-to-S phase progression, conversely, decreased cyclin D1 stability during S-phase is required for DNA replication.[82] However, overexpression of ZO-2 in canine kidney cells inhibits cell proliferation due to G_0/G_1 cell-cycle block and destabilized cyclin D1.[84] Therefore, a balance likely exists between growth suppressive and promoting properties of ZO proteins, one that is likely dependent on protein expression levels and the amount of ZO proteins that are available for translocation into the nucleus.

Other MAGUK proteins such as Dlg (discs large), Scrib (scribble), and Lgl (lethal giant larvae) regulate cell proliferation, possibly through the regulation of cell polarity.[87] Scribble is targeted by the human papillomavirus (HPV) E6 complex for ubiquitin-mediated degradation, which is associated with oropharynx, endometrial, and cervical cancer.[88] Another cell polarity protein Partner of Lim 7 1 (Pals1) has been implicated in autosomal dominant polycystic kidney disease through the regulation of PC-2 and renal cilia formation.[89] Pals1 has recently been identified in a complex with Merlin, which suppresses MAPK, and Rac1-mediated cell proliferation and is in a complex with Pals1 and Patj at tight junctions.[90] Interestingly, hDlg, MAGI-1, MAGI-2, MAGI-3, and MUPP1 proteins are targeted for proteasomal degradation, again after viral infection.[91] PDZ-LIM proteins regulate nucleo-cytoplasmic shuttling of the transcription factor Taz and are required for muscle development.[92] The PDZ family protein Mystique has potent tumor suppressor properties through suppression of NF-κB–dependent transcription.[93,94] Interestingly, Mystique is also a viral target and is implicated in human T cell leukemia virus type I (HTLV-I)-mediated tumorigenesis. Furthermore, the plaque protein Cingulin is thought to suppress proliferation through suppression of RhoA signaling and ZO-3 protein levels.[95]

In summary, TJ plaque proteins of the ZO family can directly interact with proliferative processes through the regulation of cyclin D1 protein localization, protein stability, and gene transcription. Additionally, MAGUK plaque proteins may regulate cell growth through the regulation of transmembrane protein function, or through sequestration of transcription and signaling factors.

Signaling proteins and transcription factors

TJs are potentially powerful regulators of cell signaling through their ability to sequester transcription factors and signaling molecules.[8] For example, ZO-1 functions at TJs to sequester ZONAB, a Y-box transcription factor that regulates a host of genes involved in cell replication.[68] ZONAB promotes S-phase progression by upregulating cyclin D1 gene expression, as well as the expression of proteins that are required for DNA replication and repair. Similarly, the transcription factors ASH1 and AP-1 are sequestered at TJs. ASH1 interacts with HDAC1 repression complexes that in turn regulate chromatin remodeling and gene expression, while AP-1 is a classic member of the Myc/Jun transcription factor complex. Importantly, these factors accumulate at TJs during states of low proliferation, such as cell–cell contact inhibition due to high confluence states (see Fig. 2).

Signaling molecules sequestered at the TJs include GEF-H1/Lfc, a guanine nucleotide exchange factor specific for RhoA, is involved in the regulation of paracellular permeability and cell proliferation.[96,97] Rho family GTPases cycle between active and inactive states based on their GTP or GDP bound states.[98] Therefore, activity of these factors can be controlled by TJ sequestration of proteins regulating their nucleotide binding, called GTPase-activating proteins (GAPs) and guanine nucleotide exchange factors (GEFs). It is well known that RhoA activation stimulates epithelial cell proliferation, gene expression, and differentiation.[99–101] RhoA is downregulated when epithelial cells reach confluence, resulting in inhibition of signaling pathways that stimulate proliferation.[102] Indeed, GEF-H1/Lfc directly interacts with cingulin, a TJ adaptor protein. Cingulin binding inhibits RhoA activation as illustrated by RNA interference of cingulin, which results in RhoA activation. These results indicate that developing epithelial TJs contribute to the downregulation of RhoA by inactivating GEF-H1 in a cingulin-dependent manner, providing a molecular mechanism whereby TJ formation is linked to inhibition of RhoA signaling and gene expression.[97] Similarly, the plaque protein Merlin sequesters the Rac GTPase GAP Rich1.[90] Interestingly, the TJ transmembrane protein, JAM-A is known to recruit PDZ-GEF in a dimerization-dependent fashion, although the implications of this recruitment with respect to cell proliferation have not yet been investigated.[27]

TJ structures also contain protein kinases, notably atypical protein kinase C (aPKC), which associates with TJs via the Par6/Par3 polarity complex.[8] Recent studies have determined that, in *Drosophila*, aPKC participates in the stimulation of cell proliferation after apoptosis of neighboring cells.[103] Furthermore, aPKC is linked to Hippo signaling through interactions with Lgl.[104] However, analogous regulation in vertebrates has not yet been determined. The protein phosphatases PP2A, PP1, and PTEN have been shown to both localize to TJ structures and play a role in cell growth signaling. However, these studies have focused on TJ assembly and do not directly assess the capacity of these factors to influence cell proliferation.

Concluding remarks

We have come a long way in our understanding of TJs. Originally thought to be simply static seals between the epithelial cells, TJs are emerging as complex regulators of outside–in signaling. A large number of TJ proteins have been described, and the proteins responsible for the gate-and-fence function of the epithelial sheets have been found, alongside with the proteins governing cell polarization and proliferation. Some transmembrane constituents of the TJ solely have signaling roles, such as JAM-A, which we found to regulate cell migration and morphology, and more recently proliferation, through its interaction with cytoplasmic plaque proteins. Promitogenic transcription factors and signaling molecules are regulated by intact TJs through sequestration that prevents their action in the nucleus (Fig. 2). This function for TJs is dependent on the plaque proteins and transmembrane constituents at the junctions. Indeed, during epithelial wounding events, where cell–cell contacts are altered, transmembrane and plaque constituents would potentially be unable to sequester these factors, thus leading to increased cell growth (Fig. 2). Interestingly, many of the plaque proteins discussed are targeted for ubiquitin-mediated degradation during viral infection—specifically, infections that have been shown to have a causative role in carcinogenesis. Conversely, ZO proteins act to regulate cell growth in differentiated epithelial monolayers by interfacing in the nucleus with cell cycle machinery. Together, TJs insure crosstalk between cell–cell contact structures and the nucleus, and provide a mechanism for outside–in signaling in epithelial cells. Indeed, TJs are increasingly appreciated as signaling complexes that communicate the cell–cell contact status to the cell interior, and thereby play an important role in deciding cell fate and proliferation.

Acknowledgments

This study was supported by grants from the National Institutes of Health DK59888 to A.N. and the Crohn's and Colitis Foundation of America Fellowship Award to C.T.C. and A.E.F.

Conflicts of interest

The authors declare no conflicts of interest.

References

1. Laukoetter, M.G., P. Nava & A. Nusrat. 2008. Role of the intestinal barrier in inflammatory bowel disease. *World J. Gastroenterol.* **14:** 401–407.
2. Madara, J.L. 1998. Regulation of the movement of solutes across tight junctions. *Annu. Rev. Physiol.* **60:** 143–159.
3. Meng, W. & M. Takeichi. 2009. Adherens junction: molecular architecture and regulation. *Cold Spring Harbor Persp. Biol.* **1:** 1–11.
4. Tang, V.W. 2006. Proteomic and bioinformatic analysis of epithelial tight junction reveals an unexpected cluster of synaptic molecules. *Biol. Direct* **1:** 37.
5. Umeda, K., J. Ikenouchi, S. Katahira-Tayama, *et al.* 2006. ZO-1 and ZO-2 Independently determine where claudins are polymerized in tight-junction strand formation. *Cell* **126:** 741–754.
6. Matter, K. & M.S. Balda. 2007. Epithelial tight junctions, gene expression and nucleo-junctional interplay. *J. Cell Sci.* **120:** 1505–1511.
7. Terry, S., M. Nie, K. Matter & M.S. Balda. 2010. Rho signaling and tight junction functions. *Physiology* **25:** 16–26.
8. Gonzalez-Mariscal, L., R. Tapia & D. Chamorro. 2008. Crosstalk of tight junction components with signaling pathways. *Biochim. Biophys. Acta.* **1778:** 729–756.
9. Matter, K., S. Aijaz, A. Tsapara & M.S. Balda. 2005. Mammalian tight junctions in the regulation of epithelial differentiation and proliferation. *Curr. Opin. Cell Biol.* **17:** 453–458.
10. Bauer, H., A. Traweger, J. Zweimueller-Mayer, *et al.* 2011. New aspects of the molecular constituents of tissue barriers. *J. Neural. Trans.* **118:** 7–21.
11. Aijaz, S., M.S. Balda & K. Matter. 2006. Tight junctions: molecular architecture and function. *Int. Rev. Cytol.* **248:** 261–298.
12. Anderson, J.M. & C.M. Van Itallie. 2008. Tight junctions. *Curr. Biol.* **18:** R941–R943.

13. Utech, M., M. Bruwer & A. Nusrat. 2006. Tight junctions and cell–cell interactions. *Methods Mol. Biol.* **341:** 185–195.

14. Capaldo, C.T. & A. Nusrat. 2009. Cytokine regulation of tight junctions. *Biochimica et Biophysica Acta (BBA)–Biomembranes* **1788:** 864–871.

15. Koch, S. & A. Nusrat. 2009. Dynamic regulation of epithelial cell fate and barrier function by intercellular junctions. *Ann. N. Y. Acad. Sci.* **1165:** 220–227.

16. Ivanov, A.I., C.A. Parkos & A. Nusrat. 2010. Cytoskeletal regulation of epithelial barrier function during inflammation. *Am. J. Pathol.* **177:** 512–524.

17. Williams, L.A., I. Martin-Padura, E. Dejana, *et al.* 1999. Identification and characterisation of human Junctional Adhesion Molecule (JAM). *Mol. Immunol.* **36:** 1175–1188.

18. Bergelson, J.M., J.A. Cunningham, G. Droguett, *et al.* 1997. Isolation of a common receptor for Coxsackie B viruses and adenoviruses 2 and 5. *Science* **275:** 1320–1323.

19. Raschperger, E., U. Engstrom, R.F. Pettersson & J. Fuxe. 2004. CLMP, a novel member of the CTX family and a new component of epithelial tight junctions. *J. Biol. Chem.* **279:** 796–804.

20. Hirata, K., T. Ishida, K. Penta, *et al.* 2001. Cloning of an immunoglobulin family adhesion molecule selectively expressed by endothelial cells. *J. Biol. Chem.* **276:** 16223–16231.

21. Makarova, O., M.H. Roh, C.J. Liu, S. Laurinec & B. Margolis. 2003. Mammalian Crumbs3 is a small transmembrane protein linked to protein associated with Lin-7 (Pals1). *Gene.* **302:** 21–29.

22. Laukoetter, M.G., P. Nava, W.Y. Lee, *et al.* 2007. JAM-A regulates permeability and inflammation in the intestine in vivo. *J. Exp. Med.* **204:** 3067–3076.

23. Vetrano, S., M. Rescigno, M. Rosaria Cera, *et al.* 2008. Unique Role of junctional adhesion molecule-A in maintaining mucosal homeostasis in inflammatory bowel disease. *Gastroenterology* **135:** 173–184.

24. Severson, E.A., L. Jiang, A.I. Ivanov, K.J. Mandell, A. Nusrat & C.A. Parkos. 2008. Cis-dimerization mediates function of junctional adhesion molecule A. *Mol. Biol. Cell* **19:** 1862–1872.

25. Nava, P., C.T. Capaldo, S. Koch, *et al.* 2011. JAM-A regulates epithelial proliferation through Akt/beta-catenin signalling. *EMBO Rep.* **12:** 314–320.

26. Mandell, K.J., B.A. Babbin, A. Nusrat & C.A. Parkos. 2005. Junctional adhesion molecule 1 regulates epithelial cell morphology through effects on β1 integrins and rap1 activity. *J. Biol. Chem.* **280:** 11665–11674.

27. Severson, E.A., W.Y. Lee, C.T. Capaldo, *et al.* 2009. Junctional adhesion molecule a interacts with afadin and PDZ-GEF2 to activate rap1A, regulate β1 integrin levels, and enhance cell migration. *Mole. Biol. Cell* **20:** 1916–1925.

28. Perry, J.M., X.C. He, R. Sugimura, *et al.* 2011. Cooperation between both Wnt/{beta}-catenin and PTEN/PI3K/Akt signaling promotes primitive hematopoietic stem cell self-renewal and expansion. *Genes Dev.* **25:** 1928–1942.

29. Furuse, M. 2009. Knockout animals and natural mutations as experimental and diagnostic tool for studying tight junction functions in vivo. *Biochim. Biophys. Acta.* **1788:** 813–819.

30. Daugherty, B.L., C. Ward, T. Smith, *et al.* 2007. Regulation of heterotypic claudin compatibility. *J. Biol. Chem.* **282:** 30005–30013.

31. Van Itallie, C.M. & J.M. Anderson. 2004. The role of claudins in determining paracellular charge selectivity. *Proc. Am. Thorac. Soc.* **1:** 38–41.

32. Colegio, O.R., C. Van Itallie, C. Rahner & J.M. Anderson. 2003. Claudin extracellular domains determine paracellular charge selectivity and resistance but not tight junction fibril architecture. *Am. J. Physiol. Cell Physiol.* **284:** C1346–C1354.

33. Angelow, S. & A.S. Yu. 2009. Structure-function studies of claudin extracellular domains by cysteine-scanning mutagenesis. *J. Biol. Chem.* **284:** 29205–29217.

34. Mrsny, R.J., G.T. Brown, K. Gerner-Smidt, *et al.* 2008. A key claudin extracellular loop domain is critical for epithelial barrier integrity. *Am. J. Pathol.* **172:** 905–915.

35. Piontek, J., L. Winkler, H. Wolburg, *et al.* 2008. Formation of tight junction: determinants of homophilic interaction between classic claudins. *FASEB J.* **22:** 146–158.

36. Amasheh, S., M. Fromm & D. Günzel. 2011. Claudins of intestine and nephron—a correlation of molecular tight junction structure and barrier function. *Acta. Physiol.* **201:** 133–140.

37. Tsukita, S., Y. Yamazaki, T. Katsuno & A. Tamura. 2008. Tight junction-based epithelial microenvironment and cell proliferation. *Oncogene* **27:** 6930–6938.

38. Tamura, A., Y. Kitano, M. Hata, *et al.* 2008. Megaintestine in claudin-15-deficient mice. *Gastroenterology* **134:** 523–534.

39. Nava, P., S. Koch, M.G. Laukoetter, *et al.* 2010. Interferon-[gamma] regulates intestinal epithelial homeostasis through converging [beta]-catenin signaling pathways. *Immunity* **32:** 392–402.

40. Tiwari-Woodruff, S.K., A.G. Buznikov, T.Q. Vu, *et al.* 2001. OSP/claudin-11 forms a complex with a novel member of the tetraspanin super family and beta1 integrin and regulates proliferation and migration of oligodendrocytes. *J. Cell Biol.* **153:** 295–305.

41. Haddad, N., J. El Andalousi, H. Khairallah, *et al.* 2011. The tight junction protein Claudin-3 shows conserved expression in the nephric duct and ureteric bud and promotes tubulogenesis in vitro. *Am. J. Physiol.—Renal Physiol.*

42. Zavala-Zendejas, V.E., A.C. Torres-Martinez, B. Salas-Morales, *et al.* 2011. Claudin-6, 7, or 9 overexpression in the human gastric adenocarcinoma cell line AGS increases its invasiveness, migration, and proliferation rate. *Cancer Invest.* **29:** 1–11.

43. Ikari, A., T. Sato, A. Takiguchi, *et al.* 2011. Claudin-2 knock-down decreases matrix metalloproteinase-9 activity and cell migration via suppression of nuclear Sp1 in A549 cells. *Life Sci.* **88:** 628–633.

44. Takehara, M., T. Nishimura, S. Mima, *et al.* 2009. Effect of claudin expression on paracellular permeability, migration and invasion of colonic cancer cells. *Biol. Pharm. Bull.* **32:** 825–831.

45. Yoon, C.-H., M.-J. Kim, M.-J. Park, *et al.* 2010. Claudin-1 acts through c-Abl-protein kinase Cδ (PKCδ) signaling and has a causal role in the acquisition of invasive capacity in human liver cells. *J. Biol. Chem.* **285:** 226–233.

46. Furuse, M., T. Hirase, M. Itoh, *et al.* 1993. Occludin: a novel integral membrane protein localizing at tight junctions. *J. Cell Biol.* **123:** 1777–1788.

47. Ikenouchi, J., M. Furuse, K. Furuse, *et al.* 2005. Tricellulin constitutes a novel barrier at tricellular contacts of epithelial cells. *J. Cell Biol.* **171:** 939–945.

48. Steed, E., N.T. Rodrigues, M.S. Balda & K. Matter. 2009. Identification of MarvelD3 as a tight junction-associated transmembrane protein of the occludin family. *BMC Cell Biol.* **10:** 95.

49. Sanchez-Pulido, L., F. Martin-Belmonte, A. Valencia & M.A. Alonso. 2002. MARVEL: a conserved domain involved in membrane apposition events. *Trends Biochem. Sci.* **27:** 599–601.

50. Furuse, M., K. Fujimoto, N. Sato, *et al.* 1996. Overexpression of occludin, a tight junction-associated integral membrane protein, induces the formation of intracellular multilamellar bodies bearing tight junction-like structures. *J. Cell Sci.* **109** : 429–435.

51. Saitou, M., M. Furuse, H. Sasaki, *et al.* 2000. Complex phenotype of mice lacking occludin, a component of tight junction strands. *Mol. Biol. Cell* **11:** 4131–4142.

52. Martin, T.A., M.D. Mason & W.G. Jiang. 2011. Tight junctions in cancer metastasis. *Front Biosci.* **16:** 898–936.

53. Wang, X., O. Tully, B. Ngo, *et al.* 2011. Epithelial tight junctional changes in colorectal cancer tissues. *Sci. World J.* **11:** 826–841.

54. Li, D. & R.J. Mrsny. 2000. Oncogenic Raf-1 disrupts epithelial tight junctions via downregulation of occludin. *J. Cell Biol.* **148:** 791–800.

55. Wang, Z., P. Wade, K.J. Mandell, *et al.* 2006. Raf 1 represses expression of the tight junction protein occludin via activation of the zinc-finger transcription factor slug. *Oncogene* **26:** 1222–1230.

56. Wang, Z., K.J. Mandell, C.A. Parkos, *et al.* 2005. The second loop of occludin is required for suppression of Raf1-induced tumor growth. *Oncogene* **24:** 4412–4420.

57. Runkle, E.A., J.M. Sundstrom, K.B. Runkle, *et al.* 2011. Occludin localizes to centrosomes and modifies mitotic entry. *J. Biol. Chem.* **286:** 30847–30858.

58. Zeng, F., Y. Tian, S. Shi, *et al.* 2011. Identification of mouse MARVELD1 as a microtubule associated protein that inhibits cell cycle progression and migration. *Mol. Cells* **31:** 267–274.

59. Reese, D.E., M. Zavaljevski, N.L. Streiff & D. Bader. 1999. Bves: A novel gene expressed during coronary blood vessel development. *Dev. Biol.* **209:** 159–171.

60. Smith, T.K. & D.M. Bader. 2006. Characterization of Bves expression during mouse development using newly generated immunoreagents. *Dev. Dyn.* **235:** 1701–1708.

61. Osler, M.E., M.S. Chang & D.M. Bader. 2005. Bves modulates epithelial integrity through an interaction at the tight junction. *J. Cell Sci.* **118:** 4667–4678.

62. Russ, P.K., C.J. Pino, C.S. Williams, *et al.* 2011. Bves modulates tight junction associated signaling. *PLoS One* **6:** e14563.

63. Williams, C.S., B. Zhang, J.J. Smith, *et al.* 2011. BVES regulates EMT in human corneal and colon cancer cells and is silenced via promoter methylation in human colorectal carcinoma. *J. Clin. Invest.* **121:** 4056–4069.

64. Miao, H. & B. Wang. 2009. Eph/ephrin signaling in epithelial development and homeostasis. *Int. J. Biochem. Cell Biol.* **41:** 762–770.

65. Batlle, E., J.T. Henderson, H. Beghtel, *et al.* 2002. Beta-catenin and TCF mediate cell positioning in the intestinal epithelium by controlling the expression of EphB/ephrinB. *Cell* **111:** 251–263.

66. Tanaka, M., R. Kamata & R. Sakai. 2005. Phosphorylation of ephrin-B1 via the interaction with claudin following cell–cell contact formation. *EMBO J.* **24:** 3700–3711.

67. Ozdamar, B., R. Bose, M. Barrios-Rodiles, *et al.* 2005. Regulation of the polarity protein Par6 by TGFbeta receptors controls epithelial cell plasticity. *Science* **307:** 1603–1609.

68. Balda, M.S. & K. Matter. 2009. Tight junctions and the regulation of gene expression. *Biochim. Biophys. Acta.* **1788:** 761–767.

69. Stevenson, B.R., J.D. Siliciano, M.S. Mooseker & D.A. Goodenough. 1986. Identification of ZO-1: a high molecular weight polypeptide associated with the tight junction (zonula occludens) in a variety of epithelia. *J. Cell Biol.* **103:** 755–766.

70. Itoh, M., M. Furuse, K. Morita, *et al.* 1999. Direct binding of three tight junction-associated MAGUKs, ZO-1, ZO-2, and ZO-3, with the COOH termini of claudins. *J. Cell Biol.* **147:** 1351–1363.

71. Furuse, M., M. Itoh, T. Hirase, *et al.* 1994. Direct association of occludin with ZO-1 and its possible involvement in the localization of occludin at tight junctions. *J. Cell Biol.* **127:** 1617–1626.

72. Nomme, J., A.S. Fanning, M. Caffrey, *et al.* 2011. The Src homology 3 domain is required for junctional adhesion molecule binding to the third PDZ domain of the scaffolding protein ZO-1. *J. Biol. Chem.* **286:** 43352–43360.

73. Fanning, A.S., B.J. Jameson, L.A. Jesaitis & J.M. Anderson. 1998. The tight junction protein ZO-1 establishes a link between the transmembrane protein occludin and the actin cytoskeleton. *J. Biol. Chem.* **273:** 29745–29753.

74. Wittchen, E.S., J. Haskins & B.R. Stevenson. 1999. Protein interactions at the tight junction. Actin has multiple binding partners, and ZO-1 forms independent complexes with ZO-2 and ZO-3. *J. Biol. Chem.* **274:** 35179–35185.

75. Yamamoto, T., N. Harada, Y. Kawano, *et al.* 1999. In vivo interaction of AF-6 with activated Ras and ZO-1. *Biochem. Biophys. Res. Commun.* **259:** 103–107.

76. Citi, S., F. D'Atri & D.A. Parry. 2000. Human and Xenopus cingulin share a modular organization of the coiled-coil rod domain: predictions for intra- and intermolecular assembly. *J. Struct. Biol.* **131:** 135–145.

77. Balda, M.S., M.D. Garrett & K. Matter. 2003. The ZO-1-associated Y-box factor ZONAB regulates epithelial cell proliferation and cell density. *J. Cell Biol.* **160:** 423–432.

78. Gottardi, C.J., M. Arpin, A.S. Fanning & D. Louvard. 1996. The junction-associated protein, zonula occludens-1, localizes to the nucleus before the maturation and during the remodeling of cell–cell contacts. *Proc. Natl. Acad. Sci. USA* **93:** 10779–10784.

79. Jaramillo, B.E., A. Ponce, J. Moreno, *et al.* 2004. Characterization of the tight junction protein ZO-2 localized at the nucleus of epithelial cells. *Exp. Cell Res.* **297:** 247–258.

80. Gonzalez-Mariscal, L., R. Tapia, M. Huerta & E. Lopez-Bayghen. 2009. The tight junction protein ZO-2 blocks cell cycle progression and inhibits cyclin D1 expression. *Ann. N. Y. Acad. Sci.* **1165:** 121–125.

81. Gladden, A.B. & J.A. Diehl. 2005. Location, location, location: the role of cyclin D1 nuclear localization in cancer. *J. Cell Biochem.* **96:** 906–913.

82. Sherr, C.J. 1995. D-type cyclins. *Trends Biochem. Sci.* **20:** 187–190.

83. Huerta, M., R. Munoz, R. Tapia, *et al.* 2007. Cyclin D1 is transcriptionally down-regulated by ZO-2 via an E box and the transcription factor c-Myc. *Mol. Biol. Cell* **18:** 4826–4836.

84. Tapia, R., M. Huerta, S. Islas, *et al.* 2009. Zona occludens-2 inhibits cyclin D1 expression and cell proliferation and exhibits changes in localization along the cell cycle. *Mol. Biol. Cell* **20:** 1102–1117.

85. Capaldo, C.T., S. Koch, M. Kwon, *et al.* 2011. Tight function zonula occludens-3 regulates cyclin D1-dependent cell proliferation. *Mol. Biol. Cell* **22:** 1677–1685.

86. Guo, Y., D.W. Stacey & M. Hitomi. 2002. Post-transcriptional regulation of cyclin D1 expression during G2 phase. *Oncogene* **21:** 7545–7556.

87. Humbert, P., S. Russell & H. Richardson. 2003. Dlg, Scribble and Lgl in cell polarity, cell proliferation and cancer. *Bioessays* **25:** 542–553.

88. Takizawa, S., K. Nagasaka, S. Nakagawa, *et al.* 2006. Human scribble, a novel tumor suppressor identified as a target of high-risk HPV E6 for ubiquitin-mediated degradation, interacts with adenomatous polyposis coli. *Genes Cells* **11:** 453–464.

89. Duning, K., D. Rosenbusch, M.A. Schluter, *et al.* 2010. Polycystin-2 activity is controlled by transcriptional coactivator with PDZ binding motif and PALS1-associated tight junction protein. *J. Biol. Chem.* **285:** 33584–33588.

90. Yi, C., S. Troutman, D. Fera, *et al.* 2011. A tight junction-associated Merlin-angiomotin complex mediates Merlin's regulation of mitogenic signaling and tumor suppressive functions. *Cancer Cell* **19:** 527–540.

91. Lee, C. & L.A. Laimins. 2004. Role of the PDZ domain-binding motif of the oncoprotein E6 in the pathogenesis of human papillomavirus type 31. *J. Virol.* **78:** 12366–12377.

92. Krcmery, J., T. Camarata, A. Kulisz & H.G. Simon. 2010. Nucleocytoplasmic functions of the PDZ-LIM protein family: new insights into organ development. *Bioessays* **32:** 100–108.

93. Loughran, G., N.C. Healy, P.A. Kiely, *et al.* 2005. Mystique is a new insulin-like growth factor-I-regulated PDZ-LIM domain protein that promotes cell attachment and migration and suppresses Anchorage-independent growth. *Mol. Biol. Cell* **16:** 1811–1822.

94. Qu, Z., J. Fu, P. Yan, *et al.* 2010. Epigenetic repression of PDZ-LIM domain-containing protein 2: implications for the biology and treatment of breast cancer. *J. Biol. Chem.* **285:** 11786–11792.

95. Guillemot, L. & S. Citi. 2006. Cingulin regulates claudin-2 expression and cell proliferation through the small GTPase RhoA. *Mol. Biol. Cell* **17:** 3569–3577.

96. Benais-Pont, G., A. Punn, C. Flores-Maldonado, *et al.* 2003. Identification of a tight junction–associated guanine nucleotide exchange factor that activates Rho and regulates paracellular permeability. *J. Cell Biol.* **160:** 729–740.

97. Aijaz, S., F. D'Atri, S. Citi, *et al.* 2005. Binding of GEF-H1 to the tight junction-associated adaptor cingulin results in inhibition of rho signaling and G1/S phase transition. *Develop. Cell* **8:** 777–786.

98. Bishop, A.L. & A. Hall. 2000. Rho GTPases and their effector proteins. *Biochem. J.* **348**(Pt 2): 241–255.

99. Sahai, E. & C.J. Marshall. 2002. RHO-GTPases and cancer. *Nat. Rev. Cancer* **2:** 133–142.

100. Pruitt, K. & C.J. Der. 2001. Ras and Rho regulation of the cell cycle and oncogenesis. *Cancer Lett.* **171:** 1–10.

101. Jaffe, A.B. & A. Hall. 2002. Rho GTPases in transformation and metastasis. *Advances in Cancer Research.* Vol. 84: Academic Press. New York. 57–80

102. Braga, V.M. 2002. Cell–cell adhesion and signalling. *Curr. Opin. Cell Biol.* **14:** 546–556.

103. Warner, S.J., H. Yashiro & G.D. Longmore. 2010. The Cdc42/Par6/aPKC polarity complex regulates apoptosis-induced compensatory proliferation in epithelia. *Curr. Biol.* **20:** 677–686.

104. Grzeschik, N.A., L.M. Parsons, M.L. Allott, *et al.* 2010. Lgl, aPKC, and Crumbs regulate the Salvador/Warts/Hippo pathway through two distinct mechanisms. *Curr. Biol.* **20:** 573–581.

Ann. N.Y. Acad. Sci. ISSN 0077-8923

ANNALS OF THE NEW YORK ACADEMY OF SCIENCES
Issue: *Barriers and Channels Formed by Tight Junction Proteins*

Barrier dysfunction and bacterial uptake in the follicle-associated epithelium of ileal Crohn's disease

Åsa V. Keita[1] and Johan D. Söderholm[1,2]

[1]Department of Clinical and Experimental Medicine, Division of Surgery, Faculty of Health Sciences, Linköping University, Linköping, Sweden. [2]Department of Surgery, County Council of Östergötland, Linköping, Sweden

Address for correspondence: Åsa V. Keita, Department of Clinical and Experimental Medicine, Division of Surgery, Faculty of Health Sciences, Linköping University, 58183, Linköping, Sweden. asa.keita@liu.se

The ability to control uptake across the mucosa and protect from harmful substances in the gut lumen is defined as intestinal barrier function. The etiology of Crohn's disease is unknown, but genetic, environmental, and immunological factors all contribute. The frontline between these factors lies in the intestinal barrier. The most important inflammation-driving environmental factor in Crohn's disease is the microbiota, where *Esherichia coli* strains have been assigned a key role. The first observable signs of Crohn's disease are small aphtoid ulcers over Peyer's patches and lymphoid follicles. The overlaying follicle-associated epithelium (FAE) is specialized for luminal sampling and is an entry site for antigens and bacteria. We have demonstrated increased *E. coli* uptake across the FAE in Crohn's disease, which may initiate inflammation. This short review will discuss barrier dysfunction and bacteria in the context of ileal Crohn's disease, and how the FAE might be the site of initial inflammation.

Keywords: adherent invasive *E. coli*; etiology; inflammatory bowel disease; pathogens; Peyer's patches

Introduction

The intestinal mucosa constitutes a surface area of 300–400 m^2 and is constantly exposed to and interacts with luminal content, such as antigens from ingested food, resident bacteria (10^{14} microbial cells; >400 different species), and invading viruses. The ability to control uptake and maintain the delicate balance between absorbing essential nutrients while preventing entry of, and responding to, harmful contents across the mucosa is defined as intestinal barrier function. Intestinal barrier function includes physical diffusion barriers, regulated physiologic barriers, and immunological barriers (for a detailed overview see Ref. 1).

Normally, the intestinal barrier allows only small amounts of antigens to pass the mucosa to interact with the innate and adaptive immune systems. If the control of the barrier function is broken, it can implicate enhanced passage of antigens and bacteria, which in turn may damage the mucosa leading to pathological conditions. Under normal circumstances, the host can downregulate the damage when the triggering event has resolved, whereas in a sus-

ceptible host, ongoing enhanced permeability may result in chronic inflammation, such as inflammatory bowel disease (IBD). IBD, which is primarily constituted of ulcerative colitis and Crohn's disease, has been commonly associated with intestinal barrier dysfunction.[2,3] In this review, we will mainly discuss barrier dysfunction and bacteria in Crohn's disease.

Pathophysiology and etiology of Crohn's disease

Crohn's disease is characterized by chronic, relapsing symptoms, such as diarrhea, abdominal pain, anorexia, and weight loss. There is no curing treatment, and thus Crohn's disease is a chronic disease that requires life-long medication and most often surgery. Although the etiology and the initiating factors in Crohn's disease remain unknown, an understanding of its pathogenesis is evolving. The current paradigm is that Crohn's disease represents the outcome of the interaction between the predisposition of the host (genetic and environmental), the mucosal immune response, and the microbiota in the

doi: 10.1111/j.1749-6632.2012.06502.x

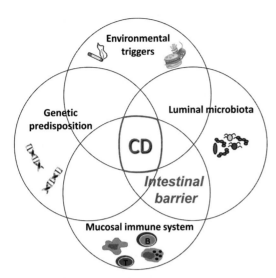

Figure 1. Pathogenesis of Crohn's disease (CD). The exact cause of CD is unknown, but evidence shows that genetic, immunological, and environmental factors all contribute to the pathogenesis of the disease. Intraluminal bacteria are considered the most important environmental triggering factor and are necessary to initiate or reactivate the disease. The frontline between the genetics, immune system, and the microbiota lies in the intestinal barrier.

intestinal lumen[4,5] (Fig. 1). Thus, Crohn's disease is initiated when an individual who is genetically predisposed is exposed to triggering environmental factors. The frontline between the genetics, the immune system, and the luminal microbiota lies in the intestinal barrier. It has been suggested that patients with Crohn's disease have impaired tolerance, for example, perturbed tolerance toward luminal antigens, probably caused by a defect in the intestinal barrier. In line with this, studies in patients revealed that T cells from Crohn's disease mucosa produce cytokines in response to dietary antigens and the patient's own microbiota.[6,7] Studies in knockout mice[8,9] have shown that induction of defects at various levels of host defense leads to chronic intestinal inflammation. In other words, there is a multiplicity of pathways leading to Crohn's disease.

Genetics and Crohn's disease

Genetic factors discovered to date include both innate and adaptive immunity as well as barrier function. The genetic background to IBD is complex; however, during the past years, a large number of genes that contribute to IBD susceptibility have successfully been identified, and today more than 100

IBD genes/loci have been confirmed.[10] As previously mentioned, a disturbed intestinal barrier function is an important factor in the pathogenesis of Crohn's disease,[2,11,12] and several of the polymorphisms that predispose to Crohn's disease are highly expressed in gut epithelia and related to barrier function. Initially, NOD2/CARD15, belonging to the NOD-like receptor family (NLRs), was identified as a susceptibility gene in Crohn's disease.[13] Further, the discovery of how toll-like receptors (TLRs) activate NF-κB and may link innate and adaptive immunity was a huge breakthrough.[14] Since this, DNA sequence variants of TLR genes in Crohn's disease have received substantial research interest.[15] A delicate balance in the regulation of NLRs and TLRs is critical for intestinal homeostasis and defense against pathogens. In addition to these receptors, there are two pathways generating the most interest at present. The first is the IL-12–IL-23 pathway, including pathogenic T cell differentiation, for example, Th17.[16] The second pathway is autophagy, and polymorphisms in the autophagy-related genes ATG16L1 and IRGM have been widely replicated as important susceptibility genes associated with Crohn's disease.[17,18] Together, NOD2, ATG16L1, and IRGM are involved in bacterial sensing and clearance, and mutations might lead to defects in the defensin secretion.[19,20] In addition, variants in the XBP1 gene were currently observed in Crohn's disease and also associated with defects in Paneth cell function.[21]

Other genes that predispose to Crohn's disease and that have been shown to be important in the maintenance of epithelial integrity, structure, and immunoregulation are IBD5 (OCTN1, 2), MDR1, and DLG5.[22] There are also several mucin genes important in barrier defense that have been linked to Crohn's disease—for example, MUC3, MUC4, MUC5B,[23] and MUC3A.[24]

It is evident that the genetics support that the intestinal mucosal barrier toward bacteria is important in the pathophysiology of Crohn's disease and that genetic variants, directly or indirectly, lead to a lack of defense, thereby initiating barrier dysfunction.

Barrier dysfunction in Crohn's disease

Studies have shown that in Crohn's disease, the barrier disturbance is a combined dysfunction of the transcellular[25,26] and the paracellular route.[11,27]

Transcytosis is the most important route for protein uptake in the intestinal epithelium and potentially contributes to inflammation and gastrointestinal disease.[28–31] Increased transcellular uptake of protein antigens has been demonstrated in microscopically normal ileum of Crohn's disease,[25,32] and this seems to be mediated by mechanisms involving tumor necrosis factor (TNF)-α.[26,33]

The paracellular route, via the junctional complexes,[34] is believed to be impermeable to protein-sized molecules, and thus under normal conditions, it constitutes an effective barrier to antigenic macromolecules. Studies of Crohn's disease mucosa have demonstrated increased small bowel permeability to medium-sized probes,[11,35,36] structural changes,[37] and leakage of the tight junctions in response to luminal stimuli.[38] Zeissig et al.[27,39] showed a reduction in tight junction strand number and manifestations of strand discontinuities in Crohn's disease. Further, Western blot analysis indicated decreased expression of occludin, claudin-5, and claudin-8 in Crohn's disease compared to controls, while claudin-2 was upregulated. Interestingly, the claudin-2 upregulation was less pronounced in active Crohn's disease compared with ulcerative colitis and was inducible by TNF-α, but not interferon-γ. Thus, TNF-α seems to play a crucial role in the barrier dysfunction seen in Crohn's disease. In addition, studies of Crohn's disease patients have shown that anti-TNF-α treatment reduces epithelial cell apoptosis and restores epithelial resistance in colonic mucosa[27] and normalizes intestinal permeability in patients.[40]

Where does it all start?

It is not fully understood how Crohn's disease is initiated. The early inflammation in Crohn's disease is, however, often located at the distal ileum,[41] where Peyer's patches are more frequent.[42] Clinical studies have shown that the first observable signs of the disease are ileal aphtoid lesions, which are well recognized by endoscopy.[41] These lesions have been shown to progress over time to larger ulcerations and structuring of the lumen.[43] It has been observed that these early lesions mainly occur at the lymphoid follicles.[41] They are found in 70% of the Crohn's disease patients,[44] and most commonly in the clusters of lymphoid follicles called the Peyer's patches of the distal ileum.[43,44] Magnifying endoscopy and scanning electron microscopy have been used to demonstrate that the aphtoid lesions of Crohn's disease are preceded by 150–200 μm–sized ultrastructural erosions of the epithelium covering the Peyer's patches, the so-called follicle-associated epithelium (FAE).[45] Taken together, these observations suggest that the lymphoid follicles are the sites of initial inflammation in ileal Crohn's disease, where the ulcerations originate from small erosions in the FAE.

Peyer's patches and function of the overlaying FAE

Peyer's patches are organized lymphoid structures that are spread throughout the human small intestine.[46] More than 300 years ago, J.K. Peyer described these aggregations of lymphoid cells in the small intestinal wall and named them "folliculi lymphatici aggregate"; consequently, there must be more than one follicle to form a Peyer's patch. In humans, they develop well before birth, though the full development of the patches as inductive sites requires acquired antigenic challenge.[47] The Peyer's patches consist of numerous follicles separated from each other by interfollicular zones characterized by high endothelial venules surrounded by densely packed lymphocytes, mainly T cells, but also dendritic cells[48] (Fig. 2). The actual follicles consist of B cell germinal centers and a marginal zone constituting of proliferating B lymphocytes and phagocytotic macrophages. Between the FAE and the follicle, the subepithelial dome (SED) is formed. The SED contains T cells and B cells and is rich in phagocyting dendritic cells, macrophages, and monocytes. It has been proposed[49] that luminal antigens and pathogens sampled by the FAE are further captured by immature dendritic cells within the SED and ferried to adjacent interfollicular T cells, where dendritic cell maturation and antigen presentation would occur.

The FAE presents a surface to the lumen that is different from the surrounding villus epithelium (VE).[50] It has phenotypic features that facilitate the recognition and adherence of microorganisms. FAE enterocytes have a network of glycocalyx, but they have lower amounts of the membrane-associated hydrolases involved in digestive functions, and the glycosylation patterns differ from those in VE enterocytes.[51–53] Moreover, 10% of the FAE contains so called membranous or microfold (M) cells, specialized in antigen sampling and transport.[54]

The intestinal mucosa is protected from luminal pathogens and toxins by immunoglobulin

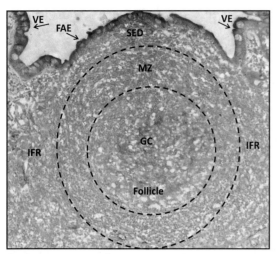 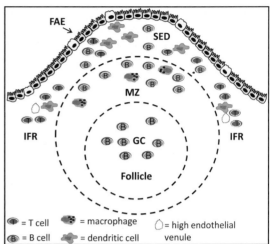

Figure 2. Structure of the FAE and underlying Peyer's patches. IFR, interfollicular region; MZ, marginal zone; GC, germinal center.

(IgA)-blocking. IgA is produced by plasma cells in the lamina propria and transported to the lumen via binding to basolateral polymeric IgA receptors. Interestingly, the entire FAE lacks these receptors and is therefore unable to transport protective IgA to the apical side.[51] It has been suggested that a lack of IgA results in increased exposure of the mucosa to bacterial proteins[55] and might lead to a breach of the intestinal barrier as a consequence of increased bacterial penetration.[56,57]

The tight junction pattern in the FAE has not been widely investigated. However, Tamagawa et al. showed[58] that a unique profile of claudin-2, claudin-3, and claudin-4 and occludin expression were noted in the tight junctions of mouse FAE. Claudin-4 was preferentially expressed at the top of the FAE dome, and its expression was stronger than in VE. Claudin-2 was weakly expressed in a restricted region of FAE near the crypt bases, and claudin-3 and occludin were found on the FAE throughout the dome. From this study, it seems likely that claudin-4 expression is preferentially associated with the dome region of FAE, the mucosal inductive site of the intestine. In that location it might correlate with the cell life cycle, help maintain the apex configuration of the dome, or be a factor favoring the uptake of antigens by the FAE.

The immune responses initiated in the SED and follicles are mainly protective and thereby important for oral tolerance, but the FAE can also function as a route of entry for pathogens, such as Yersinia[59]

and HIV.[60] It could therefore be speculated that the initiation of inflammation in the lymphoid follicles of Crohn's disease may be caused by that increased amounts of antigens and bacteria transported across the FAE, but might also be due to the presence of more aggressive bacteria in Crohn's intestine compared to healthy individuals.

Bacterial uptake across FAE in Crohn's disease

Luminal enteric bacteria are the most important inflammation-driving environmental factor in Crohn's disease.[61,62] Studies in patients with Crohn's disease have shown various changes in the luminal flora with a possible link to local inflammation,[63,64] thus the role of the luminal microflora is evident. Although no specific pathogen has been proven as a causative factor, several bacteria have been linked to Crohn's disease: for example, Mycobacterium paratuberculosis[65] and Yersinia pseudotuberculosis.[66] However, studies have been conflicting, and in a recent study,[67] the prevalence of several pathogens, including strains of Mycobacteria and Yersinia, was investigated. The pathogens were detected in specimens from Crohn's disease patients but also in control patients, suggesting that the pathogens may be associated with the disease but are not necessarily an obligatory cause.

A more confirmed role in the pathogenesis of Crohn's disease has been assigned to various

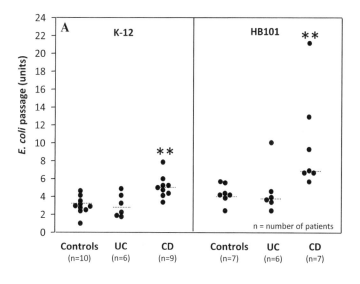

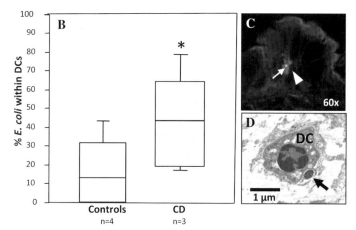

Figure 3. Uptake of nonpathogenic bacteria across the FAE in ileum from controls, ulcerative colitis (UC), or Crohn's disease (CD). (A) Permeation of chemically killed *E. coli* K-12 and live *E. coli* HB101 during 120-min exposure in Ussing chambers. Values are presented as median (25th–75th interquartile range), and comparisons were performed with the Kruskall–Wallis test, *P* = 0.002 CD versus UC and controls; 1 unit corresponds to 1.5 × 10⁶ CFU/mL. (B) *E. coli* K12 uptake (arrow) into a dendritic cell (arrow-head), squeezed in between two epithelial cells, in the FAE of a CD patient. (C) *E. coli* HB101 uptake (arrow) into a dendritic cell (DC) in CD FAE mucosa.

Esherichia coli strains. For example, studies have shown that patients with Crohn's disease have an increased number of so-called adherent-invasive *E. coli* (AIEC) in the mucosa,[68,69] and the concentration of bacteria increases progressively with the severity of the disease[62,70] (AIEC is further discussed later). We previously demonstrated[71] an increased uptake of *E. coli* K12 and HB101 across FAE in the ileum of Crohn's disease patients, compared to ileum of non-IBD controls and patients with ulcerative colitis (Fig. 3A). With confocal and electron microscopy, we showed that bacterial uptake was both transcellular and paracellular in Crohn's disease FAE, in contrast to controls, where only transcellular uptake was observed. We were also able to demonstrate that the increased uptake of bacteria was combined with an enhanced internalization of the bacteria into proinflammatory dendritic cells (Fig. 3B–D). Moreover, there was a larger mucosal release of TNF-α in Crohn's disease than in controls. This together indicates that the bacteria exert stress to the epithelium, thereby affecting paracellular permeability in the FAE of predisposed patients with Crohn's disease. It could also be speculated that the altered permeability could be due to an altered sensing of luminal bacteria. In a follow-up study,[72] the dendritic cells were characterized as mature CD83[+] and CCR7[-]. The lack of the lymph node migratory receptor CCR7[-] possibly contributes to the abnormal accumulation of dendritic cells seen in the SED of Crohn's disease. Further, the dendritic cells had greater propensity to internalize bacteria in Crohn's disease tissues compared to controls. From our studies, it could be speculated that the increased recruitment and accumulation of mature dendritic cells seen in Crohn's disease may be a consequence of the high amount of bacteria transported across the FAE.

AIEC in ileal Crohn's disease

Recently, a new pathogenic group of *E. coli* strains have been identified that not only adhere, but also invade intestinal epithelial cells and have been designated AIEC strains.[69,73] The identification of AIEC strains, both from ileum and colon of IBD patients, has been pertinent for the understanding of microbial induced intestinal inflammation in IBD. Among the *E. coli* strains isolated from the ileal mucosa of Crohn's disease patients, LF82 is the most important and by far the best characterized.[68,74] LF82 has the ability to adhere, invade to the lamina propria, and proliferate inside macrophages. Therefore, it has emerged as one of the most important luminal factors in Crohn's disease.

A possible explanation for the increased number of AIEC bacteria associated with ileal mucosa of Crohn's disease is the increased expression of CEACAM6 on the brush border of ileal enterocytes in Crohn's disease patients. CEACAM6 acts as a receptor for AIEC binding to the intestinal mucosa via type 1 pili.[75] Moreover, LF82 expression of long polar fimbriae (LPF) facilitates transport across the FAE. Chassaing *et al.*[76] recently showed that *E. coli* expressing LPF are significantly more present among *E. coli* strains of Crohn's disease mucosa than in controls. Furthermore, LPF seems to play a key role in LF82 virulence, because results showed that LPF allow the bacteria to translocate across modeled M cell monolayers and to interact with murine and human Peyer's patches. Interestingly, in a recent study by Denizot *et al.*,[77] it was shown that CEABAC10 transgenic mice (expressing human CEACAMs) infected with LF82, but not with nonpathogenic *E. coli*, showed a threefold in-

crease in intestinal permeability and disrupted mucosal integrity in a type 1 pili–dependent mechanism. Moreover, LF82 induced the expression of pore-forming claudin-2 in CEABAC10 mice and in mucosa from Crohn's disease patients. Because claudin-2 overexpression has been shown to decrease the tightness of the barrier,[78] it might be that the enhanced bacterial uptake over the FAE leads to increased claudin-2 expression altering the intestinal permeability, as seen in Crohn's disease patients. It could further be speculated that preventing type 1 pili-mediated AIEC-CEACAM6 interaction may prevent impairment of intestinal barrier function in Crohn's disease patients.

Role of defensins in bacterial uptake

One may speculate that the general mucosal barrier dysfunction seen in Crohn's disease is generated by signals initiated in FAE by interactions between microbes and innate immunity. A normal mucosal barrier function requires an intact mucus layer and epithelial cells bound together by tight junctions that prevent pathogen infiltration. Within the epithelium, there are numerous antimicrobial peptides (AMPs) with the function to kill microorganisms, attract monocytes, and potentiate macrophage opsonization. One of the most important families of AMPs are the defensins, which can be found in Paneth cells of the crypts of the small bowel (α-defensins) and throughout the colonic epithelium (β-defensins). Human α-defensin 5 (HD5), one of six human α-defensins, is stored in its pro-form inside the Paneth cells and processed to mature peptide after secretion, for example by trypsin, which is also expressed in Paneth cells.[79] Studies[80,81] have shown lower levels of HD5 and -6 in Crohn's disease patients compared to controls, and these lower levels may compromise the ability of the intestinal epithelium to respond to immune challenges by weakening both the antibacterial activity and the immune response of the epithelium. This relative lack of α-defensins in ileal Crohn's disease has recently[82] been linked to diminished mucosal mRNA levels of the Wnt pathway transcription factor TCF4, which is related to epithelial stem cell differentiation, and also to the frameshift mutation in NOD2.[19] Thus, a disturbed production of defensins may be involved in the pathogenesis of Crohn's disease and may be an important factor in the increased postoperative recurrence rate seen in patients with NOD2/CARD15

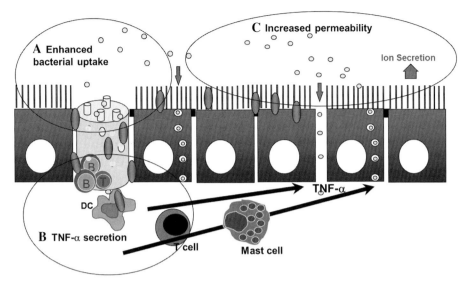

Figure 4. A simplified schema of potential mechanisms involved in barrier function during Crohn's disease. (A) In noninflamed ileal Crohn's disease mucosa, the initial mechanisms may be increased uptake of luminal bacteria at the FAE. Once the bacteria are taken up, they pass through the epithelium, and when they reach the underlying Peyer's patches (B) they are taken up by immune cells such as dendritic cells that release TNF-α—affecting mast cells, for example. Mast cells release TNF-α and other inflammatory mediators, which generate inflammation and general mucosal barrier dysfunction. Further, TNF-α affects the epithelium (C) by decreasing the sodium absorption and increasing the chloride secretion. This leads to diarrhea and increased apoptosis of epithelial cells. TNF-α further disrupts the barrier to antigens and bacteria by affecting paracellular permeability—for example, altering expression of various claudins in the tight junctions and the transcellular uptake route by increasing macropinocytosis to antigens. Consequently, an increased bacterial uptake over the FAE in uninflamed Crohn's disease leads to altered barrier function and inflamed mucosa.

polymorphisms. It might be speculated that the enhanced prevalence of AIEC in Crohn's disease mucosa could be caused by a loss of control of the bacterial attack, possibly due to dysfunctional Paneth cells and subsequently an inappropriate secretion of defensins. This, however, needs to be studied in the future.

Conclusions

To summarize, the intestinal barrier plays an essential role in innate defense by keeping the inner and the external environment separated. A small but controlled uptake of antigens takes place for immunosurveillance. However, imbalance of barrier integrity, with increased antigen and bacterial uptake, is believed to be important in the pathophysiology of several intestinal disorders, including Crohn's disease.

Figure 4 illustrates a simplified schema of the possible scenario in Crohn's disease ileal mucosa. In noninflamed mucosa, the initial mechanisms may be increased uptake of luminal bacteria at the FAE (Fig. 4A). Certain facultative pathogens like

E. coli LF82 seem to target the FAE to gain entry into the body. Once the bacteria have been taken up by the epithelium, they pass through the cells, and when they reach the SED, they get internalized into immune cells. Mast cells, macrophages, and dendritic cells release inflammatory mediators, primarily TNF-α, which generates inflammation and general mucosal barrier dysfunction (Fig. 4B). Inflamed Crohn's disease mucosa is characterized by TNF-α–driven functional alterations (Fig. 4C). Decreased sodium absorption and increased chloride secretion leads to diarrhea. Moreover, TNF-α induces increased permeability via paracellular and transcellular routes and increased apoptosis of epithelial cells. To conclude, the inflammation initiated over the Peyer's patches and FAE in Crohn's disease patients probably arises from a combination of an increased amount of bacteria at the epithelial surface, caused, for example, by either hostile properties of the same (e.g., LPF) or relative lack of defensins. The consequent increase of bacterial transport into the FAE will lead to activation of antigen-presenting and immune-orchestrating cells in the SED, thereby causing general mucosal permeability alterations, which then will propagate inflammation. The final outcome is the exacerbation of inflammatory symptoms in the patient.

Conflicts of interest

The authors declare no conflicts of interest.

References

1. Keita, A.V. & J.D. Soderholm. 2010. The intestinal barrier and its regulation by neuroimmune factors. *Neurogastroenterol Motil.* **22:** 718–733.

2. Bjarnason, I., A. MacPherson & D. Hollander. 1995. Intestinal permeability: an overview. *Gastroenterology* **108:** 1566–1581.

3. Salim, S.Y. & J.D. Soderholm. 2011. Importance of disrupted intestinal barrier in inflammatory bowel diseases. *Inflamm. Bowel Dis.* **17:** 362–381.

4. Vermeire, S., G. van Assche & P. Rutgeerts. 2007. Review article: altering the natural history of Crohn's disease—evidence for and against current therapies. *Aliment. Pharmacol. Ther.* **25:** 3–12.

5. Podolsky, D.K. 2002. Inflammatory bowel disease. *N. Engl. J. Med.* **347:** 417–429.

6. Duchmann, R. *et al.* 1995. Tolerance exists towards resident intestinal flora but is broken in active inflammatory bowel disease (IBD). *Clin. Exp. Immunol.* **102:** 448–455.

7. Macpherson, A. *et al.* 1996. Mucosal antibodies in inflammatory bowel disease are directed against intestinal bacteria. *Gut* **38:** 365–375.

8. Eckmann, L. 2006. Animal models of inflammatory bowel disease: lessons from enteric infections. *Ann. N.Y. Acad. Sci.* **1072:** 28–38.

9. Bhan, A.K. *et al.* 1999. Colitis in transgenic and knockout animals as models of human inflammatory bowel disease. *Immunol. Rev.* **169:** 195–207.

10. Cho, J.H. & S.R. Brant. 2011. Recent insights into the genetics of inflammatory bowel disease. *Gastroenterology* **140:** 1704–1712.

11. Meddings, J.B. 1997. Review article: intestinal permeability in Crohn's disease. *Aliment. Pharmacol. Ther.* **11**(Suppl. 3): 47–53.

12. Hollander, D. 1992. The intestinal permeability barrier. A hypothesis as to its regulation and involvement in Crohn's disease. *Scand. J. Gastroenterol.* **27:** 721–726.

13. Hugot, J.P. *et al.* 2001. Association of NOD2 leucine-rich repeat variants with susceptibility to Crohn's disease. *Nature* **411:** 599–603.

14. Pasare, C. & R. Medzhitov. 2005. Toll-like receptors: linking innate and adaptive immunity. *Adv. Exp. Med. Biol.* **560:** 11–18.

15. Stappenbeck, T.S. *et al.* 2011. Crohn disease: a current perspective on genetics, autophagy and immunity. *Autophagy* **7:** 355–374.

16. Barrett, J.C. *et al.* 2008. Genome-wide association defines more than 30 distinct susceptibility loci for Crohn's disease. *Nat. Genet.* **40:** 955–962.

17. McCarroll, S.A. *et al.* 2008. Deletion polymorphism upstream of IRGM associated with altered IRGM expression and Crohn's disease. *Nat. Genet.* **40:** 1107–1112.

18. Kuballa, P. *et al.* 2008. Impaired autophagy of an intracellular pathogen induced by a Crohn's disease associated ATG16L1 variant. *PLoS One* **3:** e3391.

19. Wehkamp, J. *et al.* 2004. NOD2 (CARD15) mutations in Crohn's disease are associated with diminished mucosal alpha-defensin expression. *Gut* **53:** 1658–1664.

20. Cadwell, K. *et al.* 2008. A key role for autophagy and the autophagy gene Atg16l1 in mouse and human intestinal Paneth cells. *Nature* **456:** 259–263.

21. Kaser, A. *et al.* 2008. XBP1 links ER stress to intestinal inflammation and confers genetic risk for human inflammatory bowel disease. *Cell* **134:** 743–756.

22. Vermeire, S. 2004. DLG5 and OCTN. *Inflamm. Bowel Dis.* **10:** 888–890.

23. Buisine, M.P. *et al.* 1999. Abnormalities in mucin gene expression in Crohn's disease. *Inflamm. Bowel Dis.* **5:** 24–32.

24. Kyo, K. *et al.* 2001. Associations of distinct variants of the intestinal mucin gene MUC3A with ulcerative colitis and Crohn's disease. *J. Hum. Genet.* **46:** 5–20.

25. Soderholm, J.D. *et al.* 1999. Epithelial permeability to proteins in the noninflamed ileum of Crohn's disease? *Gastroenterology* **117:** 65–72.

26. Soderholm, J.D. *et al.* 2004. Increased epithelial uptake of protein antigens in the ileum of Crohn's disease mediated by tumour necrosis factor alpha. *Gut* **53:** 1817–1824.

27. Zeissig, S. *et al.* 2004. Downregulation of epithelial apoptosis and barrier repair in active Crohn's disease by tumour necrosis factor alpha antibody treatment. *Gut* **53:** 1295–1302.

28. Sanderson, I.R. & W.A. Walker. 1993. Uptake and transport of macromolecules by the intestine: possible role in clinical disorders (an update). *Gastroenterology* **104:** 622–639.

29. Sartor, R.B. 1997. Pathogenesis and immune mechanisms of chronic inflammatory bowel diseases. *Am. J. Gastroenterol.* **92:** 5S–11S.

30. Hugot, J.P. *et al.* 2003. Crohn's disease: the cold chain hypothesis. *Lancet* **362:** 2012–2015.

31. Fiocchi, C. 1998. Inflammatory bowel disease: etiology and pathogenesis. *Gastroenterology* **115:** 182–205.

32. Schurmann, G. *et al.* 1999. Transepithelial transport processes at the intestinal mucosa in inflammatory bowel disease. *Int. J. Colorectal Dis.* **14:** 41–46.

33. Zareie, M. *et al.* 2001. Monocyte/macrophage activation by normal bacteria and bacterial products: implications for altered epithelial function in Crohn's disease. *Am. J. Pathol.* **158:** 1101–1109.

34. Farquhar, M.G. & G.E. Palade. 1963. Junctional complexes in various epithelia. *J. Cell Biol.* **17:** 375–412.

35. Hollander, D. *et al.* 1986. Increased intestinal permeability in patients with Crohn's disease and their relatives. A possible etiologic factor. *Ann. Intern. Med.* **105:** 883–885.

36. Soderholm, J.D. *et al.* 1999. Different intestinal permeability patterns in relatives and spouses of patients with Crohn's disease: an inherited defect in mucosal defense? *Gut* **44:** 96–100.

37. Marin, M.L. *et al.* 1983. A freeze fracture study of Crohn's disease of the terminal ileum: changes in epithelial tight junction organization. *Am. J. Gastroenterol.* **78:** 537–547.

38. Soderholm, J.D. *et al.* 2002. Augmented increase in tight junction permeability by luminal stimuli in the non-inflamed ileum of Crohn's disease. *Gut* **50:** 307–313.

39. Zeissig, S. *et al.* 2007. Changes in expression and distribution of claudin 2, 5 and 8 lead to discontinuous tight junctions and barrier dysfunction in active Crohn's disease. *Gut* **56:** 61–72.

40. Suenaert, P. *et al.* 2002. Anti-tumor necrosis factor treatment restores the gut barrier in Crohn's disease. *Am. J. Gastroenterol.* **97:** 2000–2004.

41. Morson, B.C. 1972. The early histological lesion of Crohn's disease. *Proc. R. Soc. Med.* **65:** 71–72.

42. Van Kruiningen, H.J. *et al.* 2002. Distribution of Peyer's patches in the distal ileum. *Inflamm. Bowel. Dis.* **8:** 180–185.

43. Rutgeerts, P. *et al.* 1984. Natural history of recurrent Crohn's disease at the ileocolonic anastomosis after curative surgery. *Gut* **25:** 665–672.

44. Rickert, R.R. & H.W. Carter. 1980. The "early" ulcerative lesion of Crohn's disease: correlative light- and scanning electron-microscopic studies. *J. Clin. Gastroenterol.* **2:** 11–19.

45. Fujimura, Y., R. Kamoi & M. Iida. 1996. Pathogenesis of aphthoid ulcers in Crohn's disease: correlative findings by magnifying colonoscopy, electron microscopy, and immunohistochemistry. *Gut* **38:** 724–732.

46. Gebert, A., H.J. Rothkotter & R. Pabst. 1996. M cells in Peyer's patches of the intestine. *Int. Rev. Cytol.* **167:** 91–159.

47. Yamanaka, T. *et al.* 2003. Microbial colonization drives lymphocyte accumulation and differentiation in the follicle-associated epithelium of Peyer's patches. *J. Immunol.* **170:** 816–822.

48. Gebert, A. 1997. M cells in the rabbit palatine tonsil: the distribution, spatial arrangement and membrane subdomains as defined by confocal lectin histochemistry. *Anat. Embryol. (Berl)* **195:** 353–358.

49. Iwasaki, A. 2007. Mucosal dendritic cells. *Annu. Rev. Immunol.* **25:** 381–418.

50. Neutra, M.R. *et al.* 1999. The composition and function of M cell apical membranes: implications for microbial pathogenesis. *Semin. Immunol.* **11:** 171–181.

51. Neutra, M.R., N.J. Mantis & J.P. Kraehenbuhl. 2001. Collaboration of epithelial cells with organized mucosal lymphoid tissues. *Nat. Immunol.* **2:** 1004–1009.

52. Kucharzik, T. *et al.* 2000. Characterization of M cell development during indomethacin-induced ileitis in rats. *Aliment. Pharmacol. Ther.* **14:** 247–256.

53. Sanders, D.S. 2005. Mucosal integrity and barrier function in the pathogenesis of early lesions in Crohn's disease. *J. Clin. Pathol.* **58:** 568–572.

54. Gebert, A. 1997. The role of M cells in the protection of mucosal membranes. *Histochem. Cell Biol.* **108:** 455–470.

55. MacDermott, R.P. & W.F. Stenson. 1988. Alterations of the immune system in ulcerative colitis and Crohn's disease. *Adv. Immunol.* **42:** 285–328.

56. Macpherson, A.J. & T. Uhr. 2004. Induction of protective IgA by intestinal dendritic cells carrying commensal bacteria. *Science* **303:** 1662–1665.

57. Macpherson, A.J., M.B. Geuking & K.D. McCoy. 2005. Immune responses that adapt the intestinal mucosa to commensal intestinal bacteria. *Immunology* **115:** 153–162.

58. Tamagawa, H. *et al.* 2003. Characteristics of claudin expression in follicle-associated epithelium of Peyer's patches: preferential localization of claudin-4 at the apex of the dome region. *Lab. Invest.* **83:** 1045–1053.

59. Schulte, R. *et al.* 2000. Translocation of Yersinia entrocolitica across reconstituted intestinal epithelial monolayers is triggered by Yersinia invasin binding to beta1 integrins apically expressed on M-like cells. *Cell Microbiol.* **2:** 173–185.

60. Fotopoulos, G. *et al.* 2002. Transepithelial transport of HIV-1 by M cells is receptor-mediated. *Proc. Natl. Acad. Sci. USA* **99:** 9410–9414.

61. Sartor, R.B. 1997. The influence of normal microbial flora on the development of chronic mucosal inflammation. *Res. Immunol.* **148:** 567–576.

62. Swidsinski, A. *et al.* 2002. Mucosal flora in inflammatory bowel disease. *Gastroenterology* **122:** 44–54.

63. Neut, C. *et al.* 2002. Changes in the bacterial flora of the neoterminal ileum after ileocolonic resection for Crohn's disease. *Am. J. Gastroenterol.* **97:** 939–946.

64. Ott, S.J. *et al.* 2004. Reduction in diversity of the colonic mucosa associated bacterial microflora in patients with active inflammatory bowel disease. *Gut* **53:** 685–693.

65. Sanderson, J.D. *et al.* 1992. Mycobacterium paratuberculosis DNA in Crohn's disease tissue. *Gut* **33:** 890–896.

66. Homewood, R. *et al.* 2003. Ileitis due to Yersinia pseudotuberculosis in Crohn's disease. *J. Infect.* **47:** 328–332.

67. Knosel, T. *et al.* 2009. Prevalence of infectious pathogens in Crohn's disease. *Pathol. Res. Pract.* **205:** 223–230.

68. Darfeuille-Michaud, A. *et al.* 2004. High prevalence of adherent-invasive Escherichia coli associated with ileal mucosa in Crohn's disease. *Gastroenterology* **127:** 412–421.

69. Martin, H.M. *et al.* 2004. Enhanced Escherichia coli adherence and invasion in Crohn's disease and colon cancer. *Gastroenterology* **127:** 80–93.

70. Kleessen, B. *et al.* 2002. Mucosal and invading bacteria in patients with inflammatory bowel disease compared with controls. *Scand. J. Gastroenterol.* **37:** 1034–1041.

71. Keita, A.V. *et al.* 2008. Increased uptake of non-pathogenic E. coli via the follicle-associated epithelium in longstanding ileal Crohn's disease. *J. Pathol.* **215:** 135–144.

72. Salim, S.Y. *et al.* 2009. CD83+CCR7- dendritic cells accumulate in the subepithelial dome and internalize translocated Escherichia coli HB101 in the Peyer's patches of ileal Crohn's disease. *Am. J. Pathol.* **174:** 82–90.

73. Boudeau, J. *et al.* 1999. Invasive ability of an Escherichia coli strain isolated from the ileal mucosa of a patient with Crohn's disease. *Infect. Immun.* **67:** 4499–509.

74. Darfeuille-Michaud, A. *et al.* 1998. Presence of adherent Escherichia coli strains in ileal mucosa of patients with Crohn's disease. *Gastroenterology* **115:** 1405–1413.

75. Barnich, N. *et al.* 2007. CEACAM6 acts as a receptor for adherent-invasive E. coli, supporting ileal mucosa colonization in Crohn disease. *J. Clin. Invest.* **117:** 1566–1574.

76. Chassaing, B. *et al.* 2011. Crohn disease–associated adherent-invasive E. coli bacteria target mouse and human Peyer's patches via long polar fimbriae. *J. Clin. Invest.* **121:** 966–975.

77. Denizot, J. *et al.* 2012 Adherent-invasive Escherichia coli induce claudin-2 expression and barrier defect in CEABAC10 mice and Crohn's disease patients. *Inflamm. Bowel Dis.* **18:** 294–304.

78. Furuse, M. *et al.* 2001. Conversion of zonulae occludentes from tight to leaky strand type by introducing claudin-2 into Madin-Darby canine kidney I cells. *J. Cell Biol.* **153:** 263–272.

79. Ghosh, D. *et al.* 2002. Paneth cell trypsin is the processing enzyme for human defensin-5. *Nat. Immunol.* **3:** 583–590.

80. Wehkamp, J. *et al.* 2005. Reduced Paneth cell alpha-defensins in ileal Crohn's disease. *Proc. Natl. Acad. Sci. USA* **102:** 18129–18134.

81. Elphick, D., S. Liddell & Y.R. Mahida 2008. Impaired luminal processing of human defensin-5 in Crohn's disease: persistence in a complex with chymotrypsinogen and trypsin. *Am. J. Pathol.* **172:** 702–713.

82. Wehkamp, J. *et al.* 2007. The Paneth cell alpha-defensin deficiency of ileal Crohn's disease is linked to Wnt/Tcf-4. *J. Immunol.* **179:** 3109–3118.

Ann. N.Y. Acad. Sci. ISSN 0077-8923

ANNALS OF THE NEW YORK ACADEMY OF SCIENCES
Issue: *Barriers and Channels Formed by Tight Junction Proteins*

The protozoan pathogen *Toxoplasma gondii* targets the paracellular pathway to invade the intestinal epithelium

Caroline M. Weight[1] and Simon R. Carding[1,2]

[1]Institute of Food Research, [2]Norwich Medical School, University of East Anglia, Norwich Research Park, Norwich, United Kingdom

Address for correspondence: Prof. Simon Carding, Gut Health and Food Safety, Institute of Food Research, Norwich Research Park, Colney Lane, Norwich NR4 7UA, UK. simon.carding@ifr.ac.uk

Toxoplasma gondii is a ubiquitous parasite found within all mammals and birds worldwide that can cause fatal infections in immunocompromised persons and fetuses. The parasite causes chronic infections by residing in long-living tissues of the muscle and brain. *T. gondii* infects the host through contaminated meat and water consumption with the gastrointestinal tract (GI tract) being the first point of contact with the host. The mechanisms by which the parasite invades the host through the GI tract are unknown, although it has been suggested that the paracellular pathway is important for parasite dissemination. Studies indicate that epithelial tight junction–associated proteins are affected by *T. gondii*, although which junctional proteins are affected and the nature of host protein–parasite interactions have not been established. We have uncovered evidence that *T. gondii* influences the cellular distribution of occludin to transmigrate the intestinal epithelium and suggest how candidate binding partners can be identified.

Keywords: *Toxoplasma gondii*; tight junctions; occludin; invasion

The history of *Toxoplasma gondii*

Toxoplasma gondii was first identified in 1908 in the gundi (*Ctenodoactylus gundi*), an African rodent, and has since been found to infect virtually all animals, including approximately one third of the human population.[1,2] *T. gondii* is an obligate intracellular parasite of the phylum Apicomplexa, which includes species of *Plasmodium*, *Cryptosporidium*, *Neospora*, and *Eimeria*. *T. gondii* has a complex life cycle in which only members of the felidae family can host the sexual phase and release oocysts containing sporozoites into the environment.[3,4] In Europe, North America, and Africa, there are three dominant clonal lineages of *T. gondii* called type I (e.g., RH and GT1), type II (e.g., ME49), and type III (e.g., VEG). The lineages differ in prevalence, virulence, migratory capacity within the host, and ability to convert to the bradyzoite cyst phase. Expansion of these lineages coincided with the domestication of the cat and other changes in human agricultural practices 10,000 years ago.[5] Multiple infections of different strains can result in the production of a large number of recombinant and atypical forms, which are highly prevalent in South America where ancient or exotic strains, such as COUG, MAS, and CAST, are also found.[6–8]

Life cycle

Humans and other animals become infected through congenital transmission, transfusion, and transplantation of infected blood and organs, but more commonly, via ingestion of contaminated water and undercooked meat containing bradyzoite cysts, the slow replicating form, and ingestion of contaminated soil containing sporozoites from feline feces (Fig. 1).[9,10] When individuals of the feline family become infected with *T. gondii*, the bradyzoites or sporozoites are released into the small intestine and multiply by schizogony (multiple nuclei and cytoplasm divisions) to produce merozoites.[11] This is thought to initiate the sexual phase of the life cycle whereby gametogenesis leads to the rupture of infected epithelial cells. The oocysts are shed in the feline feces and sporulation occurs in the

doi: 10.1111/j.1749-6632.2012.06534.x
Ann. N.Y. Acad. Sci. 1258 (2012) 135–142 © 2012 New York Academy of Sciences.

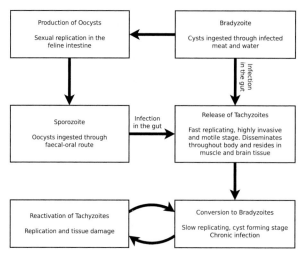

Figure 1. The life cycle of *T. gondii*. Sporozoites generated in the felidae gut and bradyzoites are ingested by animals. Conversion to tachyzoites leads to activation of the immune system and dissemination throughout the body. Once tachyzoites reach long-lasting tissues, they convert back to bradyzoites to establish a chronic infection. Reactivation occurs following immune suppression and results in inflammation.

environment within five days.[11] The oocysts can survive and remain highly infectious in the environment for many years.[12] Bradyzoites and sporozoites released from oocysts infect cells in the small intestine where they convert to fast replicating and highly invasive tachyzoites. Asexual replication takes place by endodyogeny whereby two daughter cells form within one parasite.[13] Tachyzoites disseminate throughout the body and reside in long-lasting tissues and organs such as the muscle and brain. Here, due to pressure from the immune system, parasites convert back into bradyzoites and remain dormant in cysts, which are invisible to immune attack, until reactivation.[14] Interconversion between stages occurs in response to changes in temperature, pH, chemical stress, and stress from the immune system, and can occur spontaneously.[15,16] Reactivation can occur following host stress such as illness and causes clinical symptoms such as encephalitis and chorioretinitis.[2] *T. gondii* is one of the most common secondary opportunistic infections to occur in immunocompromised individuals where reduced numbers of T cells results in mass systemic inflammation and pathology.[17] With the exception of the immunocompromised and the unborn child, *T. gondii* causes a relatively asymptomatic infection associated with typical symptoms of fever in both humans and animals. This includes members of the felidae family.

In experimental infection of rodents, it has been shown that *T. gondii* rapidly infects host cells, taking less than 40 seconds to infect small intestinal epithelial cells, to being present in the lamina propria after one hour, transported to lymph nodes after two hours, converting to tachyzoites after 18 hours, and reaching the brain within six days of initial contact with the host.[18–22] Parasites infect toward the distal half of the intestine and only tachyzoites are thought to migrate through the body to infect the brain and muscles.[18,23]

Invasion strategies of *T. gondii*

Within all species, *T. gondii* preferentially attaches to, and invades, cells that are in mid-S phase of the cell cycle, during the upregulation of surface receptors.[24] *T. gondii* influences the cell cycle to remain in S phase by deregulating molecules such as cyclin E1, cyclin dependent kinase 4 (CDK4), and CDK6.[25] Once inside cells, parasites remain below the cell surface above the nucleus within a parasitophorous vacuole, and can replicate six hours after infection.[16,26,27] However, in the small intestine of humans and other animals, *T. gondii* has only been found within cells of the villi and, to date, it is not known whether parasites that infect enterocytes can exit through the basal domain before replicating (transcellular pathway).[11]

Once the epithelial barrier has been breached, *T. gondii* can exploit the mobility of immune cells, such as dendritic cells, to disseminate to adjacent lymph nodes in a Trojan horse–type mechanism.[28] The process of dendritic cells invasion involves parasitic cyclophilin-18, which is a mimetic of CCL5 that binds to its receptor on dendritic cells, recruiting them to sites of infection, causing them to exhibit a higher migratory capacity than normal. This is strain dependent, favoring type II and type III lineages over type I, which tend to migrate in an extracellular manner via virulence factors instead of going through cells.[29]

Migration is most common in type I strains and tachyzoites that move between epithelial cells, via the paracellular pathway, and do so without affecting the integrity of the monolayer as measured by transepithelial electrical resistance and permeability to the soluble tracer dextran.[30] However, *T. gondii* does appear to upregulate intercellular junction–associated proteins such as intercellular adhesion molecule-1 (ICAM-1), which is an adherens junction protein, in MDCK II and BeWo cells.[30] These proteins, which contain catenins (such as α, β, γ) and cadherins (such as E, N, P), link to cytoskeletal microfilaments and consist of calcium-dependent proteins that contribute to cell–cell adhesion, through interactions with one another and with tight junction proteins.[31–33] ICAM-1 is expressed on leukocytes and endothelial cells and binding to its ligand (leukocyte function-associated antigen-1) activates leukocyte transmigration via actin-cytoskeletal rearrangements.[34,35] It could be predicted that *T. gondii* takes advantage of junctional proteins to initiate signaling cascades, which recruit lymphocytes providing the parasite with an opportunity to infect immune cells for use as a transport vehicle for reaching lymph nodes. It is therefore interesting to note that dendritic cells, T lymphocytes, and intraepithelial lymphocytes express tight junction proteins such as occludin.[36–38] This mechanism may not induce changes in the epithelial barrier function and so the integrity of the monolayer remains intact. The rapid dissemination out of the gastrointestinal tract before activation of an immune response is paramount for the establishment of a chronic infection.

If *T. gondii* exploits the paracellular pathway as a mechanism of entry into the host, then tight junction proteins, and in particular the transmembrane

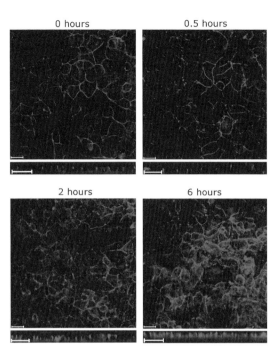

Figure 2. Bradyzoites alter the distribution of occludin. Murine-derived small intestinal epithelial cells (m-IC$_{cl2}$ cells) were infected with bradyzoites for 0 to 6 hours. After fixing with acetone, cells were immunostained for occludin and visualized by confocal laser-scanning microscopy. XZ images are shown underneath each XY image. Scale bar = 20 μm.

proteins in the first instance, may be affected. Sporozoites and bradyzoites have been shown to leave host cells without lysing them, and to infect neighboring cells, suggesting that invasion via the lateral membrane is possible.[39,40]

We have assessed the effects of *T. gondii* exposure on a selection of junctional-associated proteins in the murine small intestinal epithelium. The distribution of occludin is altered within the first hour of infection, being less uniform at the tight junctions and becoming more concentrated within the cytoplasm and at lateral membranes.[41] These *in vivo* observations have recently been confirmed *in vitro* using m-IC$_{cl2}$ cells, a small intestinal epithelial cell line, in which striking changes in occludin distribution were seen within minutes of exposure to bradyzoites, leading to an increase in occludin expression at two and six hours (Fig. 2).[42] This could be explained by increased synthesis or redistribution of occludin to tight junction complexes and may also be associated with membrane ruffling.[27] Within 30 minutes, bradyzoites are able to transmigrate through small intestinal epithelial cells grown on cell culture inserts, and after two hours is accompanied by an increase in permeability to 3 kDa dextran.[42] This suggests that barrier function is affected by bradyzoites and it could be predicted that other tight junction associated proteins are also affected as a consequence of changes in occludin signaling.

Host cell invasion and the molecules involved

To date, only a few cell surface receptors involved in *T. gondii* attachment and invasion have been identified. These molecules are mainly associated with the invasion of cells other than intestinal epithelial cells such as fibroblasts, but it is the studies on enterocytes and gastrointestinal immune cells derived from humans and rodents that provide the most useful information when considering the initial contact with the host during a natural infection. The small intestinal epithelial cell lines that have been used to study *T. gondii* infection to date are the murine MODE-K and m-IC$_{cl2}$ cells and the rat IEC-6 cell line, although they focus primarily on the immune response to infection rather than the mechanism of infection itself.[43–45] The human colonic carcinoma cell line HT29/B6 infected with *T. gondii* shows changes in mannitol and chloride flux, which may be caused by the paracellular movement of parasites affecting signaling pathways within the tight junction complex, or by secondary effects from *T. gondii* that alter tight junction permeability.[46]

To identify the parasite-derived and host-derived molecules involved in attachment and invasion, a number of experimental procedures have been employed. For example, observations of parasites clustering around the edges of lateral cells using confocal and electron microscopy led Barragan and colleagues to focus on host molecules associated with the lateral domain and used recombinant ICAM-1, as well as blocking antibodies to ICAM-1, to demonstrate the involvement of this adhesion protein in the transmigration of, but not cellular invasion by, the parasite.[30] Having established this effect, they immunoprecipitated parasite preparations with ICAM-1 and identified by immunoblotting the binding to MIC2, a microneme adhesion protein from *T. gondii*. It was not possible in this study to determine if binding to ICAM-1 is sufficient to facilitate parasite invasion or if other host

proteins are bound and utilized by the parasite to affect cell entry.

The generation of cells containing reduced levels of a particular protein, using interference RNA (shRNA or siRNA) techniques *in vitro* or knock out parasite models, would provide further evidence of an individual protein's involvement during attachment and/or invasion. Including recombinant proteins and peptides of the cellular junctional complex in pull down experiments combined with mass spectrometry analysis would enhance this approach and could also identify novel potential binding partners.

The direct interaction of recombinant host–parasite proteins could be investigated by flow cytometric analyses through the incorporation of fluorescently tagged recombinant proteins. The use of atomic force microscopy may also shed light on the strength of these interactions and would be useful for quantifying a particular molecule's involvement. This might give an indication of the different types and numbers of surface receptors involved when compared to whole parasites and intact cell monolayer's interactions. This approach may yield different results according to the cell type used as the parasite is capable of infecting almost any cell type, so one may predict a different range of molecules used and may even depend on parasite strain or life stage.

The initial stage of invasion is attachment to the cell surface. The surface of a tachyzoite is covered in glycosylphosphatidylinositol (GPI)-anchored proteins, which include surface antigen (SAG) 1 and SAG1-related sequences.[47] SAG1 is the most abundant protein in tachyzoites and is thought to contribute to virulence and adherence to the surface of different cell types via host glucosamine receptors and via laminin binding to β_1 integrin receptor $\alpha_6\beta_1$, which have been confirmed using macrophage, fibroblasts, and ovarian cells in antibody blocking assays.[48–50] GPIs from *T. gondii* can also bind to galectin-like molecules on the cells surface.[51] Galectins are involved in the formation of the microneme MIC1-MIC6 protein complex, which is secreted by *T. gondii* during invasion.[52] Sialic acid residues and sulfated glycosaminoglycans (GAGs) on host cells also mediate parasite binding, an example of which is the binding of SAG3 to heparin sulfated proteoglycans.[53–56]

Following attachment, the parasite reorients itself so that the apical cytoskeletal complex (which is made of a tubulin polymer and is called a conoid) is in contact with the host membrane.[57,58] The parasite moves into the cell by conoid extension in an anterior to posterior fashion using an inherent actin–myosin motor (gliding motility).[22,59–62]

Actin polymerization is essential for parasite motility to occur and actin filaments must be expressed in a polarized manner to initiate directed gliding motility into cells.[63] Actin polymerization within the host may also assist in internalization and establishment of the moving junction during invasion.[64] Myosin light chain kinase induces the contraction of myosin through calmodulin-dependent increases in calcium levels within the parasite, and to initiate microneme binding.[62] This process shows that active penetration of the tachyzoite is required for invasion to occur and is different to bradyzoite invasion or phagocytosis by macrophages as no membrane ruffling occurs and microfilaments are not reorganized.[21,22] Although *T. gondii* actively invades cells, host dynamin (a GTPase involved in vesicle transport), endocytosis pathways, and membrane tubule formation are also necessary for the parasite to internalize, as illustrated by reduced invasion levels in Dynasore-treated cells.[65] This indicates that host proteins, like cytoplasmic tight junction proteins, could also be exploited by the parasite during invasion owing to the range of different functions these proteins have on cellular development, motility, polarity, and signaling cascades.[66]

To identify the potential molecules involved in cellular invasion or transmigration, it is necessary to consider the types of molecules secreted and upregulated by *T. gondii* during infection. In general, tachyzoites have few microneme and rhoptry proteins, but numerous dense granules (secreted from specialized organelles from the apical cytoplasmic complex), whereas bradyzoites have many microneme and rhoptry proteins, but few dense granules.[11] Tachyzoites and bradyzoites both express SAG3 and dense granule proteins 1–7, whereas bradyzoites exclusively express the SAG1-related sequence bradyzoite specific recombinant 4, and bradyzoite antigen 1 (BAG1), which has homology to small heat shock proteins (HSP).[47]

Micronemes and rhoptries

Microneme proteins are homologs of vertebrate adhesive proteins and contain thrombospondin type I repeats, integrin-like domains, epidermal growth factor-like domains, and lectin-like domains, consistent with binding to cell surface receptors.[52,67–69] Following increases in parasite intracellular calcium, microneme proteins such as apical membrane antigen 1 (AMA1) are secreted from the apical complex, which aid in attachment to the host membrane.[70–72] Impairment of AMA1 signaling causes decreased invasion, providing evidence that these molecules are important for host–parasite cell surface interactions.[70] Microneme proteins, such as the MIC2–M2AP complex, are involved in invasion and motility and directly participate in host cell binding.[73] MIC6 and MIC8 act as escorts for other MIC protein complexes through binding to cytoskeletal molecules such as adolase to provide a gateway between the actin–myosin motor and the cell surface.[68,74] MIC4 is complexed with MIC1 and MIC6, and contains cysteine residues that form disulfide bonds within six conserved apple domains.[75] Binding to cells indicates that MIC1 and MIC4 proteins act as a bridge between the parasite and host cells.[67,75] These are potential molecules that could interact with junctional proteins.

Rhoptries are another type of specialized organelle associating with microneme proteins to initiate the moving junction and formation of the parasitophorous vacuole.[76] AMA1 complexes with rhoptry neck proteins RON2, RON4, and Ts4705, with RON2 using AMA1 to attach to the moving junction.[77] This complex binds and integrates with the host membrane and thereby providing a gateway for secreted RON4, RON5, and RON8 to commence formation of the parasitophorous vacuole, progressively eliminating host proteins from the invaginated surface.[70,77,78] RON8 also adheres to an as yet unidentified cell cytoskeletal molecule, which could be a junction-associated protein.[78] Straub and colleagues recently demonstrated that RON8 contributes to virulence and is required for attachment and invasion, as deficient parasites were less able to do so in addition to not binding with cell cytoskeletal molecules.[79] ROP18 is associated with the parasitophorous vacuole, has active serine/threonine kinase activity, and is involved in controlling parasite replication.[80] ROP18 shows high sequence divergence between strains, which correlates to high type I virulence.[81] ROP16 is also a serine/threonine kinase that enters the host cell nucleus where it induces variable levels of interleukin (IL)-12 secretion from macrophages, depending on the strain of *T.*

gondii.[82] Rab11 and Toxofilin have also been identified to be rhoptry proteins.[76] The GTPase Rab11 controls cholesterol recycling while Toxofilin binds to parasite G-actin and may play a role in assisting the actin–myosin motor during invasion.[83,84] Parasite protein phosphatases have been identified that translocate to the cell nucleus upon infection and target rhoptry proteins or host proteins.[85–87] This suggests that phosphorylation and dephosphorylation events are important during invasion and growth of *T. gondii* inside cells, which may directly or indirectly affect, or be affected by, junction-associated proteins.[86]

Conclusions

We propose that *T. gondii* uses the tight junction complex of proteins as a gateway for entering enterocytes and disseminating into the intestinal mucosa. Our own studies suggest that occludin is affected, and experiments are currently underway to identify potential *T. gondii*–derived proteins involved in this process.

Conflicts of interest

The authors declare no conflicts of interest.

References

1. Sukthana, Y. 2006. Toxoplasmosis: beyond animals to humans. *Trends Parasitol.* **22:** 137–142.
2. Weiss, L.M. & J.P. Dubey. 2009. Toxoplasmosis: a history of clinical observations. *Int. J. Parasitol.* **39:** 895–901.
3. Hill, D. & J.P. Dubey. 2002. *Toxoplasma gondii*: transmission, diagnosis and prevention. *Clin. Microbiol. Infect.* **8:** 634–640.
4. Elmore, S.A., J.L. Jones, P.A. Conrad, *et al.* 2010. *Toxoplasma gondii*: epidemiology, feline clinical aspects, and prevention. *Trends Parasitol.* **26:** 190–196.
5. Su, C., D. Evans, R.H. Cole, *et al.* 2003. Recent expansion of Toxoplasma through enhanced oral transmission. *Science* **299:** 414–416.
6. Sibley, L.D., A. Khan, J.W. Ajioka & B.M. Rosenthal. 2009. Genetic diversity of *Toxoplasma gondii* in animals and humans. *Philos. Trans. R. Soc. Lond. B. Biol. Sci.* **364:** 2749–2761.
7. Lindström, I., N. Sundar, J. Lindh, *et al.* 2008. Isolation and genotyping of *Toxoplasma gondii* from Ugandan chickens reveals frequent multiple infections. *Parasitology* **135:** 39–45.
8. Dubey, J.P., G.V. Velmurugan, C. Rajendran, *et al.* 2011. Genetic characterisation of *Toxoplasma gondii* in wildlife from North America revealed widespread and high prevalence of the fourth clonal type. *Int. J. Parasitol.* **41:** 1139–1147.
9. Derouin, F., H. Pelloux; ESCMID Study Group on Clinical Parasitology 2008. Prevention of toxoplasmosis in transplant patients. *Clin. Microbiol. Infect.* **14:** 1089–1101.
10. Kijlstra, A. & E. Jongert. 2008. Control of the risk of human toxoplasmosis transmitted by meat. *Int. J. Parasitol.* **38:** 1359–1370.
11. Dubey, J.P., D.S. Lindsay & C.A. Speer. 1998. Structures of *Toxoplasma gondii* tachyzoites, bradyzoites, and sporozoites and biology and development of tissue cysts. *Clin. Microbiol. Rev.* **11:** 267–299.
12. Black, M.W. & J.C. Boothroyd. 2000. Lytic cycle of *Toxoplasma gondii*. *Microbiol. Mol. Biol. Rev.* **64:** 607–623.
13. Sheffield, H.G. & M.L. Melton. 1968. The fine structure and reproduction of *Toxoplasma gondii*. *J. Parasitol.* **54:** 209–226.
14. Scharton-Kersten, T.M., T.A. Wynn, E.Y. Denkers, *et al.* 1996. In the absence of endogenous IFN-γ, mice develop unimpaired IL-12 responses to *Toxoplasma gondii* while failing to control acute infection. *J. Immunol.* **157:** 4045–4054.
15. Soête, M., D. Camus & J.F. Dubremetz. 1994. Experimental induction of bradyzoite-specific antigen expression and cyst formation by the RH strain of *Toxoplasma gondii in vitro*. *Exp. Parasitol.* **78:** 361–370.
16. Jerome, M.E., J.R. Radke, W. Bohne, *et al.* 1998. *Toxoplasma gondii* bradyzoites form spontaneously during sporozoite-initiated development. *Infect. Immun.* **66:** 4838–4844.
17. Nissapatorn, V. 2009. Toxoplasmosis in HIV/AIDS: a living legacy. *Southeast Asian J. Trop. Med. Public Health* **40:** 1158–1178.
18. Dubey, J.P. 1997. Bradyzoite-induced murine toxoplasmosis: stage conversion, pathogenesis, and tissue cyst formation in mice fed bradyzoites of different strains of *Toxoplasma gondii*. *J. Eukaryot Microbiol.* **44:** 592–602.
19. Buzoni-Gatel, D., H. Debbabi, M. Moretto, *et al.* 1999. Intraepithelial lymphocytes traffic to the intestine and enhance resistance to *Toxoplasma gondii* oral infection. *J. Immunol.* **162:** 5846–5852.
20. Courret, N., S. Darche, P. Sonigo, *et al.* 2006. CD11c- and CD11b-expressing mouse leukocytes transport single *Toxoplasma gondii* tachyzoites to the brain. *Blood* **107:** 309–316.
21. Morisaki, J.H., J.E. Heuser & L.D. Sibley. 1995. Invasion of *Toxoplasma gondii* occurs by active penetration of the host cell. *J. Cell Sci.* **108:** 2457–2464.
22. Hirai, K., K. Hirato & R. Yanagawa. 1966. A cinematographic study of the penetration of cultured cells by *Toxoplasma gondii*. *Jpn J. Vet. Res.* **14:** 81–90.
23. Dubey, J.P. 1998. Advances in the life cycle of *Toxoplasma gondii*. *Int. J. Parasitol.* **28:** 1019–1024.
24. Grimwood, J., J.R. Mineo & L.H. Kasper. 1996. Attachment of *Toxoplasma gondii* to host cells is host cell cycle dependent. *Infect. Immun.* **64:** 4099–4104.
25. Molestina, R.E., N. El-Guendy & A.P. Sinai. 2008. Infection with *Toxoplasma gondii* results in dysregulation of the host cell cycle. *Cell Microbiol.* **10:** 1153–1165.
26. Woodmansee, D.B. 2003. Kinetics of the initial rounds of cell division of *Toxoplasma gondii*. *J. Parasitol.* **89:** 895–898.
27. Sasono, P.M. & J.E. Smith. 1998. *Toxoplasma gondii*: an ultrastructural study of host-cell invasion by the bradyzoite stage. *Parasitol. Res.* **84:** 640–645.
28. Aliberti, J., J.G. Valenzuela, V.B. Carruthers, *et al.* 2003. Molecular mimicry of a CCR5 binding-domain in the microbial activation of dendritic cells. *Nat. Immunol.* **4:** 485–490.
29. Lambert, H., P.P. Vutova, W.C. Adams, *et al.* 2009. The *Toxoplasma gondii*-shuttling function of dendritic cells is linked to the parasite genotype. *Infect. Immun.* **77:** 1679–1688.

30. Barragan, A., F. Brossier & L.D. Sibley. 2005. Transepithelial migration of *Toxoplasma gondii* involves an interaction of intercellular adhesion molecule 1 (ICAM-1) with the parasite adhesin MIC2. *Cell Microbiol.* **7:** 561–568.

31. Hinck, L., I.S. Näthke, J. Papkoff & W.J. Nelson. 1994. Dynamics of cadherin/catenin complex formation: novel protein interactions and pathways of complex assembly. *J. Cell Biol.* **125:** 1327–1340.

32. Knudsen, K.A. & M.J. Wheelock. 1992. Plakoglobin, or an 83-kD homologue distinct from β-catenin, interacts with E-cadherin and N-cadherin. *J. Cell Biol.* **118:** 671–679.

33. Itoh, M., K. Morita & S. Tsukita. 1999. Characterization of ZO-2 as a MAGUK family member associated with tight as well as adherens junctions with a binding affinity to occludin and α catenin. *J. Biol. Chem.* **274:** 5981–5986.

34. Etienne-Manneville, S., J.B. Manneville, P. Adamson, *et al.* 2000. ICAM-1-coupled cytoskeletal rearrangements and transendothelial lymphocyte migration involve intracellular calcium signaling in brain endothelial cell lines. *J. Immunol.* **165:** 3375–3383.

35. Marlin, S.D. & T.A. Springer. 1987. Purified intercellular adhesion molecule-1 (ICAM-1) is a ligand for lymphocyte function-associated antigen 1 (LFA-1). *Cell* **51:** 813–819.

36. Alexander, J.S., T. Dayton, C. Davis, *et al.* 1998. Activated T-lymphocytes express occludin, a component of tight junctions. *Inflammation* **22:** 573–582.

37. Inagaki-Ohara, K., A. Sawaguchi, T. Suganuma, *et al.* 2005. Intraepithelial lymphocytes express junctional molecules in murine small intestine. *Biochem. Biophys. Res. Commun.* **331:** 977–983.

38. Rescigno, M., M. Urbano, B. Valzasina, *et al.* 2001. Dendritic cells express tight junction proteins and penetrate gut epithelial monolayers to sample bacteria. *Nat. Immunol.* **2:** 361–367.

39. Speer, C.A., J.P. Dubey, J.A. Blixt & K. Prokop. 1997. Time lapse video microscopy and ultrastructure of penetrating sporozoites, types 1 and 2 parasitophorous vacuoles, and the transformation of sporozoites to tachyzoites of the VEG strain of *Toxoplasma gondii*. *J. Parasitol.* **83:** 565–574.

40. Dzierszinski, F., M. Nishi, L. Ouko & D.S. Roos. 2004. Dynamics of *Toxoplasma gondii* differentiation. *Eukaryot. Cell* **3:** 992–1003.

41. Dalton, J.E., S.M. Cruickshank, C.E. Egan, *et al.* 2006. Intraepithelial γδ+ lymphocytes maintain the integrity of intestinal epithelial tight junctions in response to infection. *Gastroenterology* **131:** 818–829.

42. Weight, C. 2011. The interactions of *Toxoplasma gondii* with epithelial tight junctions. Ph.D. thesis, University of East Anglia.

43. Mennechet, F.J.D., L.H. Kasper, N. Rachinel, *et al.* 2002. Lamina propria CD4+ T lymphocytes synergize with murine intestinal epithelial cells to enhance proinflammatory response against an intracellular pathogen. *J. Immunol.* **168:** 2988–2996.

44. Gopal, R., D. Birdsell & F.P. Monroy. 2011. Regulation of chemokine responses in intestinal epithelial cells by stress and *Toxoplasma gondii* infection. *Parasite Immunol.* **33:** 12–24.

45. Dimier, I.H. & D.T. Bout. 1993. Rat intestinal epithelial cell line IEC-6 is activated by recombinant interferon-gamma to inhibit replication of the coccidian *Toxoplasma gondii*. *Eur. J. Immunol.* **23:** 981–983.

46. Kowalik, S., W. Clauss & H. Zahner. 2004. *Toxoplasma gondii*: changes of transepithelial ion transport in infected HT29/B6 cell monolayers. *Parasitol. Res.* **92:** 152–158.

47. Manger, I.D., A.B. Hehl & J.C. Boothroyd. 1998. The surface of Toxoplasma tachyzoites is dominated by a family of glycosylphosphatidylinositol-anchored antigens related to SAG1. *Infect. Immun.* **66:** 2237–2244.

48. Furtado, G.C., Y. Cao & K.A. Joiner. 1992b. Laminin on *Toxoplasma gondii* mediates parasite binding to the β1 integrin receptor α6β1 on human foreskin fibroblasts and Chinese hamster ovary cells. *Infect. Immun.* **60:** 4925–4931.

49. Furtado, G.C., M. Slowik, H.K. Kleinman & K.A. Joiner. 1992a. Laminin enhances binding of *Toxoplasma gondii* tachyzoites to J774 murine macrophage cells. *Infect. Immun.* **60:** 2337–2342.

50. Mineo, J.R., R. McLeod, D. Mack, *et al.* 1993. Antibodies to *Toxoplasma gondii* major surface protein (SAG-1, P30) inhibit infection of host cells and are produced in murine intestine after peroral infection. *J. Immunol.* **150:** 3951–3964.

51. Debierre-Grockiego, F., S. Niehus, B. Coddeville, *et al.* 2010. Binding of *Toxoplasma gondii* glycosylphosphatidylinositols to galectin-3 is required for their recognition by macrophages. *J. Biol. Chem.* **285:** 32744–32750.

52. Saouros, S., B. Edwards-Jones, M. Reiss, *et al.* 2005. A novel galectin-like domain from *Toxoplasma gondii* micronemal protein 1 assists the folding, assembly, and transport of a cell adhesion complex. *J. Biol. Chem.* **280:** 38583–38591.

53. Monteiro, V.G., C.P. Soares & W. de Souza. 1998. Host cell surface sialic acid residues are involved on the process of penetration of *Toxoplasma gondii* into mammalian cells. *FEMS Microbiol. Lett.* **164:** 323–327.

54. Jacquet, A., L. Coulon, J.D. Nève, *et al.* 2001. The surface antigen SAG3 mediates the attachment of *Toxoplasma gondii* to cell-surface proteoglycans. *Mol. Biochem. Parasitol.* **116:** 35–44.

55. Carruthers, V.B., S. Håkansson, O.K. Giddings & L.D. Sibley. 2000. *Toxoplasma gondii* uses sulfated proteoglycans for substrate and host cell attachment. *Infect. Immun.* **68:** 4005–4011.

56. Friedrich, N., J.M. Santos, Y. Liu, *et al.* 2010. Members of a novel protein family containing microneme adhesive repeat domains act as sialic acid-binding lectins during host cell invasion by apicomplexan parasites. *J. Biol. Chem.* **285:** 2064–2076.

57. Hu, K., D.S. Roos & J.M. Murray. 2002. A novel polymer of tubulin forms the conoid of *Toxoplasma gondii*. *J. Cell. Biol.* **156:** 1039–1050.

58. Mordue, D.G., N. Desai, M. Dustin & L.D. Sibley. 1999. Invasion by *Toxoplasma gondii* establishes a moving junction that selectively excludes host cell plasma membrane proteins on the basis of their membrane anchoring. *J. Exp. Med.* **190:** 1783–1792.

59. Håkansson, S., H. Morisaki, J. Heuser & L.D. Sibley. 1999. Time-lapse video microscopy of gliding motility in *Toxoplasma gondii* reveals a novel, biphasic mechanism of cell locomotion. *Mol. Biol. Cell.* **10:** 3539–3547.

60. Dobrowolski, J.M., V.B. Carruthers & L.D. Sibley. 1997. Participation of myosin in gliding motility and host cell invasion by *Toxoplasma gondii*. *Mol. Microbiol.* **26:** 163–173.

61. Dobrowolski, J.M. & L.D. Sibley. 1996. Toxoplasma invasion of mammalian cells is powered by the actin cytoskeleton of the parasite. *Cell* **84:** 933–939.

62. Pezzella-D'Alessandro, N., H.L. Moal, A. Bonhomme, *et al.* 2001. Calmodulin distribution and the actomyosin cytoskeleton in *Toxoplasma gondii*. *J. Histochem. Cytochem.* **49:** 445–454.

63. Wetzel, D.M., S. Håkansson, K. Hu, *et al.* 2003. Actin filament polymerization regulates gliding motility by apicomplexan parasites. *Mol. Biol. Cell* **14:** 396–406.

64. Gonzalez, V., A. Combe, V. David, *et al.* 2009. Host cell entry by apicomplexa parasites requires actin polymerization in the host cell. *Cell Host Microbe* **5:** 259–272.

65. Caldas, L.A., M. Attias & W. de Souza. 2009. Dynamin inhibitor impairs *Toxoplasma gondii* invasion. *FEMS Microbiol. Lett.* **301:** 103–108.

66. Aijaz, S., M.S. Balda & K. Matter. 2006. Tight junctions: molecular architecture and function. *Int. Rev. Cytol.* **248:** 261–298.

67. Fourmaux, M.N., A. Achbarou, O. Mercereau-Puijalon, *et al.* 1996. The MIC1 microneme protein of *Toxoplasma gondii* contains a duplicated receptor-like domain and binds to host cell surface. *Mol. Biochem. Parasitol.* **83:** 201–210.

68. Meissner, M., M. Reiss, N. Viebig, *et al.* 2002. A family of transmembrane microneme proteins of *Toxoplasma gondii* contain EGF-like domains and function as escorters. *J. Cell. Sci.* **115:** 563–574.

69. Wan, K.L., V.B. Carruthers, L.D. Sibley & J.W. Ajioka. 1997. Molecular characterisation of an expressed sequence tag locus of *Toxoplasma gondii* encoding the micronemal protein MIC2. *Mol. Biochem. Parasitol.* **84:** 203–214.

70. Alexander, D.L., J. Mital, G.E. Ward, *et al.* 2005. Identification of the moving junction complex of *Toxoplasma gondii*: a collaboration between distinct secretory organelles. *PLoS Pathog.* **1:** e17.

71. Carruthers, V.B. & L.D. Sibley. 1999. Mobilization of intracellular calcium stimulates microneme discharge in *Toxoplasma gondii*. *Mol. Microbiol.* **31:** 421–428.

72. Lovett, J.L. & L.D. Sibley. 2003. Intracellular calcium stores in *Toxoplasma gondii* govern invasion of host cells. *J. Cell. Sci.* **116:** 3009–3016.

73. Huynh, M.H., K.E. Rabenau, J.M. Harper, *et al.* 2003. Rapid invasion of host cells by Toxoplasma requires secretion of the MIC2-M2AP adhesive protein complex. *EMBO J.* **22:** 2082–2090.

74. Zheng, B., A. He, M. Gan, *et al.* 2009. MIC6 associates with aldolase in host cell invasion by *Toxoplasma gondii*. *Parasitol Res.* **105:** 441–445.

75. Brecht, S., V.B. Carruthers, D.J. Ferguson, *et al.* 2001. The toxoplasma micronemal protein MIC4 is an adhesin composed of six conserved apple domains. *J. Biol. Chem.* **276:** 4119–4127.

76. Bradley, P.J., C. Ward, S.J. Cheng, *et al.* 2005. Proteomic analysis of rhoptry organelles reveals many novel constituents for host-parasite interactions in *Toxoplasma gondii*. *J. Biol. Chem.* **280:** 34245–34258.

77. Lamarque, M., S. Besteiro, J. Papoin, *et al.* 2011. The RON2-AMA1 interaction is a critical step in moving junction-dependent invasion by apicomplexan parasites. *PLoS Pathog.* **7:** e1001276.

78. Straub, K.W., S.J. Cheng, C.S. Sohn & P.J. Bradley. 2009. Novel components of the Apicomplexan moving junction reveal conserved and coccidia-restricted elements. *Cell Microbiol.* **11:** 590–603.

79. Straub, K.W., E.D. Peng, B.E. Hajagos, *et al.* 2011. The moving junction protein RON8 facilitates firm attachment and host cell invasion in *Toxoplasma gondii*. *PLoS Pathog.* **7:** e1002007.

80. Hajj, H.E., M. Lebrun, S.T. Arold, *et al.* 2007. ROP18 is a rhoptry kinase controlling the intracellular proliferation of *Toxoplasma gondii*. *PLoS Pathog.* **3:** e14.

81. Taylor, S., A. Barragan, C. Su, *et al.* 2006. A secreted serine-threonine kinase determines virulence in the eukaryotic pathogen *Toxoplasma gondii*. *Science* **314:** 1776–1780.

82. Saeij, J.P.J., S. Coller, J.P. Boyle, *et al.* 2007. Toxoplasma co-opts host gene expression by injection of a polymorphic kinase homologue. *Nature* **445:** 324–327.

83. Hölttä-Vuori, M., K. Tanhuanpää, W. Möbius, *et al.* 2002. Modulation of cellular cholesterol transport and homeostasis by Rab11. *Mol. Biol. Cell.* **13:** 3107–3122.

84. Poupel, O., H. Boleti, S. Axisa, *et al.* 2000. Toxofilin, a novel actin-binding protein from *Toxoplasma gondii*, sequesters actin monomers and caps actin filaments. *Mol. Biol. Cell.* **11:** 355–368.

85. Gilbert, L.A., S. Ravindran, J.M. Turetzky, *et al.* 2007. *Toxoplasma gondii* targets a protein phosphatase 2C to the nuclei of infected host cells. *Eukaryot. Cell* **6:** 73–83.

86. Jan, G., V. Delorme, N. Saksouk, et al. 2009. A Toxoplasma type 2C serine-threonine phosphatase is involved in parasite growth in the mammalian host cell. *Microbes Infect.* **11:** 935–945.

87. Delorme, V., A. Garcia, X. Cayla & I. Tardieux. 2002. A role for *Toxoplasma gondii* type 1 ser/thr protein phosphatase in host cell invasion. *Microbes Infect.* **4:** 271–278.

Ann. N.Y. Acad. Sci. ISSN 0077-8923

ANNALS OF THE NEW YORK ACADEMY OF SCIENCES

Issue: *Barriers and Channels Formed by Tight Junction Proteins*

Ion transport and barrier function are disturbed in microscopic colitis

Christian Barmeyer,[1] Irene Erko,[2] Anja Fromm,[2] Christian Bojarski,[1] Kristina Allers,[1] Verena Moos,[1] Martin Zeitz,[1] Michael Fromm,[3] and Jörg-Dieter Schulzke[2]

[1]Department of Gastroenterology, Infectiology and Rheumatology, [2]Department of Gastroenterology, Infectiology and Rheumatology, Division of Nutritional Medicine, [3]Institute of Clinical Physiology, Charité, Campus Benjamin Franklin, Freie Universität and Humboldt-Universität, Berlin, Germany

Address for correspondence: Jörg-Dieter Schulzke, Department of Gastroenterology, Infectiology and Rheumatology, Division of Nutritional Medicine, Charité, Campus Benjamin Franklin, Hindenburgdamm 30, 12203 Berlin, Germany. joerg.schulzke@charite.de

In this paper, we identify mechanisms of watery diarrhea in microscopic colitis (MC). Biopsies from the sigmoid colon of patients with collagenous colitis and treated lymphocytic colitis were analyzed in miniaturized Ussing chambers for electrogenic sodium transport and barrier function with one-path impedance spectroscopy. Cytometric bead arrays (CBA) served to analyze cytokine profiles. In active MC, electrogenic sodium transport was diminished and epithelial resistance decreased. CBA revealed a Th1 cytokine profile featuring increased IFN-γ, TNF-α, and IL-1β levels. After four weeks of steroid treatment with budesonide, electrogenic sodium transport recovered while epithelial barrier defects remained. Diarrhea in MC results at least in part from a combination of impaired electrogenic sodium transport and barrier defects. From a therapeutic perspective it can be postulated that the functional importance of loss of ions may be higher than that caused by barrier impairment.

Keywords: microscopic colitis; ENaC; epithelial barrier; ion transport; Th1 cytokine profile

Introduction

Microscopic colitis (MC) is a chronic inflammatory bowel disease that typically affects patients between the fifth and seventh decades of life.[1] The two major disease entities are collagenous colitis (CC) and lymphocytic colitis (LC). CC exhibits many similarities to LC, but differs in the appearance of a collagenous band that acts as an additional diffusion barrier; the cardinal symptom is a severe chronic watery diarrhea. Usually, disease onset is slow, with the development of a chronic recurring pattern in later stages that can last for years. Because of a lack of knowledge of the pathophysiologic causality, therapy is currently directed toward alleviating symptoms.

A symptom of many diseases, diarrhea is often characterized as being the result of different mechanisms. Whereas diarrhea in cholera is based on a secretory mechanism—a consequence of an enterotoxin-based abnormal anion secretion of mainly chloride and bicarbonate—malabsorptive diarrhea in cystic fibrosis is due to malabsorption of solutes or a reduced absorptive surface, as in celiac disease or short bowel syndrome. Leak-flux diarrhea, as another mechanism, occurs during disturbances of the epithelial barrier function, leading to an increased passive loss of osmotically active solutes into the gut lumen, for example, in clostridium difficile infection.[2,3] Often, a coaction of different mechanisms, such as a combination of malabsorption and leak flux, or even a combination of leak flux, malabsorption, and anion secretion, can be observed, for example, as in ulcerative colitis, Crohn's disease, or CC.[4–7]

For MC, however, not much is known about the inherent molecular mechanisms of the resulting diarrhea. Only a few reports exist suggesting abnormal colonic fluid absorption, such as reduced active and passive sodium and chloride absorption.[8,9] In the

doi: 10.1111/j.1749-6632.2012.06631.x

first comprehensive study of the molecular mechanisms of diarrhea in MC, we reported reduced electroneutral NaCl transport as the key mechanism in CC.[4]

Thus, the aim of the study here was to use a combination of electrophysiological and molecular approaches to identify further diarrheal mechanisms, and the signaling events responsible for these changes, as well as therapy effects in MC. These studies were performed in patients with active CC, LC, and in patients with LC after treatment.

Materials and methods

Patients and tissue preparation

Human sigmoid colon was obtained from patients with active CC, LC, and inactive-while-treated LC with an oral therapy of budesonide (9 mg daily) initiated at least four weeks previously. Patients who underwent lower GI endoscopy for prevention of colon cancer served as controls. To confirm the presence of CC and LC, tissues were investigated with H&E and CD3 staining (mean intraepithelial lymphocyte count in active CC/LC, 54/100 enterocytes; in treated LC, 30/100 enterocytes; in controls, 18/100 enterocytes). The median stool frequency in active CC/LC was 5.7 per day, in treated LC 1.7 per day. Immediately after removal from the colon, the samples were placed in cold bathing solution and mounted into Ussing chambers after about 45 minutes. The study was approved by the local ethics committee and written informed consent was obtained from each patient.

Solutions and drugs

The bathing solution consisted of (in mmol/L): 140.0 Na^+, 123.8 Cl^-, 5.4 K^+, 1.2 Ca^{2+}, 1.2 Mg^{2+}, 2.4 HPO_4^{2-}, 0.6 $H_2PO_4^-$, 21.0 HCO_3^-, 10.0 D(+)-glucose, 10.0 D(+)-mannose, 2.5 glutamine, and 0.5 β-OH-butyrate. The solution was gassed with a mixture of 95% O_2 and 5% CO_2. Temperature was kept at 37 °C, and pH was 7.4. To prevent bacterial growth, 50 mg/L azlocillin and 4 mg/L tobramycin were added to the bathing solution.

Aldosterone, amiloride, bumetanide, theophylline, carbachol, and PGE_2 were obtained from Sigma Aldrich (St. Louis, MO).

Electrophysiologic experiments

Data acquisition of electrogenic sodium transport via the epithelial sodium channel (ENaC) has been performed as described before.[10,11] After mounting

the tissue into a miniaturized Ussing chamber as described by Stockmann *et al.*, spontaneous electrogenic chloride secretion was inhibited by serosal addition of 10^{-5} M bumetanide.[12] Then, 3×10^{-9} M aldosterone was added to both sides of the epithelium. Eight hours after steroid addition amiloride (10^{-4} M) was added to the mucosal compartment in a concentration that completely blocks the aldosterone-induced apical sodium conductivity to quantify electrogenic sodium absorption (J_{Na}).

After obtaining the data on electrogenic sodium transport, chloride secretion was stimulated, as a proof of tissue viability. Therefore, PGE_2 (10^{-6} M) and carbachol (10^{-4} M) were added to the serosal side and theophylline (10^{-2} M) to both the mucosal and serosal side of the chamber. An increase in short circuit current reflected viability of the tissue.

One-path impedance spectroscopy

Impedance spectroscopy was performed to determine epithelial barrier function. As described earlier, by one-path impedance spectroscopy, the epithelial (R^{epi}) and subepithelial (R^{sub}) portion of the total wall resistance (R^t) can be differentiated.[13] In brief, after mounting the tissue into the Ussing chamber, 48 discrete frequencies of an effective sine-wave–alternating current of 35 μA/cm^2 were applied ranging from 1.3 Hz to 65 kHz. Resulting voltage changes were detected by phase-sensitive amplifiers. After correction for the resistance of the bathing solution and the frequency behavior of the measuring device for each frequency, calculated complex impedance values were plotted in a Nyquist diagram and fitted by least square analysis. Because of the frequency-dependent electrical characteristics of the epithelium, R^t could be obtained at minimum and R^{sub} at maximum frequency. $R^{epi} = R^t - R^{sub}$, where R^{epi} indicates epithelial resistance.

Cytometric bead arrays

Amounts of IFN-γ, TNF-α, IL-1β, IL-8, and IL-13 were determined using the cytometric bead array (CBA) system (BD™, Heidelberg, Germany). Experiments were carried out following the manufacturer's instructions. In brief, biopsies from the sigmoid colon were incubated in 1 mL RPMI 1640 medium containing 10% fetal calf serum (Gibco-BRL, Berlin, Germany), 100 U/mL penicillin, 100 μg/mL streptomycin, 50 μg/mL gentamicin, and 2.5 μg/mL amphotericin (Seromed Biochrom KG, Berlin, Germany) at 37 °C in a humidified 5%

carbon dioxide/95% oxygen atmosphere for 48 h, and cytokines produced *in situ* were quantified in the supernatants with a CBA (Becton Dickinson, Heidelberg, Germany) according to the manufacturer's protocol. Samples were measured on a FACS Canto II (BD[TM]) and analyzed by FCAP array software (BD[TM]). The levels of IFN-γ, TNF-α, IL-1β, IL-8, and IL-13 (in pg/g tissue) were calculated using a specific standard curve.

Statistics
Results are given as means $\pm$ SEM. Significance was tested by the two-tailed Student's *t*-test. $P < 0.05$ was considered significant.

Results

Electrogenic sodium transport
Electrogenic sodium transport (J_{Na}) via the ENaC is the major mechanism in the distal colon to prevent loss of sodium and water. ENaC is essential for the fine regulation of the feces. J_{Na} can be determined as the drop in I_{SC} after the addition of amiloride after an eight-hour incubation period with aldosterone. After the addition of amiloride, controls exhibited a J_{Na} of 15.5 ± 3.2 μmol/h/cm^2 ($n = 5$, Fig. 1). In contrast, in patients with CC J_{Na} was markedly reduced to 4.7 ± 2.1 μmol/h/cm^2 ($n = 4$). Similar results have been obtained for LC (unpublished data). These findings indicate a severe defect in electrogenic sodium absorption in untreated MC, resulting in a reduced capability to prevent loss of solutes and water.

By contrast, in a subset of patients with treated LC of at least four weeks after initiation of the budesonide therapy J_{Na} had almost returned to normal again (11.7 ± 4.0 μmol/h/cm^2; $n = 3$, Fig. 1). Although these are only preliminary data in a small number of patients, a positive effect of budesonide on the recovery of electrogenic sodium transport can be assumed.

Epithelial barrier function
The data are presented in Table 1. In controls R^{epi} was 41 ± 2 $\Omega \cdot$cm^2. In contrast, in CC R^{epi} was reduced by about 25% to 31 ± 1 $\Omega \cdot$cm^2. These results indicate a modest impairment of mucosal barrier function in CC and confirm observations of Bürgel *et al.*, who previously described a mucosal barrier disturbance in CC.[4] For LC, similar data have been obtained (unpublished data).

In the group of LC patients treated with budesonide, R^{epi} was 27 ± 4 $\Omega \cdot$cm^2 and thus did not exhibit an increase after at least four weeks of therapy. No significant differences between the three groups were detected for R^{sub}.

When comparing true active transport rates in inflamed and noninflamed conditions, transport rates become underestimated by a factor that is given by the ratio of R^t over R^{epi}.[14,15] These factors were 1.4 ± 0.1 in controls, 1.7 ± 0.2 in CC, and 2.2 ± 0.2 in treated LC (Table 1). Because these factors differed significantly, electrogenic sodium transport and active chloride secretion were compared in all groups only after this correction.

Activation of anion secretion
Before data can be interpreted, the viability of the tissues used for the experiments had to be ensured. Thus, we stimulated and quantified active anion secretion after having obtained the above mentioned data on absorption and barrier function.

Stimulation of cAMP-induced active anion secretion was carried out by the simultaneous addition of PGE$_2$ and theophylline, whereas intracellular Ca-dependent anion secretion was induced by addition of carbachol. The results are shown in Figure 2. Both the combination of PGE$_2$/theophylline and carbachol activated active anion secretion. As demonstrated in Figure 2, there were no marked differences between controls, LC, and CC, indicating that viability of the epithelium was preserved in all three groups.

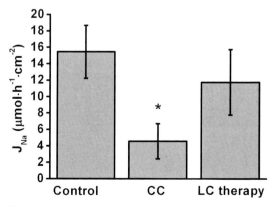

Figure 1. Amiloride-sensitive electrogenic sodium transport (J_{Na}) in controls ($n = 5$), CC ($n = 4$), and treated LC four weeks after the beginning of budesonide therapy ($n = 3$). *$P < 0.05$ versus control.

Table 1. Impedance spectroscopy data from controls and patients with collagenous colitis (CC) and treated lymphocytic colitis (LC therapy)

	R^t ($\Omega\cdot cm^2$)	R^{epi} ($\Omega\cdot cm^2$)	R^{sub} ($\Omega\cdot cm^2$)	R^t/R^{epi}	n
Control	60 ± 6	41 ± 2	19 ± 4	1.4 ± 0.1	5
CC	52 ± 4	$31 \pm 1^*$	21 ± 5	1.7 ± 0.2	3
LC therapy	58 ± 8	$27 \pm 4^*$	31 ± 5	$2.2 \pm 0.2^*$	3

NOTE: Data represent means $\pm$ SEM.
$^*P < 0.05$ versus control.

Cytokine profile in MC

The etiology of MC is unknown, although circumstantial evidence suggests a response to a luminal antigen.[16-18] Thus, CBA were performed to determine the cytokine profile of MC. The levels of IFN-γ, TNF-α, and IL-1β were markedly increased compared to controls, whereas levels of IL-8 and IL-13 did not reveal any difference between both groups (Fig. 3). These data indicate a predominantly Th1-based cytokine pattern in MC.

It should be mentioned, however, that IL-13 showed a tendency toward an increase in MC but the scatter of data was too large to reach statistical significance with the numbers examined.

Discussion

MC is a chronic diarrheal disease with unknown etiology. The mechanisms leading to diarrhea are not well understood. Only a few reports exist that reported impaired electroneutral sodium and chloride transport based on *in vitro* and *in vivo* measurements,[4,8,9] but most of these studies lack information about the specific transport processes.

In our study, we demonstrate for the first time that electrogenic sodium transport mediated by the epithelial sodium channel ENaC is impaired in CC; we previously demonstrated that electroneutral NaCl-absorption is seriously impaired in CC.[4,8,9] Here, we found that the ENaC transport system was neither spontaneously activated nor could it be activated by $3\cdot10^{-9}$ M aldosterone. Thus, electrogenic sodium transport is severely disturbed, which adds to sodium loss in the distal colon. This finding is of importance because electrogenic sodium transport is thought to be a compensatory mechanism for the prevention of loss of ions and water, in addition to electroneutral NaCl-absorption in the proximal colon. This is also supported by the observation of a marked increase of ENaC activity in NHE3-deficient mice.[19] Our finding of recovered electrogenic sodium transport in a small subset of symptom-free patients with treated LC could be additional evidence, although data interpretation should be done with caution due to the limited number of patients evaluated.

Similar observations of disturbed electrogenic sodium transport have been made in ulcerative colitis and Crohn's disease,[6,20] and preliminary data suggest the same for LC (unpublished data). The findings reported in Amasheh *et al.* could demonstrate that impaired electrogenic sodium transport in ulcerative colitis is a result of reduced transcription of the γ-ENaC subunit and that this effect has been provoked by TNF-α.[20] As demonstrated for Crohn's disease this might additionally be due to activation of ERK1/2.[6] Experiments in LC suggest that this signaling pathway might apply to LC as well (unpublished data).

In addition, impedance spectroscopy analysis revealed an impaired barrier function in patients with CC and the small group of patients with treated LC. Epithelial resistance was markedly reduced, indicating increased permeability for solutes and water. A morphologic correlate for impaired barrier function with increased permeability is the tight junction, which consists of multiple proteins, including occludin and the claudin multigene family. Alterations in epithelial resistance usually reflect alterations in the formation of tight junction proteins. This has been demonstrated for Crohn's disease where the *sealing* claudins 5 and 8 are downregulated and the *pore-forming* claudin-2 is upregulated, resulting in reduced epithelial resistance.[7] Similar results have been reported recently for CC, where reduced epithelial resistance due to reduced expression of occludin and sealing claudins were observed.[4]

To demonstrate tissue viability, active anion secretion was stimulated and quantified after the end of each experiment. No significant differences were observed between controls, LC, and CC after the addition of PGE_2/theophylline and carbachol at the end of each experiment, indicating that viability of the epithelium was preserved throughout the entire experiment. However, it should be noted that endogenous chloride secretion was inhibited at the beginning of the experiment with bumetanide. The

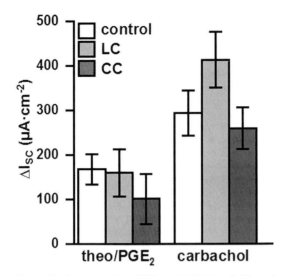

Figure 2. ΔI_{SC} after the addition of PGE_2/theophyllin and carbachol at the end of each experiment. ΔI_{SC} reflects the increase of the short circuit current after the addition of PGE_2/theophylline and carbachol. ΔI_{SC} therefore indicates the viability of the epithelium in the Ussing chamber.

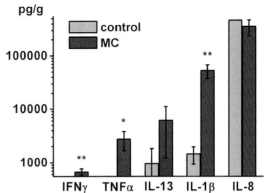

Figure 3. Cytometric bead arrays from controls and patients with microscopic colitis (predominantly LC). Data represent means ± SEM. *$P < 0.05$ versus control, **$P < 0.01$ versus control.

obtained data therefore might represent only residual anion secretion (e.g., bicarbonate secretion) instead of cAMP-induced or Ca-dependent chloride secretion. Independent of that, the data demonstrate that tissue viability was intact throughout the experiment.

As an additional question, we addressed the predominant cytokine profile in MC. We found a prevailing T helper cell type 1 (Th1)–like immune response with high concentrations of IFN-γ, TNF-α, and IL-1β that points toward a response to a luminal antigen rather than an allergic reaction. In fact, there is circumstantial evidence that the immune response occurs as a response to a luminal antigen. This evidence includes the decreasing gradient of histological changes from right to left colon,[18] the induction of histological changes similar to MC in conditions, with the ingestion of known or suspected luminal antigens, including Brainerd diarrhea[16] and celiac disease,[21] and rapid clinical and histological remission after fecal stream diversion with ileostomy.[17] Furthermore, our data confirm the findings of a previous report by Tagkalidis *et al.*, who found increased mucosal mRNA levels of IL-15, IFN-γ, TNF-α, and inducible nitric oxide synthase in MC, whereas mRNA levels of T helper cell type 2 (Th2) cytokines, including IL-2 and IL-4, were too low to be accurately quantified.[22]

Taken together, diarrheal mechanisms in MC are characterized by (1) impaired electrogenic sodium transport due to impaired ENaC function, possibly due to increased TNF-α concentrations, and (2) a barrier defect leading to increased permeability and therefore causing a leak flux of solutes and water. However, the observation that LC patients after at least four weeks of budesonide treatment exhibited normal bowel movement, and that only electrogenic sodium transport was restored but not the epithelial barrier indicates that impaired ion transport might play a larger functional role than barrier impairment. Nevertheless, these data have to be interpreted with caution because the number of patients in the treated LC group was small.

Conflicts of interest

The authors declare no conflicts of interest.

References

1. Fernandez-Banares, F., A. Salas, M. Forne, *et al.* 1999. Incidence of collagenous and lymphocytic colitis: a 5-year population-based study. *Am. J. Gastroenterol.* **94:** 418–423.

2. Hecht, G., C. Pothoulakis, J.T. LaMont & J.L. Madara. 1988. Clostridium difficile toxin A perturbs cytoskeletal structure and tight junction permeability of cultured human intestinal epithelial monolayers. *J. Clin. Invest.* **82:** 1516–1524.

3. Moore, R., C. Pothoulakis, J.T. LaMont, *et al.* 1990. C. difficile toxin A increases intestinal permeability and induces Cl- secretion. *Am. J. Physiol.* **259:** G165–G172.

4. Burgel, N., C. Bojarski, J. Mankertz, *et al.* 2002. Mechanisms of diarrhea in collagenous colitis. *Gastroenterology* **123:** 433–443.

5. Schmitz, H., C. Barmeyer, M. Fromm, *et al.* 1999. Altered tight junction structure contributes to the impaired

epithelial barrier function in ulcerative colitis. *Gastroenterology* **116:** 301–309.

6. Zeissig, S., T. Bergann, A. Fromm, *et al.* 2008. Altered ENaC expression leads to impaired sodium absorption in the non-inflamed intestine in Crohn's disease. *Gastroenterology* **134:** 1436–1447.

7. Zeissig, S., N. Burgel, D. Gunzel, *et al.* 2007. Changes in expression and distribution of claudin 2, 5 and 8 lead to discontinuous tight junctions and barrier dysfunction in active Crohn's disease. *Gut* **56:** 61–72.

8. Bo-Linn, G.W., D.D. Vendrell, *et al.* 1985. An evaluation of the significance of microscopic colitis in patients with chronic diarrhea. *J. Clin. Invest.* **75:** 1559–1569.

9. Protic, M., N. Jojic, D. Bojic, *et al.* 2005. Mechanism of diarrhea in microscopic colitis. *World J. Gastroenterol.* **11:** 5535–5539.

10. Amasheh, S., H.J. Epple, J. Mankertz, *et al.* 2000. Differential regulation of ENaC by aldosterone in rat early and late distal colon. *Ann. N. Y. Acad. Sci.* **915:** 92–94.

11. Epple, H.J., S. Amasheh, J. Mankertz, *et al.* 2000. Early aldosterone effect in distal colon by transcriptional regulation of ENaC subunits. *Am. J. Physiol. Gastrointest Liver Physiol.* **278:** G718–G724.

12. Stockmann, M., A.H. Gitter, D. Sorgenfrei, *et al.* 1999. Low edge damage container insert that adjusts intestinal forceps biopsies into Ussing chamber systems. *Pflugers Arch.* **438:** 107–112.

13. Gitter, A.H., J.D. Schulzke, D. Sorgenfrei & M. Fromm 1997. Ussing chamber for high-frequency transmural impedance analysis of epithelial tissues. *J. Biochem. Biophys. Methods.* **35:** 81–88.

14. Fromm, M., J.D. Schulzke & U. Hegel 1985. Epithelial and subepithelial contributions to transmural electrical resistance of intact rat jejunum, in vitro. *Pflugers Arch.* **405:** 400–402.

15. Tai, Y.H. & C.Y. Tai. 1981. The conventional short-circuiting technique under-short-circuits most epithelia. *J. Membr. Biol.* **59:** 173–177.

16. Bryant, D.A., E.D. Mintz, N.D. Puhr, *et al.* 1996. Colonic epithelial lymphocytosis associated with an epidemic of chronic diarrhea. *Am. J. Surg. Pathol.* **20:** 1102–1109.

17. Jarnerot, G., C. Tysk, J. Bohr & S. Eriksson. 1995. Collagenous colitis and fecal stream diversion. *Gastroenterology* **109:** 449–455.

18. Wang, N., J.A. Dumot, E. Achkar, *et al.* 1999. Colonic epithelial lymphocytosis without a thickened subepithelial collagen table: a clinicopathologic study of 40 cases supporting a heterogeneous entity. *Am. J. Surg. Pathol.* **23:** 1068–1074.

19. Schultheis, P.J., L.L. Clarke, P. Meneton, *et al.* 1998. Renal and intestinal absorptive defects in mice lacking the NHE3 Na+/H+ exchanger. *Nat. Genet.* **19:** 282–285.

20. Amasheh, S., C. Barmeyer, C.S. Koch, *et al.* 2004. Cytokine-dependent transcriptional down-regulation of epithelial sodium channel in ulcerative colitis. *Gastroenterology* **126:** 1711–1720.

21. Wolber, R., D. Owen & H. Freeman 1990. Colonic lymphocytosis in patients with celiac sprue. *Hum. Pathol.* **21:** 1092–1096.

22. Tagkalidis, P.P., P.R. Gibson & P.S. Bhathal. 2007. Microscopic colitis demonstrates a T helper cell type 1 mucosal cytokine profile. *J. Clin. Pathol.* **60:** 382–387.

Ann. N.Y. Acad. Sci. ISSN 0077-8923

ANNALS OF THE NEW YORK ACADEMY OF SCIENCES
Issue: *Barriers and Channels Formed by Tight Junction Proteins*

Enteropathogenic *E. coli* effectors EspG1/G2 disrupt tight junctions: new roles and mechanisms

Lila G. Glotfelty and Gail A. Hecht

Section of Digestive Diseases and Nutrition, Department of Medicine, University of Illinois at Chicago, Chicago, Illinois

Address for correspondence: Gail A. Hecht, Department of Medicine, Section of Digestive Diseases and Nutrition, University of Illinois at Chicago, 840 S Wood Street, CSB, Chicago, IL 60612. gahecht@uic.edu

Enteropathogenic *E. coli* (EPEC) infection is a major cause of infantile diarrhea in the developing world. Using a type-three secretion system, bacterial effector proteins are transferred to the host cell cytosol where they affect multiple physiological functions, ultimately leading to diarrheal disease. Disruption of intestinal epithelial cell tight junctions is a major consequence of EPEC infection and is mediated by multiple effector proteins, among them EspG1 and its homologue EspG2. EspG1/G2 contribute to loss of barrier function via an undefined mechanism that may be linked to their disruption of microtubule networks. Recently new investigations have identified additional roles for EspG. Sequestration of active ADP-ribosylating factor (ARF) proteins and promotion of p21-activated kinase (PAK) activity as well as inhibition of Golgi-mediated protein secretion have all been linked to EspG. In this review, we examine the functions of EspG1/G2 and discuss potential mechanisms of EspG-mediated tight junction disruption.

Keywords: EPEC; tight junctions; EspG; microtubules; ARF; PAK

Introduction

Enteropathogenic *Escherichia coli* (EPEC) infection is a major cause of diarrhea in the developing world. Infants are primarily affected by this pathogen and may experience repeated infections.[1,2] This leads to prolonged intervals during early childhood development without the benefit of micronutrients, sufficient glucose, and other nutritive factors absorbed via the intestine.[3] EPEC belongs to a family of related attaching and effacing (A/E) pathogens, including enterohemorrhagic *E. coli* (EHEC) and the murine pathogen *Citrobacter rodentium*. A/E pathogens use a type-three secretion system (TTSS) to transfer bacterial effector proteins into host epithelial cells via a needle-like structure.[4–6] This "molecular syringe" spans the bacterial and host cell membranes and delivers the effectors directly into the host cytosol, working in concert with multiple chaperone proteins.

EPEC houses a 35-kb pathogenicity island called the locus of enterocyte effacement (LEE).[7,8] Commensal *E. coli* lack this chromosomally encoded region. Some bacterial effector proteins are encoded within the LEE, while others are non-LEE encoded (Nle). EPEC effectors have multiple deleterious effects on host cells.[9,10] EPEC-induced diarrhea is largely attributable to inhibition of absorption by targeting the Na^+/H^+ exchanger (NHE3), the Na^+/glucose cotransporter (SGLT1), as well as the anion exchanger downregulated-in-adenoma (DRA) in human small intestinal epithelium.[11–15] The EPEC effector EPEC-secreted protein F (EspF) inhibits NHE3 and SGLT1 activity, while EspG is responsible for decreasing the activity of DRA. In addition to the effects on intestinal transporters, multiple effectors, including EspF, MAP, NleA, and EspG, contribute to disruption of intestinal epithelial tight junctions (TJ), which enhances the diarrhea phenotype by preventing the formation of ion concentration gradients across the epithelium (Fig. 1).[16–18]

Epithelial TJs are disrupted in multiple disease states underscoring the importance of maintaining epithelial barriers. Intestinal barrier defects were first described over 25 years ago in patients with

doi: 10.1111/j.1749-6632.2012.06563.x

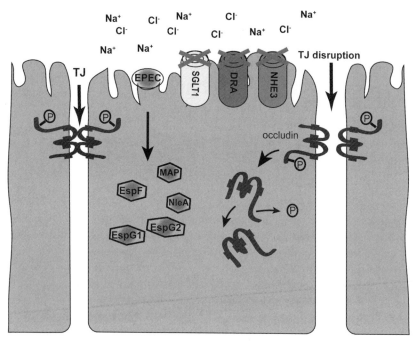

Figure 1. Mechanisms of EPEC-induced diarrhea. After attachment to epithelial cells, EPEC inject effector proteins (red hexagons) into host cell cytosol. Several effectors mediate tight junction disruption by driving occludin (shown in blue) dephosphorylation and localization to the cytosol. Inhibition of the Na$^+$/glucose cotransporter (SGLT1), the anion exchanger downregulated-in-adenoma (DRA), and Na$^+$/H$^+$ exchanger (NHE3) leads to an increase in luminal sodium and chloride concentrations and ensuing net water loss.

Crohn's disease, and since then have been linked to TNF-induced expression of myosin light chain kinase MLCK and caveolin-dependent endocytosis of occludin.[19,20–23] Interleukin-13 (IL-13)–induced upregulation of claudin-2 has been identified as the primary cause of barrier dysfunction in ulcerative colitis.[24–26] The intestine is not the only location where barrier defects cause severe clinical consequences. The blood–testis barrier (BTB) is partially maintained via TJs. Differential expression and mislocalization of TJ proteins has been correlated with increased BTB permeability leading to increased levels of anti-sperm antibody, a possible factor in male infertility.[27–30] In lung epithelial tissues, environmental factors may decrease expression of TJ proteins and play a role in the development of asthma.[31–33]

In general, tight junction maintenance, disruption, and assembly can be regulated at the transcriptional level, by post-translational protein modification, phosphorylation for example, and by protein half-life. Endocytosis mechanisms involving cytoskeletal changes also play a role in TJ regulation.

TJs are highly dynamic structures that undergo continual maintenance through a process that requires actin.[34–36] Cells transfected to stably express fluorescent-tagged occludin, claudin-1, and ZO-1 were treated with the actin-perturbing drug latrunculin A and imaged and transepithelial electrical resistance (TER) was measured simultaneously in real time. Actin depolymerization induced the relocalization of TJ proteins to the cytosol with a corresponding drop in TER, suggesting that actin is required for TJ maintenance. Evidence suggests that microtubules also contribute to TJ maintenance as their disruption by nocodazole impedes significantly the movement of occludin-containing vesicles.[37] Real-time tracking of GFP-occludin revealed that the fraction of mobile cytosolic occludin was significantly reduced in the presence of nocodazole. Colchicine, an inhibitor of microtubule assembly, also disrupts epithelial cell barrier function although it is not clear if this is due to the mislocalization of TJ proteins or another mechanism.[38]

These studies indicate that cytoskeletal components play an important role in TJ homeostasis, making EPEC effectors that interfere with the cytoskeleton of high interest with regard to their effects on TJs. The EPEC effector EspG1 and its homologue EspG2 have previously been demonstrated to disrupt microtubules. Recent investigations have identified another role for EspG, that of a GTPase pathway regulator with potential downstream effects on the actin cytoskeleton and on inhibition of vesicle trafficking. Effects of EspG on the Golgi apparatus include the inhibition of protein secretion and possible interference with proper membrane trafficking. Although these studies did not specifically investigate how the various roles ascribed to EspG impact TJs, several mechanisms can be envisioned. In this review, we examine the current literature regarding the identified activities of EspG and discuss the potential mechanisms of EspG-mediated TJ disruption.

EPEC infection disrupts TJ structure and function

EPEC infection has been demonstrated to perturb TJs in both *in vitro* and *in vivo* models. Simonovic *et al.* demonstrated that EPEC infection of intestinal epithelial cells (IEC) shifts occludin localization from the plasma membrane to an intracellular compartment.[39] There was a strong correlation between decreased occludin phosphorylation and movement into the cytosol. There was also a corresponding drop in TER, suggesting interdependence of TJ structure and barrier function. Both the structural and functional alterations normalized following eradication of infection with gentamicin, highlighting the restorative nature of TJs. Infection of IEC with an EPEC mutant lacking the effector EspF reduced the shift in occludin localization and the decrease in TER. Claudin-1 and ZO-1 localization are also altered in EPEC-infected IEC.[40]

Occludin localization and TER are similarly affected in the mouse model of EPEC infection.[41] After one day of EPEC infection, ileal, and colonic tissues exhibited substantial relocalization of occludin to the cytosol as well as a significant drop in TER. Neither of these phenomena was observed following infection with an EspF mutant. In contrast to studies done in IEC, ZO-1 localization was not altered in the infected murine intestine.[41] In a separate study, claudin-1 was shown to move from the membrane to the cytosol as well.[42]

EPEC infection also induces a number of transcriptional and post-translational changes that are important for pathogenesis.[39,43–46]

EspG plays a role in host colonization and disrupts intestinal epithelial TJ function

While EspF and MAP have been implicated in disruption of TJs, EspG has also been shown to contribute to loss of TJ function.[16–18] In EPEC, EspG1 is LEE-encoded and its homolog EspG2 is encoded in the espC pathogenicity island.[18,47–49] EspG1 was first described as a type-three secreted homolog of the *Shigella flexneri* effector VirA harboring 21% identity and 40% similarity.[47] EspG is also highly conserved across other attaching and effacing pathogens, namely EHEC O157:H7 and the murine pathogen *Citrobacter rodentium.* EspG2 is 42% identical and 62% similar to EspG and 20% identical and 38% similar to VirA.[47] Functionally, EspG1 can complement an *S. flexneri* VirA mutant.[47] An EPEC EspG1/G2 double mutant can be complemented by EspG1, EspG2, or VirA, indicating a high level of functional conservation.[18,50] Structurally EspG and VirA are similar. Both contain a centrally placed six-stranded sheet, a four-stranded sheet forming the amino terminal domain, and both also contain a helical domain.[51,52]

Relatively few studies have investigated the *in vivo* function of EspG, but in a rabbit model of infection, deletion of *espG* from rabbit EPEC greatly reduced colonization compared to wild type. A study performed in mice demonstrated similar findings.[47,53] While neither study examined TJ structure or function *in vivo*, reduced colonization suggests that EspG plays an important role in EPEC pathogenesis.

EspG also contributes to loss of TJ function during EPEC infection. Tomson *et al.* reported that infection of either T84 or Caco-2 IEC with ∆*espG1/G2* attenuated the drop in TER, compared to wild-type EPEC.[18] Five hours after infection with wild-type EPEC, TER dropped approximately 70%, while infection with ∆*espG1/G2* reduced TER by only 30%. Complementation of the mutant with *espG* restored the reduction in TER to levels comparable to wild type. A separate study reported the decrease in TER to be comparable between wild type and ∆*espG1/G2* although the different results are possibly attributable to the use of a nonintestinal cell line.[54] While ectopic expression of EspG1 in Madin-Darby canine kidney epithelial cells in the absence

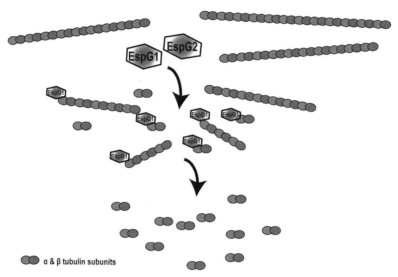

Figure 2. EspG disrupts microtubules via an unknown mechanism. EspG disrupts microtubules via an unknown mechanism. EPEC EspG has been demonstrated to bind directly to α-tubulin. The presence of EspG alone in the absence of other effectors induces depolymerization of microtubules into tubulin subunits.

of other EPEC effectors failed to decrease TER, paracellular permeability to 4 kDa dextran was increased significantly, indicating that EspG may have an impact on the mechanisms that regulate the selective permeability of TJs.[54] The phenotypes attributable to EspG are specific as this effector was shown to play no role in induction of the inflammatory cascade or pedestal formation.[18,55]

EspG disrupts microtubule networks

EspG was first identified as a homolog of the *Shigella flexneri* effector VirA.[47] VirA disrupts microtubule networks and extensive literature has been published regarding a similar function for EspG1/G2 (Fig. 2).[47,56] It was first established that IEC microtubule networks were absent under attached EPEC microcolonies and that this phenomenon was dependent on a functional TTSS.[48] Infection of IEC with various EPEC mutants identified EspG1/G2 as the relevant effector proteins. EspG1 was also spatially associated with microtubule loss as tagged bacterial protein localized directly under microcolonies in areas devoid of microtubules.[48]

A separate study examined the state of microtubules during EPEC infection of IEC revealing extensive loss of microtubule networks at five hours after wild-type infection but no change in cells infected with ΔespG1.[18] Eradication of infection with

gentamicin allowed for recovery of microtubules, which was dependent on new protein synthesis as treatment with cycloheximide blocked microtubule formation. This finding is consistent with the reported degradation of α-tubulin by EspG.[18]

Transient expression of EspG in mammalian cells, which allows for assessment of EspG effects in the absence of other bacterial effectors, confirmed that this protein alone is sufficient to depolymerize microtubules in the absence of other EPEC effectors.[18] Furthermore, gel overlay assays using purified His-tagged EHEC EspG demonstrated that it complexes directly with tubulin heterodimers in the absence of additional cofactors.[53] The same study used a bacterial two-hybrid assay to determine that EspG binds directly to the α-tubulin subunit. In another reductionist system, purified EspG depolymerized microtubules in solution, underscoring the effect of this protein on the cytoskeleton.[57]

While the role of microtubules in TJ maintenance has been only superficially investigated, two studies suggest that these structures are required for TJ homeostasis. Subramanian *et al.* transiently expressed GFP-occludin in epithelial cells and tracked its movement.[37] Individual vesicle tracks were aligned and the distance traveled was measured over time. Approximately 75 ±10% of vesicles traveled >6 μm/min. Microtubule disruption with nocodazole reduced the percentage of mobile vesicles to 13%. Although this study did not examine directionality, it suggests that occludin traffics on microtubules, supporting the speculation that EspG-mediated microtubule disruption impedes the movement of TJ proteins.

Calcium chelation studies investigated the requirement for microtubules in TJ disassembly.[58] Chelation of calcium from epithelial monolayers rapidly disrupts TJs, driving TJ proteins into the cytosol. Microtubule disruption by nocodazole prevented the mislocalization of TJ proteins following calcium chelation in two different intestinal epithelial cell lines; β-catenin, occludin, and ZO-1 all remained junction-associated even after 60 min of chelation. Cells harboring intact microtubule networks, however, exhibited an increase in cytosolic TJ proteins. While this study investigated only the movement of TJ proteins from the plasma membrane to the cytosol, it supports the theory that microtubules participate in the movement of TJ proteins. We speculate that EspG-mediated microtubule disruption may interfere with TJ protein movement, thus preventing TJ restoration and contributing to EPEC-induced barrier loss.

Additional data link EspG-induced microtubule disruption to remodeling of the actin cytoskeleton. A microtubule-bound RhoA-specific guanine nucleotide exchange factor, GEF-H1, was shown to be released and activated following infection with wild-type EPEC, but not after infection with an EspG1/G2 mutant.[57] This suggests that EspG1/G2-induced microtubule depolymerization activates GEF-H1 and induces the ensuing downstream effects on the actin cytoskeleton.[54,57] As predicted, RhoA was found to be active after wild-type EPEC infection, but not after infection with an EspG1/G2 mutant.[54] RhoA mediates its downstream effects via Rho-associated kinase (ROCK). ROCK phosphorylates multiple targets including the myosin phosphatase target subunit (MYPT1) of myosin light chain phosphatase.[59,62] Phosphorylation of MYPT1 inactivates the phosphatase resulting in increased phosphorylated myosin light chain (MLC). MLC activity is regulated by the opposing actions of MLCK and MYPT1. EPEC induces phosphorylation of MLC causing contraction of the perijunctional actomyosin ring and a decrease in TER, which is attenuated by inhibition of MLCK.[39,63-66] The mechanism by which MLCK is activated by EPEC is not fully defined however. We speculate that EspG-mediated

microtubule disruption may lead to GEF-H1 release, Rho/ROCK activation and inhibition of MYPT1. Inhibition of MYPT1 combined with activation of MLCK might contribute to greatly increased levels of phospho-MLC and ensuing TJ disruption.

We note here that one study used transient transfection of GFP-GEF-H1 to investigate its association with microtubules.[57] This raises the possibility that protein overexpression is responsible for the association with microtubules and that endogenous GEF-H1 is not as strongly associated with microtubules, as a separate study showed that native GEF-H1 is associated with TJs and not microtubules in polarized epithelia.[67] If indeed GEF-H1 is not associated with microtubules, EspG-mediated microtubule disruption may affect the TJ simply by preventing the trafficking of TJ proteins.[37] TJ homeostasis requires the correct targeting of recycled and newly synthesized proteins to these structures. Interruption of TJ protein delivery could have substantial effects on both TJ structure and function. Further work is needed to clarify the effect of EspG-mediated microtubule disruption on the TJ.

EspG acts as a regulator of GTPase signaling

Two recent investigations revealed previously unrecognized activities for EspG. Selyunin *et al.* demonstrated that EHEC EspG binds to both and p21-activated kinase (PAK) proteins.[52] Structural analysis of crystallized EspG in complex with ARF6 revealed that upon binding, ARF6 adopts a conformation nearly identical to that of the protein in its active GTP-bound state. In GST-pulldown assays, EspG selectively bound to GTP-loaded ARF1 and ARF6. In the GDP-bound (inactive) conformation, the EspG–binding region of ARF is inaccessible to EspG. The EspG–ARF interaction also blocked the access of GTPase activating proteins (GAPs), thus preventing hydrolysis of the ARF-GTP γ-phosphate and disrupting the normal guanine nucleotide cycle. These data suggest that EspG preferentially binds and sequesters ARF proteins in the active GTP-bound conformation. An EspG ARF-binding mutant had no effect on GAP activity, indicating the specificity of EspG's stearic hindrance.

ARF proteins are required for the recycling of vesicles to the plasma membrane and for organization of vesicle trafficking.[68-70] EspG was shown to localize to the Golgi apparatus and disrupt Golgi

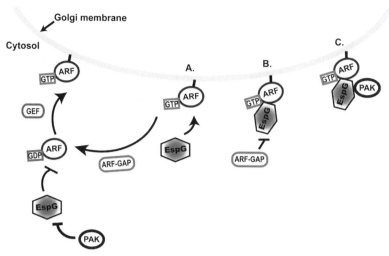

Figure 3. EspG regulates the GTPase cycle. (A) Active, GTP-bound ARF is membrane bound. GTPase activating proteins (GAPs) stimulate GTP hydrolysis and release from the membrane. Guanine nucleotide exchange factors (GEFs) catalyze the exchange of GDP for GTP and the reactivation of ARF. EspG is unable to bind to inactive GDP-bound ARF and preferentially binds GTP-bound ARF. PAK2 will not bind to EspG independently of ARF. (B) The formation of the ARF$_{GTP}$-EspG complex renders GAPs unable to stimulate GTP hydrolysis. (C) Once the ARF$_{GTP}$-EspG complex has formed, PAK is able to bind EspG.

stacks, suggesting that EspG interferes with vesicle trafficking. An EspG ARF-binding mutant did not induce Golgi disruption. Interestingly, Golgi dispersal is a well-established indicator of microtubule disruption, however in this study microinjection of 10 nM recombinant EspG into rat kidney cells did not disrupt microtubule networks.[71–75] The absence of this phenotype may be due to differences in protocols, including the use of different cell types (rat kidney versus IEC), microinjection versus infection or transfection and/or the time points examined. Previous ectopic expression studies were performed using either stable or transient transfection of *espG* and microtubules were examined at later time points. Microtubules were not examined in cells transfected with *espG* in this publication, only EspG microinjected cells.

Most interesting was the demonstration that, in contrast to the sequestration of active ARF proteins by EspG, this effector protein also binds p21-activated kinase 2 (PAK2) and promotes its activity. In the auto-inhibited homodimer conformation, the region of PAK that interacts with EspG has three functions—it blocks substrate binding, it blocks the catalytic site, and it stabilizes the homodimer structure. In the presence of EspG, PAK2 activity is increased 7.6-fold, suggesting the reversal of these inhibitions. PAK and ARF proteins bind EspG

on adjacent nonoverlapping surfaces. In addition, PAK binding is entirely dependent on the formation of the EspG-ARF complex as in the absence of ARF interaction, PAK2 fails to associate with EspG (Fig. 3). The PAK family of proteins transduce signals from the Rac1 and Cdc42 GTPases, regulating cytoskeletal dynamics, and cell motility.[76] Of note, both Rac1 and CdC42 have also been determined to play a role in TJ disruption.[77,78] By linking GTPase inhibition and Rac1/Cdc42 signal transduction, EspG may serve as a "catalytic scaffold" that permits EPEC to interfere with host cell processes in a specific location within the cell, namely at the Golgi membranes. A separate study also reported that EspG binds to the Rac/Cdc42-binding site of PAK1 leading to the conclusion that by imitating a small GTPase, EspG permits EPEC to bypass host cell GTPases and enable PAK-dependent actin remodeling regardless of the status of native Rac/Cdc42.[51] In summary, while the impact of EspG-PAK1 binding on TJs is currently not known, these studies identify cytoskeletal remodeling as a major function of EspG and suggest that this effector plays multiple roles in EPEC pathogenesis.

Another role for EspG may be linked to its localization to the Golgi apparatus.[52] Ectopic expression of EspG from EPEC, and EHEC, EspG2 from EPEC, and VirA from *S. flexneri*–induced Golgi disruption. Expression of these proteins significantly inhibited Golgi-based secretion as determined by the release of a reporter protein. Another EPEC effector, NleA, also inhibits the secretory pathway by binding to SEC24 and inhibiting COPII anterograde trafficking.[79] Comparison of the activity levels of NleA to EspG from EPEC and EHEC, EspG2 from EPEC, and VirA from *S. flexneri* revealed that all EspG and VirA inhibit protein secretion to a significantly greater extent than NleA. A yeast two-hybrid screen identified Golgi matrix protein GM130 as a binding partner for EPEC EspG. GM130 is a Golgi-associated protein that interacts with vesicle docking protein p115 as well as Rab1 and syntaxin 5.[80] The link between the EspG-GM130 interaction and EspG-mediated interruption of protein transport is currently unclear. Transient expression of EHEC EspG in this study was not observed to induce dramatic fragmentation of microtubules. Instead, "thickening" of tubular structures was observed, as well as membrane ruffling, indicating that EspG may have hitherto undefined effects on both microtubule and actin cytoskeletal components.

Disruption of Golgi stacks, via binding to GM130 or sequestration of ARF proteins, may be another mechanism by which EspG contributes to barrier dysfunction. As TJ proteins move through the Golgi, they are glycosylated, and directed to specific cell membrane domains.[81] Disruption of the Golgi via a GM130-dependent mechanism or by sequestration of active ARF may prevent newly synthesized TJ proteins from reaching the plasma membrane and contributing to TJ reconstruction. The EPEC effector NleA has already been shown to contribute to disruption of TJ function mediated by inhibition of Golgi trafficking.[17]

These data open new avenues of investigation into the many potential roles of EspG1/G2 in EPEC pathogenesis and provide new insight into how they may contribute to EPEC-induced TJ disruption, potentially by disrupting microtubules, by as-yet undetermined mechanisms related to vesicle trafficking or cytoskeletal remodeling or by blocking secretion of TJ proteins.

Conclusion

Loss of TJ structure and function during EPEC infection is due to multiple effectors, each of which may have myriad effects on the host cell. For example, EspF has been termed the bacterial "Swiss army knife" and has been shown to associate with mitochondria, promote apoptosis, regulate host membrane interactions, promote invasion of host IEC and inhibit macrophage uptake in addition to disrupting TJs.[82–87] While microtubule disruption was initially thought to be the sole function of EspG1/G2, new work uncovering inhibition of vesicle trafficking, promotion of cytoskeletal remodeling and inhibition of Golgi secretion as additional functions of these proteins paints a more complex picture. In the ongoing hunt for therapeutic targets, EPEC effectors with broad scopes of capabilities invite further investigation; disabling one potent effector with a large range of functions could lead to greatly reduced physiological disturbance. The expanding repertoire of activities of EspG1/G2 identified demonstrates that further investigation is needed to fully understand their role in EPEC-induced loss of epithelial barrier function.

Acknowledgments

This work was supported by Grants DK50694, DK58964, DK067887, and a VA Merit to GH and DK091151 to LG.

Conflicts of interest

The authors declare no conflicts of interest.

References

1. Fagundes-Neto, U., M.R. Kallas & F.R. Patricio. 1997. Morphometric study of the small bowel mucosa in infants with diarrhea due to enteropathogenic Escherichia coli strains. *Hepato-gastroenterology.* **44:** 1051–1056.
2. Hill, S.M., A.D. Phillips & J.A. Walker-Smith. 1991. Enteropathogenic Escherichia coli and life threatening chronic diarrhoea. *Gut* **32:** 154–158.
3. Fagundes-Neto, U. *et al.* 1996. Nutritional impact and ultrastructural intestinal alterations in severe infections due to enteropathogenic Escherichia coli strains in infants. *J. Am. Col. Nutri.* **15:** 180–185.
4. Jarvis, K.G. *et al.* 1995. Enteropathogenic Escherichia coli contains a putative type III secretion system necessary for the export of proteins involved in attaching and effacing lesion formation. *Proc. Natl. Acad. Sci. USA* **92:** 7996–8000.
5. Jarvis, K.G. & J.B. Kaper. 1996. Secretion of extracellular proteins by enterohemorrhagic Escherichia coli via a putative type III secretion system. *Infect. Immun.* **64:** 4826–4829.
6. Wolff, C. *et al.* 1998. Protein translocation into host epithelial cells by infecting enteropathogenic Escherichia coli. *Mol. Microbiol.* **28:** 143–155.
7. Karaolis, D.K. *et al.* 1997. Cloning of the RDEC-1 locus of enterocyte effacement (LEE) and functional analysis of the phenotype on HEp-2 cells. *Ad. Exp. Medi. Biol.* **412:** 241–245.

8. McDaniel, T.K. *et al.* 1995. A genetic locus of entero-cyte effacement conserved among diverse enterobacterial pathogens. *Proc. Natl. Acad. Sci. USA* **92:** 1664–1668.

9. Dean, P. & B. Kenny. 2009. The effector repertoire of en-teropathogenic E. coli: ganging up on the host cell. *Curr. Opin. Microbiol.* **12:** 101–109.

10. Dean, P., M. Maresca & B. Kenny. 2005. EPEC's weapons of mass subversion. *Curr. Opin. Microbiol.* **8:** 28–34.

11. Dean, P. *et al.* 2006. Potent diarrheagenic mechanism me-diated by the cooperative action of three enteropathogenic Escherichia coli-injected effector proteins. *Proc. Natl. Acad. Sci. USA* **103:** 1876–1881.

12. Guttman, J.A. *et al.* 2007. Aquaporins contribute to diar-rhoea caused by attaching and effacing bacterial pathogens. *Cell Microbiol.* **9:** 131–141.

13. Hecht, G. *et al.* 2004. Differential regulation of Na+/H+ exchange isoform activities by enteropathogenic E. coli in human intestinal epithelial cells. *Am. J. Physiol. Gastrointest. Liver Physiol.* **287:** G370–G378.

14. Hodges, K. *et al.* 2008. The enteropathogenic Escherichia coli effector protein EspF decreases sodium hydrogen exchanger 3 activity. *Cell Microbiol.* **10:** 1735–1745.

15. Gill, R.K. *et al.* 2007. Mechanism underlying inhibition of intestinal apical Cl–/OH– exchange following infection with enteropathogenic E. coli. *J. Clin. Invest.* **117:** 428–437.

16. Dean, P. & B. Kenny. 2004. Intestinal barrier dysfunction by enteropathogenic Escherichia coli is mediated by two effec-tor molecules and a bacterial surface protein. *Mol. Microbiol.* **54:** 665–675.

17. Thanabalasuriar, A. *et al.* 2010. The bacterial virulence factor NleA is required for the disruption of intestinal tight junc-tions by enteropathogenic Escherichia coli. *Cell Microbiol.* **12:** 31–41.

18. Tomson, F.L. *et al.* 2005. Enteropathogenic Escherichia coli EspG disrupts microtubules and in conjunction with Orf3 enhances perturbation of the tight junction barrier. *Mol. Microbiol.* **56:** 447–464.

19. Pearson, A.D. *et al.* 1982. Intestinal permeability in children with Crohn's disease and coeliac disease. *Br. Medi. J. Clin.Res.* **285:** 20–21.

20. Ukabam, S.O., J.R. Clamp & B.T. Cooper. 1983. Abnor-mal small intestinal permeability to sugars in patients with Crohn's disease of the terminal ileum and colon. *Digestion* **27:** 70–74.

21. Blair, S.A. *et al.* 2006. Epithelial myosin light chain kinase ex-pression and activity are upregulated in inflammatory bowel disease. *Lab. Invest.* **86:** 191–201.

22. Ye, D., I. Ma & T.Y. Ma. 2006. Molecular mechanism of tumor necrosis factor-alpha modulation of intestinal epithelial tight junction barrier. *Am. J. Physiol. Gastrointest. Liver Physiol.* **290:** G496–G504.

23. Marchiando, A.M. *et al.* 2010. Caveolin-1-dependent oc-cludin endocytosis is required for TNF-induced tight junc-tion regulation in vivo. *J. Cell Biol.* **189:** 111–126.

24. Fuss, I.J. *et al.* 2004. Nonclassical CD1d-restricted NK T cells that produce IL-13 characterize an atypical Th2 response in ulcerative colitis. *J. Clin. Invest.* **113:** 1490–1497.

25. Heller, F. *et al.* 2005. Interleukin-13 is the key effector Th2 cytokine in ulcerative colitis that affects epithelial tight junc-

tions, apoptosis, and cell restitution. *Gastroenterology* **129:** 550–564.

26. Rosen, M.J. *et al.* 2011. STAT6 activation in ulcerative colitis: a new target for prevention of IL-13-induced colon epithelial cell dysfunction. *Inflammat. Bowel Dis.* **17:** 2224–2234.

27. Cheng, C.Y. *et al.* 2011. Environmental toxicants and male reproductive function. *Spermatogenesis* **1:** 2–13.

28. Wang, X.W. *et al.* 2010. Mechanisms involved in the blood-testis barrier increased permeability induced by EMP. *Toxi-cology* **276:** 58–63.

29. Wong, E.W. *et al.* 2010. Regulation of blood-testis bar-rier dynamics by TGF-beta3 is a Cdc42-dependent protein trafficking event. *Proc. Natl. Acad. Sci. USA* **107:** 11399–11404.

30. Wong, E.W.P. *et al.* 2010. 11.09 – Cell Junctions in the Testis as Targets for Toxicants. *In* Comprehensive Toxicology (Second Edition). Editor-in-Chief: Charlene, A.M., Ed.: 167–188. Elsevier. Oxford.

31. Tai, H.Y. *et al.* 2006. Pen ch 13 allergen induces secretion of mediators and degradation of occludin protein of human lung epithelial cells. *Allergy* **61:** 382–388.

32. Vermeer, P.D. *et al.* 2009. MMP9 modulates tight junction integrity and cell viability in human airway epithelia. *Am. J. Physiol. Lung Cel. Mol. Physiol.* **296:** L751–762.

33. Vinhas, R. *et al.* 2011. Pollen proteases compromise the air-way epithelial barrier through degradation of transmem-brane adhesion proteins and lung bioactive peptides. *Allergy* **66:** 1088–1098.

34. Morimoto, S. *et al.* 2005. Rab13 mediates the continuous endocytic recycling of occludin to the cell surface. *J. Biol. Chem.* **280:** 2220–2228.

35. Shen, L., C.R. Weber & J.R. Turner. 2008. The tight junction protein complex undergoes rapid and continuous molecular remodeling at steady state. *J. Cell Biol.* **181:** 683–695.

36. Shen, L. & J.R. Turner. 2005. Actin depolymerization dis-rupts tight junctions via caveolae-mediated endocytosis. *Mol. Biol. Cell* **16:** 3919–3936.

37. Subramanian, V.S. *et al.* 2007. Tight junction targeting and intracellular trafficking of occludin in polarized epithelial cells. *Am. J. Physiol. Cell Physiol.* **293:** C1717–C1726.

38. Banan, A. *et al.* 2000. Oxidant-induced intestinal barrier disruption and its prevention by growth factors in a human colonic cell line: role of the microtubule cytoskeleton. *Free Radi. Biol. Medi.* **28:** 727–738.

39. Simonovic, I. *et al.* 2000. Enteropathogenic Escherichia coli dephosphorylates and dissociates occludin from intestinal epithelial tight junctions. *Cell Microbiol.* **2:** 305–315.

40. Muza-Moons, M.M., E.E. Schneeberger & G.A. Hecht. 2004. Enteropathogenic Escherichia coli infection leads to appear-ance of aberrant tight junctions strands in the lateral mem-brane of intestinal epithelial cells. *Cell Microbiol.* **6:** 783–793.

41. Shifflett, D.E. *et al.* 2005. Enteropathogenic E. coli disrupts tight junction barrier function and structure in vivo. *Lab. Invest.* **85:** 1308–1324.

42. Zhang, Q. *et al.* 2010. Redistribution of Tight Junction Pro-teins During EPEC Infection In Vivo. *Inflammation*

43. Bonazzi, M. *et al.* 2011. Clathrin phosphorylation is required for actin recruitment at sites of bacterial adhesion and inter-nalization. *J. Cell Biol.* **195:** 525–536.

44. de Grado, M. *et al.* 2001. Enteropathogenic Escherichia coli infection induces expression of the early growth response factor by activating mitogen-activated protein kinase cascades in epithelial cells. *Infect. Immun.* **69:** 6217–6224.

45. Hardwidge, P.R. *et al.* 2004. Proteomic analysis of the intestinal epithelial cell response to enteropathogenic Escherichia coli. *J. Biol. Chem.* **279:** 20127–20136.

46. Bhatt, S., T. Romeo & D. Kalman. 2011. Honing the message: post-transcriptional and post-translational control in attaching and effacing pathogens. *Tre. Microbiol.* **19:** 217–224.

47. Elliott, S.J. *et al.* 2001. EspG, a novel type III system-secreted protein from enteropathogenic Escherichia coli with similarities to VirA of Shigella flexneri. *Infect. Immun.* **69:** 4027–4033.

48. Shaw, R.K. *et al.* 2005. Enteropathogenic Escherichia coli type III effectors EspG and EspG2 disrupt the microtubule network of intestinal epithelial cells. *Infect. Immun.* **73:** 4385–4390.

49. Mellies, J.L. *et al.* 2001. espC pathogenicity island of enteropathogenic Escherichia coli encodes an enterotoxin. *Infect. Immun.* **69:** 315–324.

50. Smollett, K. *et al.* 2006. Function and distribution of EspG2, a type III secretion system effector of enteropathogenic Escherichia coli. *Microbes Infect.* **8:** 2220–2227.

51. Germane, K.L. & B.W. Spiller. 2011. Structural and functional studies indicate that the EPEC effector, EspG, directly binds p21-activated kinase. *Biochemistry* **50:** 917–919.

52. Selyunin, A.S. *et al.* 2011. The assembly of a GTPase-kinase signalling complex by a bacterial catalytic scaffold. *Nature* **469:** 107–111.

53. Hardwidge, P.R. *et al.* 2005. Modulation of host cytoskeleton function by the enteropathogenic Escherichia coli and Citrobacter rodentium effector protein EspG. *Infect. Immun.* **73:** 2586–2594.

54. Matsuzawa, T., A. Kuwae & A. Abe. 2005. Enteropathogenic Escherichia coli type III effectors EspG and EspG2 alter epithelial paracellular permeability. *Infect. Immun.* **73:** 6283–6289.

55. Savkovic, S.D. *et al.* 2001. EPEC-activated ERK1/2 participate in inflammatory response but not tight junction barrier disruption. *Am. J. Physiol. Gastrointest. Liver Physiol.* **281:** G890–G898.

56. Yoshida, S. *et al.* 2006. Microtubule-severing activity of Shigella is pivotal for intercellular spreading. *Science* **314:** 985–989.

57. Matsuzawa, T. *et al.* 2004. Enteropathogenic Escherichia coli activates the RhoA signaling pathway via the stimulation of GEF-H1. *EMBO J.* **23:** 3570–3582.

58. Ivanov, A.I. *et al.* 2006. Microtubules regulate disassembly of epithelial apical junctions. *BMC Cell Biol.* **7:** 12.

59. Amano, M. *et al.* 1997. Formation of actin stress fibers and focal adhesions enhanced by Rho-kinase. *Science* **275:** 1308–1311.

60. Chang, Y.C. *et al.* 2008. GEF-H1 couples nocodazole-induced microtubule disassembly to cell contractility via RhoA. *Mol. Biol. Cell* **19:** 2147–2153.

61. Enomoto, T. 1996. Microtubule disruption induces the formation of actin stress fibers and focal adhesions in cultured cells: possible involvement of the rho signal cascade. *Cell Stru. Fun.* **21:** 317–326.

62. Kimura, K. *et al.* 1996. Regulation of myosin phosphatase by Rho and Rho-associated kinase (Rho-kinase). *Science* **273:** 245–248.

63. Manjarrez-Hernandez, H.A. *et al.* 1996. Phosphorylation of myosin light chain at distinct sites and its association with the cytoskeleton during enteropathogenic Escherichia coli infection. *Infect. Immun.* **64:** 2368–2370.

64. Yuhan, R. *et al.* 1997. Enteropathogenic Escherichia coli-induced myosin light chain phosphorylation alters intestinal epithelial permeability. *Gastroenterology* **113:** 1873–1882.

65. Zolotarevsky, Y. *et al.* 2002. A membrane-permeant peptide that inhibits MLC kinase restores barrier function in vitro models of intestinal disease. *Gastroenterology* **123:** 163–172.

66. Shen, L. *et al.* 2006. Myosin light chain phosphorylation regulates barrier function by remodeling tight junction structure. *J. Cell Sci.* **119:** 2095–2106.

67. Guillemot, L. *et al.* 2008. Paracingulin regulates the activity of Rac1 and RhoA GTPases by recruiting Tiam1 and GEF-H1 to epithelial junctions. *Mol. Biol. Cell* **19:** 4442–4453.

68. D'Souza-Schorey, C. *et al.* 1998. ARF6 targets recycling vesicles to the plasma membrane: insights from an ultrastructural investigation. *J. Cell Biol.* **140:** 603–616.

69. Franco, M. *et al.* 1999. EFA6, a sec7 domain-containing exchange factor for ARF6, coordinates membrane recycling and actin cytoskeleton organization. *EMBO J.* **18:** 1480–1491.

70. Kahn, R.A. 2009. Toward a model for Arf GTPases as regulators of traffic at the Golgi. *FEBS letters* **583:** 3872–3879.

71. Cole, N.B. *et al.* 1996. Golgi dispersal during microtubule disruption: regeneration of Golgi stacks at peripheral endoplasmic reticulum exit sites. *Mol. Biol. Cell* **7:** 631–650.

72. Ho, W.C. *et al.* 1989. Reclustering of scattered Golgi elements occurs along microtubules. *Eur. J. Cell Biol.* **48:** 250–263.

73. Rogalski, A.A. & S.J. Singer. 1984. Associations of elements of the Golgi apparatus with microtubules. *J. Cell Biol.* **99:** 1092–1100.

74. Thyberg, J. & S. Moskalewski. 1999. Role of microtubules in the organization of the Golgi complex. *Exp. Cell Res.* **246:** 263–279.

75. Wehland, J. *et al.* 1983. Role of microtubules in the distribution of the Golgi apparatus: effect of taxol and microinjected anti-alpha-tubulin antibodies. *Proc. Natl. Acad. Sci. USA* **80:** 4286–4290.

76. Bokoch, G.M. 2003. Biology of the p21-activated kinases. *Ann. Rev. Biochem.* **72:** 743–781.

77. Bruewer, M. *et al.* 2004. RhoA, Rac1, and Cdc42 exert distinct effects on epithelial barrier via selective structural and biochemical modulation of junctional proteins and F-actin. *Am. J. Physiol. Cell Physiol.* **287:** C327–335.

78. Jou, T.S., E.E. Schneeberger & W.J. Nelson. 1998. Structural and functional regulation of tight junctions by RhoA and Rac1 small GTPases. *J. Cell Biol.* **142:** 101–115.

79. Kim, J. *et al.* 2007. The bacterial virulence factor NleA inhibits cellular protein secretion by disrupting mammalian COPII function. *Cell Host. Microbe.* **2:** 160–171.

80. Clements, A. *et al.* 2011. EspG of enteropathogenic and enterohemorrhagic E. coli binds the Golgi matrix protein

GM130 and disrupts the Golgi structure and function. *Cell Microbiol.* **13:** 1429–1439.

81. Gut, A. *et al.* 1998. Carbohydrate-mediated Golgi to cell surface transport and apical targeting of membrane proteins. *EMBO J.* **17:** 1919–1929.

82. Holmes, A. *et al.* The EspF effector, a bacterial pathogen's Swiss army knife. *Infect. Immun.* **78:** 4445–4453.

83. Alto, N.M. *et al.* 2007. The type III effector EspF coordinates membrane trafficking by the spatiotemporal activation of two eukaryotic signaling pathways. *J. Cell Biol.* **178:** 1265–1278.

84. McNamara, B.P. *et al.* 2001. Translocated EspF protein from enteropathogenic Escherichia coli disrupts host intestinal barrier function. *J. Clin. Invest.* **107:** 621–629.

85. Nagai, T., A. Abe & C. Sasakawa. 2005. Targeting of enteropathogenic Escherichia coli EspF to host mitochondria is essential for bacterial pathogenesis: critical role of the 16th leucine residue in EspF. *J. Biol. Chem.* **280:** 2998–3011.

86. Nougayrede, J.P. & M.S. Donnenberg. 2004. Enteropathogenic Escherichia coli EspF is targeted to mitochondria and is required to initiate the mitochondrial death pathway. *Cell Microbiol.* **6:** 1097–1111.

87. Tahoun, A. *et al.* 2011. Comparative analysis of EspF variants in inhibition of Escherichia coli phagocytosis by macrophages and inhibition of E. coli translocation through human- and bovine-derived M cells. *Infect. Immun.* **79:** 4716–4729.

Ann. N.Y. Acad. Sci. ISSN 0077-8923

ANNALS OF THE NEW YORK ACADEMY OF SCIENCES

Issue: *Barriers and Channels Formed by Tight Junction Proteins*

Abnormal intestinal permeability in Crohn's disease pathogenesis

Christopher W. Teshima,[1] Levinus A. Dieleman,[1] and Jon B. Meddings[2]

[1]Division of Gastroenterology, University of Alberta, Edmonton, Alberta, Canada. [2]Department of Medicine, University of Calgary, Calgary, Alberta, Canada

Address for correspondence: Dr. Jon Meddings, 7th Floor, TRW Building, 3280 Hospital Drive NW, Calgary, AB Canada, T2N 4Z6. meddings@ucalgary.ca

Increased small intestinal permeability is a longstanding observation in both Crohn's disease patients and in their healthy, asymptomatic first-degree relatives. However, the significance of this compromised gut barrier function and its place in the pathogenesis of the disease remains poorly understood. The association between abnormal small intestinal permeability and a specific mutation in the *NOD2* gene, which functions to modulate both innate and adaptive immune responses to intestinal bacteria, suggests a common, genetically determined pathway by which an abnormal gut barrier could result in chronic intestinal inflammation. Furthermore, rodent colitis models show that gut barrier defects precede the development of inflammatory changes. However, it remains possible that abnormal permeability is simply a consequence of mucosal inflammation. Further insight into whether abnormal barrier function is the cause or consequence of chronic intestinal inflammation will be crucial to understanding the role of intestinal permeability in the pathogenesis of Crohn's disease.

Keywords: Crohn's disease; experimental colitis; intestinal permeability; microbiome; pathogenesis

Introduction

The barrier function of the intestine is essential for the normal homeostasis of the gut and mucosal immune system. Abnormalities in this barrier function, as reflected by increased intestinal permeability, have long been observed in patients with Crohn's disease (CD) and in a proportion of their healthy, asymptomatic first-degree relatives. While the etiology of CD remains obscure, it is believed that the interplay among the patient's genetic background, host immune system, intestinal microflora, and other environmental factors determine an abnormal immune response that leads to disease. Whether increased intestinal permeability is a consequence of this inflammatory response or a pathophysiologic determinant of disease has not yet been determined. Since it is not usually possible to examine the intestinal permeability of CD patients before the onset of their disease, attention has instead been placed on their first-degree relatives who are the cohort at highest risk for developing inflam-

matory bowel disease (IBD). Determining whether abnormal intestinal permeability is the result of an ongoing inflammatory response or whether the impaired barrier function of the gut contributes to the development of the disease will be a necessary piece of the puzzle in our understanding of CD pathogenesis.

Gut barrier

The gastrointestinal tract is a complex interface between the external environment and the immune system, establishing a dynamic barrier that enables the absorption of dietary nutrients and the exclusion of potentially harmful compounds from the intestinal lumen, while permitting the sampling of luminal antigens as part of immune surveillance. The ability to selectively restrict the movement of these compounds across the epithelial cell layer is referred to as the barrier function of the gut. Passage of molecules across the gut lining may follow transcellular or paracellular routes, the former relying on endocytosis and transmembrane

doi: 10.1111/j.1749-6632.2012.06612.x

channels or transporters and the latter on tight junctions that bridge the apicolateral border of the epithelial cells.[1] These tight junctions are an intricate complex of numerous proteins spanning the plasma membrane and linking to the internal cytoskeleton that regulate the paracellular passage of molecules in a dynamic fashion that is responsive to numerous stimuli.[2] This gut barrier function is reflected by the measurement of intestinal permeability.

Measurement of intestinal permeability

The permeability of the paracellular pathway can be safely and easily measured *in vivo* by using small, nontoxic, noncharged, water-soluble compounds that are orally administered, neither metabolized nor stored in the body after being absorbed, constitutively cleared by the kidney, and easily quantifiable in urine.[3] Typically, this has been achieved using Cr-EDTA or with small saccharide probes. Whereas Cr-EDTA is stable throughout the entire gastrointestinal tract, the saccharide probes are differentially metabolized in varying locations of the gut, which allows for regional characterization of gut permeability.[4] For instance, sucrose can be used to assess gastroduodenal permeability because it is quickly metabolized by sucrase-isomaltase after entering the duodenum.[5] Therefore, the urinary excretion of intact sucrose only reflects its absorption proximal to this region. Similarly, the probes used to assess the small intestine, such as lactulose, rhamnose, cellobiose, and mannitol, are all degraded by colonic bacteria. Since the surface area of the small intestine is exponentially greater than that of the stomach, any amount of gastric absorption can be ignored, and therefore the fractional urinary excretion of these sugars will only reflect small intestinal permeability.[3] Throughout the small intestine, the tips of villi contain numerous tiny channels through which mannitol can pass freely, whereas disaccharides such as lactulose are excluded.[6] However, lactulose can move across intermediate-sized channels at the base of the villi. These channels are in fact paracellular pathways formed by tight junctions that regulate the movement of particles on the basis of molecular size and charge. Since the villi essentially represent the surface area of the entire small intestine, the measurement of the urinary excretion of lactulose relative to mannitol, expressed as a ratio, reflects the permeability of the

small intestine relative to the surface area available.[2] Thus, small intestinal permeability as measured by the lactulose-to-mannitol ratio is not confounded by the loss of intestinal mucosa or alterations in transcellular transport, but is only determined by the leakiness of the paracellular tight junction channels.

Increased intestinal permeability in CD

Abnormal intestinal barrier function has long been a recognized feature of inflammatory bowel disease. Increased permeability has been clearly demonstrated in a significant proportion of patients with CD.[7–11] Patients with active CD have increased permeability that decreases with the induction of remission,[12] while anti-TNF-α therapy with infliximab has been shown to normalize intestinal permeability.[13] Furthermore, increased intestinal permeability in patients in clinical remission predicts a high risk of early relapse.[14–18]

The molecular mechanisms responsible for this increase in intestinal permeability were examined by a study that took biopsy samples from the sigmoid colon of CD patients and studied them using Ussing chambers.[19] The barrier function of the sample tissues was determined by using impedance testing and conductance scanning, while the tight junction structure was analyzed by freeze-fracture electron microscopy and confocal laser scanning with immunohistochemistry. Samples from patients with active CD showed downregulation of the "sealing" tight junction proteins occludin, claudin-5, and claudin-8, with redistribution of claudin-5 and claudin-8 off the tight junctions, and showed upregulation of the pore-forming tight junction protein claudin-2. Furthermore, the upregulation of claudin-2 was inducible by TNF-α. The freeze fracture analysis revealed reduced numbers and discontinuous tight junction strands, a finding thought to correlate with easier passage and uptake of food antigens and bacterial lipopolysaccharides.[20] In addition, there was also significantly increased epithelial apoptosis in active CD.[19] Finally, there was a spatially uniform increase in transepithelial conductivity despite the presence of only focal epithelial defects (e.g., mucosal erosions) in the biopsy samples. This suggests that intestinal permeability was more broadly affected than simply the areas of epithelial injury.

Increased intestinal permeability in relatives of CD patients

Increased intestinal permeability has been proposed as a possible pathogenic mechanism for the development of CD since abnormally elevated permeability was recognized in both patients and in a subset of their healthy relatives 25 years ago.[21] Subsequent studies confirmed the presence of increased intestinal permeability in 10–20% of healthy, asymptomatic first-degree relatives of CD patients.[22–25] This was an interesting finding given that first-degree relatives have the greatest risk for the development of the disease,[26,27] with a 17–35 times greater likelihood than the general population.[28] Furthermore, the proportion of relatives with abnormal permeability was increased significantly in families with multiple cases of CD compared to relatives of sporadic CD cases.[29] Together, these observations supported the hypothesis that a genetic defect compromising the integrity of the intestinal barrier serves as an early insult preceding the development of intestinal inflammation and the disease phenotype.

However, determining whether abnormal permeability is an important factor in the pathogenesis of CD or whether this is simply the result of inflammation and gut injury has proven a challenging question. Certainly, the case report of a young, symptom-free girl who was found to have increased permeability while having a negative workup that included X-rays, upper endoscopy and colonoscopy, and whose brother had CD, and who then went on to develop clinical CD 7 years later, is suggestive of a genetic link between abnormal permeability and the subsequent development of the disease.[30]

The possibility of a genetic basis for abnormal intestinal permeability among healthy relatives of CD patients was further supported by the finding of an association between mutations in the *NOD2* gene, already known to confer significant risk for the development of ileal disease especially,[31] and abnormal intestinal permeability among both CD patients and their first-degree relatives.[29,32] In particular, the frameshift mutation 3020insC in the *NOD2* gene was significantly more prevalent among both CD patients and their first-degree relatives compared to controls (22%, 23%, and 2%, respectively).[32] Moreover, increased intestinal permeability was seen in 40% of *NOD2* 3020insC carriers but in only 15% of those with the wild-type gene, and the median intestinal permeability measurement was significantly higher in first-degree relatives with the 3020insC *NOD2* mutation. These findings further suggest a link between the genetic susceptibility for CD and a genetic determination of abnormal intestinal permeability in CD patients and their first-degree relatives.

Biologic plausibility for a genetically determined increase in intestinal permeability in the pathogenesis of CD

There is evidence from animal models of IBD that abnormal permeability may precede the development of the disease. In the IL-10 knockout mouse, abnormal intestinal permeability can be detected well before the development of a Crohn's-like colitis.[33] Not only do these animals demonstrate increased intestinal permeability long before they develop colitis, but if the increase in small intestinal permeability is prevented pharmacologically, the colitis is significantly attenuated.[34] These data suggest that increased small intestinal permeability is causally related to subsequent inflammation and that it is not simply an epiphenomenon. Furthermore, in the SAMP1/YitFc mouse model of CD, small intestinal permeability becomes abnormal weeks to months before evidence of disease onset, appearing immediately after weaning.[35] While implicating increased small intestinal permeability in the pathogenesis of the colitis observed in these animals may not be immediately intuitive, there are many possible mechanisms by which permeability alterations in the small intestine could affect inflammation at distant sites in the colon. The small intestine is the largest site of interaction between the innate and adaptive immune systems and the external environment, and alterations in intestinal permeability may lead to sustained inflammatory responses that disrupt homeostatic mechanisms, some of which may result in increases in circulating proinflammatory cytokines or modification of regulatory T cell populations that may have effects elsewhere in the mucosal immune system, including the colon.[36,37] Furthermore, increased intestinal permeability results in significant increases in secretory IgA secretion into the small intestine, which can affect the bacterial load within the intestinal microflora.[38] While speculative, it is possible that this change in secretory IgA may also alter the microbial flora in the colon. Thus, it is entirely consistent with these animal models that the gut permeability defects are not simply the result of intestinal inflammation that is the disease phenotype, but rather may be a determining factor.

The association between abnormal permeability and the 3020insC frameshift mutation in the *NOD2* gene is particularly interesting given the status of *NOD2* as one of the most well-characterized and significant susceptibility loci conferring risk for the development of CD.[31] The *NOD2* protein recognizes the peptidoglycan muramyl dipeptide from the bacterial cell wall and modulates both the innate and the adaptive immune responses.[39] Human intestinal cells transfected with the 3020insC *NOD2* mutation overexpressed the *NOD2* protein but demonstrated reduced antibacterial defenses[40] due to decreased expression of α-defensins from Paneth cells.[41,42] In addition, muramyl dipeptide stimulation of *NOD2* induces autophagy, which is important for limiting bacterial replication and for antigen presentation, and together with costimulation of TLR2 receptors on dendritic cells promotes Th17 cell differentiation.[43,44] Chronic activation of NOD2 by muramyl dipeptide is also important for the induction of both self-tolerance and of cross-tolerance to IL-1β and to TLR2 and TLR4 ligands.[31] This muramyl dipeptide-induced immune tolerance and the costimulation of Th17 cell differentiation is lost or significantly reduced in *NOD2*-deficient mice and in humans with the 3020insC *NOD2* mutation. Recently, it has been shown that *NOD2* is important for the development and regulation of the intestinal microflora, since *NOD2*-deficient mice have a significantly increased bacterial load in the terminal ileum and in their stool, as well as have an altered microbial composition to their gut flora.[45] Furthermore, *NOD2* genotype also influences the composition of the intestinal microflora in humans.[45] These findings imply that mutations in *NOD2* may contribute to the development of CD by altering the normal innate and adaptive immune responses to the intestinal microflora or by modifying the composition of the flora itself. How exactly a *NOD2* mutation would result in increased small intestinal permeability in humans is not entirely clear, unless the permeability defect is secondary to inflammatory and other immune-mediated mechanisms triggered by increased anti-

genic exposure due to compromised antibacterial defenses.

However, this is obviously only part of the story, since not all individuals with a *NOD2* mutation develop CD, and more than half of relatives with the 3020insC *NOD2* mutation continue to have normal intestinal permeability.[32] In fact, as genome-wide association studies have begun to shed light on the numerous genetic risk loci involved in IBD, we have come to realize the enormous complexity of the role of genetics and its contribution to the pathogenesis of the disease. Almost 100 genetic risk loci have now been identified, but over 50% of these are also associated with other autoimmune or inflammatory conditions.[31] The fact that the monozygotic twin concordance is less than 50% in CD[46] speaks to the important role of nongenetic, and likely environmental, factors in disease pathogenesis. Therefore, it is likely that genetic determination does not entirely account for the changes in intestinal permeability seen in CD patients and in a subset of their first-degree relatives.

Role of inflammation in determining abnormal permeability

The finding of abnormal intestinal permeability in both CD patients and in their healthy, asymptomatic first-degree relatives, and the association between increased permeability in both patients and in family members with mutations in the *NOD2* gene that has a role in epithelial bacterial defense and immune tolerance, does not reveal whether intestinal inflammation is an intervening step leading to the development of abnormal permeability, or whether a compromised gut barrier with increased permeability drives a subsequent inflammatory response. However, the impact of inflammation on intestinal permeability has been directly examined by *in vitro* studies using intestinal epithelial cell monolayers that showed increased tight junction permeability in response to treatment with the proinflammatory cytokines TNF-α and IFN-γ and decreased tight junction permeability with the anti-inflammatory cytokine TGF-β.[47–49] These changes occurred rapidly and were mediated through modulation of cytoskeleton pathways, predominantly via increased expression and activity of myosin light-chain kinase.[1] Furthermore, patients with active CD treated with anti-TNF-α antibody therapy showed improvements in gut barrier function, although this seemed to relate to a reduction in epithelial apoptosis.[50] Taken together, these findings implicate a role for inflammation in the alteration of intestinal permeability, although this is likely only partly true. Previously, we have shown that CD relatives demonstrate an exaggerated response in the development of abnormal intestinal permeability caused by injury from nonsteroidal anti-inflammatory drugs (NSAIDs).[51] However, subsequent work has shown that while anti-TNF-α antibody treatment can restore normal intestinal permeability in patients with CD, the hyperresponsiveness to NSAID exposure remains unchanged.[52] This implies that the gut barrier function in CD is likely altered in ways more than just simply by inflammation.

Unanswered questions about abnormal intestinal permeability

While much insight has been achieved, the pathogenesis of CD remains unclear, although much evidence suggests that it develops from an abnormal immune response to the intestinal microflora in individuals with a susceptible genetic predisposition. Abnormal intestinal permeability is now a well-recognized finding among a subset of healthy, asymptomatic first-degree relatives of CD patients, individuals who have the greatest risk for developing the disease. However, how this abnormal gut permeability fits into our understanding of CD pathogenesis continues to be uncertain. It is possible that increased intestinal permeability may be an early step, leading to excessive antigen exposure to the mucosal immune system, which results in an exaggerated immune response in a genetically predisposed individual. Conversely, compromised mucosal immune defenses caused by genetic defects could lead to abnormal bacterial exposure to the mucosal immune system, resulting in chronic inflammation that disrupts the integrity of the gut barrier, and leading to increased intestinal permeability.

The difficulty with our current understanding of gut barrier function in CD relatives is that while increased intestinal permeability is a marker of abnormal tight junction function, it may also be a marker of mild intestinal disease. Thus, observing abnormal small intestinal permeability in a population at high risk of developing Crohn's can be explained by either hypothesis. Answering this question remains an important component in the understanding of CD

pathogenesis and may prove helpful in directing future efforts. If abnormal permeability is recognized to be a consequence of inflammatory changes early in the disease, it may provide a useful tool for the early identification and possible treatment of at-risk first-degree relatives of CD patients. If, however, abnormal permeability is found to be a risk factor for the subsequent development of inflammation, it may create the opportunity to attempt innovative therapies such as regulating the functional state of epithelial tight junctions and/or altering the intestinal microflora that affect it.

Efforts are currently under way to address this question in a large, prospective study of asymptomatic first-degree relatives of CD patients. Intestinal permeability measurements are being performed using the lactulose:mannitol test followed by endoscopic assessment of the small intestine using video capsule endoscopy to examine for ulcers, erosions, or other inflammatory changes. In addition, blood and stool samples are being collected to perform genetic tests and to measure other inflammatory markers, respectively. This study will likely determine if the abnormal small intestinal permeability found in healthy, first-degree relatives is associated with macroscopic mucosal abnormalities suggestive of subclinical CD or whether increased intestinal permeability exists despite an endoscopically normal small bowel.

Conflicts of interest

The authors declare no conflicts of interest.

References

1. Salim, S.Y. & J.D. Söderholm. 2011. Importance of disrupted intestinal barrier in inflammatory bowel diseases. *Inflamm. Bowel Dis.* **1:** 362–381.
2. Arrieta, M.C., L. Bistritz & J.B. Meddings. 2006. Alterations in intestinal permeability. *Gut* **5:** 1512–1520.
3. Teshima, C.W. & J.B. Meddings. 2008. The measurement and clinical significance of intestinal permeability. *Curr. Gastroenterol. Rep.* **1:** 443–449.
4. Meddings, J.B. & I. Gibbons. 1998. Discrimination of site-specific alterations in gastrointestinal permeability in the rat. *Gastroenterology* **11:** 83–92.
5. Meddings, J.B., L.R. Sutherland, N.I. Byles, *et al.* 1993. Sucrose: a novel permeability marker for gastroduodenal disease. *Gastroenterology* **10:** 1619–1626.
6. Fihn, B.M., A. Sjöqvist & M. Jodal. 2000. Permeability of the rat small intestinal epithelium along the villus-crypt axis: effects of glucose transport. *Gastroenterology* **11:** 1029–1036.

7. Jenkins, R.T., D.B. Jones, R.L. Goodacre, *et al.* 1987. Reversibility of increased intestinal permeability to 51Cr-EDTA in patients with gastrointestinal inflammatory diseases. *Am. J. Gastroenterol.* **8:** 1159–1164.

8. Jenkins, R.T., J.K. Ramage, D.B. Jones, *et al.* 1988. Small bowel and colonic permeability to 51Cr-EDTA in patients with active inflammatory bowel disease. *Clin. Invest. Med.* **1:** 151–155.

9. Pironi, L., M. Miglioli, E. Ruggeri, *et al.* 1990. Relationship between intestinal permeability to [51Cr]EDTA and inflammatory activity in asymptomatic patients with Crohn's disease. *Dig. Dis. Sci.* **3:** 582–588.

10. Adenis, A., J.F. Colombel, P. Lecouffe, *et al.* 1992. Increased pulmonary and intestinal permeability in Crohn's disease. *Gut* **3:** 678–682.

11. Wyatt, J., G. Oberhuber, S. Pongratz, *et al.* 1997. Increased gastric and intestinal permeability in patients with Crohn's disease. *Am. J. Gastroenterol.* **9:** 1891–1896.

12. Sanderson, I.R., P. Boulton, I. Menzies, *et al.* 1987. Improvement of abnormal lactulose/rhamnose permeability in active Crohn's disease of the small bowel by an elemental diet. *Gut* **2:** 1073–1076.

13. Suenaert, P., V. Bulteel, L. Lemmens, *et al.* 2002. Anti-tumor necrosis factor treatment restores the gut barrier in Crohn's disease. *Am. J. Gastroenterol.* **9:** 2000–2004.

14. Wyatt, J., H. Vogelsang, W. Hübl, *et al.* 1993. Intestinal permeability and the prediction of relapse in Crohn's disease. *Lancet* **34:** 1437–1439.

15. Hilsden, R.J., J.B. Meddings, J. Hardin, *et al.* 1999. Intestinal permeability and postheparin plasma diamine oxidase activity in the prediction of Crohn's disease relapse. *Inflamm. Bowel Dis.* **5:** 85–91.

16. D'Incà, R., V. Di Leo, G. Corrao, *et al.* 1999. Intestinal permeability test as a predictor of clinical course in Crohn's disease. *Am. J. Gastroenterol.* **9:** 2956–2960.

17. Arnott, I.D., K. Kingstone & S. Ghosh. 2000. Abnormal intestinal permeability predicts relapse in inactive Crohn disease. *Scand. J. Gastroenterol.* **3:** 1163–1169.

18. Tibble, J.A., G. Sigthorsson, S. Bridger, *et al.* 2000. Surrogate markers of intestinal inflammation are predictive of relapse in patients with inflammatory bowel disease. *Gastroenterology* **11:** 15–22.

19. Zeissig, S., N. Bürgel, D. Günzel, *et al.* 2007. Changes in expression and distribution of claudin 2, 5 and 8 lead to discontinuous tight junctions and barrier dysfunction in active Crohn's disease. *Gut* **5:** 61–72.

20. Schulzke, J.D., S. Ploeger, M. Amasheh, *et al.* 2009. Epithelial tight junctions in intestinal inflammation. *Ann. N. Y. Acad. Sci.* **116:** 294–300.

21. Hollander, D., C.M. Vadheim, E. Brettholz, *et al.* 1986. Increased intestinal permeability in patients with Crohn's disease and their relatives. A possible etiologic factor. *Ann. Intern. Med.* **10:** 883–885.

22. May, G.R., L.R. Sutherland & J.B. Meddings. 1993. Is small intestinal permeability really increased in relatives of patients with Crohn's disease? *Gastroenterology* **10:** 1627–1632.

23. Peeters, M., B Geypens, D. Claus, *et al.* 1997. Clustering of increased small intestinal permeability in families with Crohn's disease. *Gastroenterology* **11:** 802–807.

24. Secondulfo, M., L. de Magistris, R. Fiandra, *et al.* 2001. Intestinal permeability in Crohn's disease patients and their first degree relatives. *Dig. Liver Dis.* **3:** 680–685.

25. Fries, W., M.C. Renda, M.A. Lo Presti, *et al.* 2005. Intestinal permeability and genetic determinants in patients, first-degree relatives, and controls in a high-incidence area of Crohn's disease in Southern Italy. *Am. J. Gastroenterol.* **10:** 2730–2736.

26. Orholm, M., P. Munkholm, E. Langholz, *et al.* 1991. Familial occurrence of inflammatory bowel disease. *N. Engl. J. Med.* **32:** 84–88.

27. Monsén, U., O. Bernell, C. Johansson, *et al.* 1991. Prevalence of inflammatory bowel disease among relatives of patients with Crohn's disease. *Scand. J. Gastroenterol.* **2:** 302–306.

28. Fielding, J.F. 1986. The relative risk of inflammatory bowel disease among parents and siblings of Crohn's disease patients. *J. Clin. Gastroenterol.* **8:** 655–657.

29. D'Incà, R., V. Annese, V. Di Leo, *et al.* 2006. Increased intestinal permeability and NOD2 variants in familial and sporadic Crohn's disease. *Aliment. Pharmacol. Ther.* **2:** 1455–1461.

30. Irvine, E.J. & J.K. Marshall. 2000. Increased intestinal permeability precedes the onset of Crohn's disease in a subject with familial risk. *Gastroenterology* **11:** 1740–1744.

31. Khor, B., A. Gardet & R.J. Xavier. 2011. Genetics and pathogenesis of inflammatory bowel disease. *Nature* **47:** 307–317.

32. Buhner, S., C. Buning, J. Genschel, *et al.* 2006. Genetic basis for increased intestinal permeability in families with Crohn's disease: role of CARD15 3020insC mutation? *Gut* **5:** 342–347.

33. Meddings, J. 2008. What role does intestinal permeability have in IBD pathogenesis? *Inflamm. Bowel Dis.* **14:** S138–S139.

34. Arrieta, M.C., K. Madsen, J. Doyle, *et al.* 2009. Reducing small intestinal permeability attenuates colitis in the IL10 gene-deficient mouse. *Gut* **5:** 41–48.

35. Olson, T.S., B.K. Reuter, K.G.E. Scott, *et al.* 2006. The primary defect in experimental ileitis originates from a non-hematopoietic source. *J. Exp. Med.* **20:** 541–552.

36. Macdonald, T.T., I. Monteleone, M.C. Fantini, *et al.* 2011. Regulation of homeostasis and inflammation in the intestine. *Gastroenterology* **14:** 1768–1775.

37. Maloy, K.J. & F. Powrie. 2011. Intestinal homeostasis and its breakdown in inflammatory bowel disease. *Nature* **47:** 298–306.

38. Santaolalla, R. & M.T. Abreu. 2012. Innate immunity in the small intestine. *Curr. Opin. Gastroenterol.* **2:** 124–129.

39. Shaw, M.H., N. Kamada, N. Warner, *et al.* 2011. The ever-expanding function of NOD: autophagy, viral recognition, and T cell activation. *Trends Immunol.* **3:** 73–79.

40. Hisamatsu, T., M. Suzuki, H.C. Reinecker, *et al.* 2003. CARD15/NOD2 functions as an antibacterial factor in human intestinal epithelial cells. *Gastroenterology* **12:** 993–1000.

41. Wehkamp, J., J. Harder, M. Weichenthal, *et al.* 2004. NOD2 (CARD15) mutations in Crohn's disease are associated with diminished mucosal alpha-defensin expression. *Gut* **5:** 1658–1664.

42. Wehkamp, J., N.H. Salzman, E. Porter, *et al.* 2005. Reduced Paneth cell alpha-defensins in ileal Crohn's disease. *Proc. Natl. Acad. Sci. USA* **10:** 18129–18134.

43. Cooney, R., J. Baker, O. Brain, *et al.* 2010. NOD2 stimulation induces autophagy in dendritic cells influencing bacterial handling and antigen presentation. *Nat. Med.* **1:** 90–97.

44. Travassos, L.H., L.A.M. Carneiro, M. Ramjeet, *et al.* 2010. Nod1 and Nod2 direct autophagy by recruiting ATG16L1 to the plasma membrane at the site of bacterial entry. *Nat. Immunol.* **1:** 55–62.

45. Rehman, A., C. Sina, O. Gavrilova, *et al.* 2011. Nod2 is essential for temporal development of intestinal microbial communities. *Gut* **6:** 1354–1362.

46. Halfvarson, J., L. Bodin, C. Tysk, *et al.* 2003. Inflammatory bowel disease in a Swedish twin cohort: a long-term follow-up of concordance and clinical characteristics. *Gastroenterology* **12:** 1767–1773.

47. Thjodleifsson, B., G. Sigthorsson, N. Cariglia, *et al.* 2003. Subclinical intestinal inflammation: an inherited abnormality in Crohn's disease relatives? *Gastroenterology* **12:** 1728–1737.

48. Pham, M., S.T. Leach, D.A. Lemberg, *et al.* 2010. Subclinical intestinal inflammation in siblings of children with Crohn's disease. *Dig. Dis. Sci.* **5:** 3502–3507.

49. Nusrat, A., J.R. Turner & J.L. Madara. 2000. Molecular physiology and pathophysiology of tight junctions. IV. Regulation of tight junctions by extracellular stimuli: nutrients, cytokines, and immune cells. *Am. J. Physiol. Gastrointest. Liver Physiol.* **27:** G851–G857.

50. Howe, K.L., C. Reardon, A. Wang, *et al.* 2005. Transforming growth factor-beta regulation of epithelial tight junction proteins enhances barrier function and blocks enterohemorrhagic Escherichia coli O15:H7-induced increased permeability. *Am. J. Pathol.* **16:** 1587–1597.

51. Utech, M., M. Brüwer & A. Nusrat. 2006. Tight junctions and cell-cell interactions. *Methods Mol. Biol.* **34:** 185–195.

52. Zeissig, S., C. Bojarski, N. Buergel, *et al.* 2004. Downregulation of epithelial apoptosis and barrier repair in active Crohn's disease by tumour necrosis factor alpha antibody treatment. *Gut* **5:** 1295–1302.

53. Hilsden, R.J., J.B. Meddings & L.R. Sutherland. 1996. Intestinal permeability changes in response to acetylsalicylic acid in relatives of patients with Crohn's disease. *Gastroenterology* **11:** 1395–1403.

54. Suenaert, P., V. Bulteel, S. Vermeire, *et al.* 2005. Hyperresponsiveness of the mucosal barrier in Crohn's disease is not tumor necrosis factor-dependent. *Inflamm. Bowel Dis.* **1:** 667–673.

Ann. N.Y. Acad. Sci. ISSN 0077-8923

ANNALS OF THE NEW YORK ACADEMY OF SCIENCES
Issue: *Barriers and Channels Formed by Tight Junction Proteins*

A dynamic paracellular pathway serves diuresis in mosquito Malpighian tubules

Klaus W. Beyenbach

Department of Biomedical Sciences, Cornell University, Ithaca, New York

Address for correspondence: Prof. Dr. Klaus W. Beyenbach, Professor of Physiology, Department of Biomedical Sciences, VRT 804, Cornell University, Ithaca, NY 14853. KWB1@CORNELL.EDU

Female mosquitoes gorge on vertebrate blood, a rich nutrient source for developing eggs, but gorging meals increase the risk of predation. Mosquitoes are quick to reduce the flight payload with a potent diuresis. Diuretic peptides of the insect kinin family induce a tenfold reduction in the paracellular resistance of Malpighian tubules and increase the paracellular permeation of Cl^-, the counterion of the transepithelial secretion of Na^+ and K^+. As a result, the transepithelial secretion of NaCl and KCl and water increases. Insect kinins signal the opening of the paracellular pathway via G protein–coupled receptors and the elevation of intracellular $[Ca^{2+}]$, which leads to the reorganization of the cytoskeleton associated with the septate junction (SJ). The reorganization may affect the septate junctional proteins that control the barrier and permselectivity properties of the paracellular pathway. The proteins involved in the embryonic formation of the SJ and in epithelial polarization are largely known for ectodermal epithelia, but the proteins that form and mediate the dynamic functions of the SJ in Malpighian tubules remain to be determined.

Keywords: septate junction; paracellular permeability; paracellular chloride transport; leucokinin; aedeskinin; insect kinins; septa; integral membrane proteins; scaffolding proteins; actin; adducin; actin-depolymerizing factor; cytoskeleton

Introduction

Only female mosquitoes take blood meals and only during the reproductive period.[1] Blood is a convenient source of salt, water, ATP, iron, and nutrients for developing eggs. However, the blood meal challenges the mosquito with excess salt and water, hemolymph dilution and hypertension, and a huge flight payload. For example, when taking a blood meal, the yellow fever mosquito *Aedes aegypti* more than doubles her body weight in less than two minutes.[2] Such an increase in weight reduces lift and flight speed, which increases the risk of predation and shortens the flight range for finding an aqueous habitat to deposit her eggs.[3] Getting rid of the unwanted salt and water of the blood meal is therefore the immediate priority. Accordingly, the mosquito begins to urinate at high rates even before she has finished her meal.[4] The initial diuresis lasts about 20 minutes. The urine of this diuresis consists largely of Na^+, Cl^-, and water.[2] Thereafter, K^+ starts to ap-

pear in the urine, which reflects the digestions of red blood cells.

The formation of urine in mosquitoes begins in the distal blind end of the Malpighian (renal) tubules by the process of epithelial secretion. The cations Na^+ and K^+ take transport pathways through large principal cells of the tubule (Fig. 1). Chloride, the counterion of transepithelial Na^+ and K^+ secretion, can take transcellular routes through principal and stellate cells and a paracellular route between the epithelial cells (Fig. 1).

Central to the transepithelial secretion of NaCl and KCl (and consequently water) is the V-type H^+ ATPase located at the apical, brush-border membrane of principal cells (Fig. 1). It energizes both transepithelial cation and anion transport. In particular, the V-type H^+ ATPase provides the electrochemical potential for driving the extrusion of Na^+ and K^+ across the apical membrane via presumably nH^+/cation exchange mediated by the NHA family of transporters.[10,11] By generating a lumen-positive

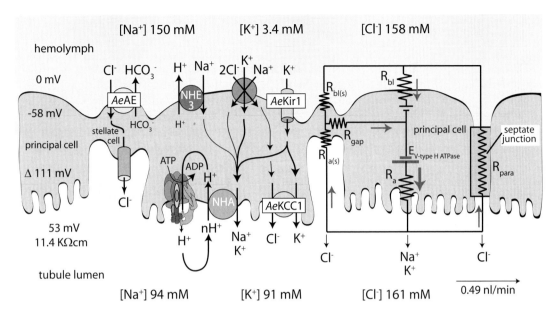

Figure 1. Mechanisms of transepithelial NaCl and KCl secretion in the Malpighian tubules of the yellow fever mosquito *Aedes aegypti*. Transepithelial secretion of Na$^+$ and K$^+$ is active in generating a lumen-positive voltage of 53 mV; transepithelial secretion of Cl$^-$ is passive. The transport of all three ions is energized by the V-type H$^+$ ATPase located at the apical membrane of principal cells. The electrical circuit illustrates electrical coupling of transcellular cation transport and paracellular anion transport. The mosquito anion exchanger *Ae*AE, the KCl cotransporter *Ae*KCC1, and the inwardly rectifying K$^+$ channel *Ae*Kir1 were recently cloned by our group.[5–7] E, electromotive force; R, electrical resistance; a, apical membrane; bl, basolateral membrane; para, paracellular pathway; (s), stellate cells. Although the tubule epithelium expresses the genes encoding subunits of the Na/K ATPase,[8] we measured no significant ouabain-sensitive ATPase activity in *Aedes* Malpighian tubules.[9]

transepithelial voltage, the V-type H$^+$ ATPase also provides the electrochemical potential that favors the passive transport of Cl$^-$ from hemolymph to the tubule lumen via a paracellular pathway permeable to Cl$^-$.[12] Moreover, exporting protons, the V-type H$^+$ ATPase generates current that must return to the cytoplasmic face of the proton pump. The current is carried by Cl$^-$ passing through the paracellular pathway and/or stellate cells (Fig. 1). Current is carried across the basolateral membrane by K$^+$ passing through an inward-rectifying K$^+$ (Kir) channel, such as *Ae*Kir1, which we have recently cloned from *Aedes* Malpighian tubules.[7] Na$^+$ can enter principal cells across the basolateral membrane via exchange transport with H$^+$ mediated by NHE3 cloned in the laboratory of Gill[13] and via cotransport with K$^+$ and Cl$^-$ presumably carried by NKCC.[14] Cl$^-$ can enter the epithelium via the bumetanide-sensitive NKCC of principal cells and the SLC4-like SCL4 Cl/HCO$_3$ anion exchanger (AE) of stellate cells.[5] Cl$^-$ can exit principal cells via a SCL12-like K, Cl cotransporter (KCC)[6] and stellate cells via Cl$^-$ channels we have identified in apical membranes of these cells.[15] Un-

der control conditions, Cl$^-$ is likely to take both transcellular and paracellular pathways for getting into the tubule lumen. However, under diuretic conditions triggered by leucokinin or aedeskinin, the paracellular pathway is the dominant route for moving Cl$^-$ across the epithelium in *Aedes* Malpighian tubules.[16–19]

Role of diuretic hormones

Using HPLC methods, our laboratory has purified a protein from mosquito heads that triggers the renal excretion of NaCl and water in the yellow fever mosquito *Aedes aegypti*.[20] The protein was named mosquito natriuretic peptide (MNP) because it selectively increases the transepithelial secretion of NaCl (and not KCl) by stimulating active transepithelial transport of Na$^+$.[20] MNP was found to be released into the hemolymph of the mosquito upon taking a blood meal.[21] MNP elevates intracellular concentrations of cyclic AMP, and cAMP mimics the effect of MNP in the Malpighian tubules by stimulating transepithelial Na$^+$ secretion.[22] Electrophysiological studies show that MNP and cAMP activate a Na$^+$ conductance in the basolateral membrane of principal cells of the Malpighian tubules,[23] but the channel or electrogenic carrier providing this conductance is unknown. The Coast laboratory has gone on to identify MNP as the calcitonin-like Anoga-DH31 in *Anopheles gambiae*.[24] Anoga-DH31 selectively activates the transepithelial secretion of Na$^+$ in the Malpighian tubules of *Anopheles* and *Aedes* using cAMP as second messenger.

MNP, Anoga-DH31, and cAMP all have a common electrophysiological signature in isolated Malpighian tubules: they hyperpolarize the transepithelial voltage in *Anopheles* and *Aedes* Malpighian tubules as Na$^+$ enters principal cells across the basolateral membrane.[24–27] In contrast, other HPLC fractions of mosquito (*Aedes*) heads depolarize the transepithelial voltage, consistent with the presence of other diuretic peptides and other mechanisms of action at the level of the Malpighian tubules.[20] Leucokinin turned out to be such a depolarizing peptide of the transepithelial voltage. Leucokinin was known to stimulate the contractions of the hindgut in the cockroach *Leucophaea maderae*,[28] which suggested excretory activity not only in the hindgut but also upstream in the Malpighian tubules. Indeed, leucokinin from the cockroach was found to have diuretic activity in the mosquito Malpighian tubules[29] consistent with the wide distribution of kinins in insects. Leucokinin and other insect kinins are now known to stimulate fluid secretion in the house cricket,[30] locust,[31] corn earworm,[32] fruit fly,[33] and housefly.[34] Kinins have been found in every order of the winged insects, which includes 97% of all insects.

The mechanism of action of kinins in the Malpighian tubules of *Aedes aegypti*

Isolated distal segments of the Malpighian tubules of the yellow fever mosquito spontaneously secrete fluid *in vitro* for several hours. Secreted fluid is approximately isosmotic with the peritubular bathing Ringer, and Na$^+$, K$^+$, and Cl$^-$ are the major electrolytes in secreted fluid, accounting for most of the osmotic pressure.[25] The transepithelial voltage is lumen positive by 59 mV, reflecting the active transepithelial secretion of cations in a secretory epithelium (Fig. 2A). The transepithelial resistance is 18.5 KΩcm, indicating a moderately tight epithelium.[17,25,26] When leucokinin-VIII is added to the peritubular medium, the

transepithelial voltage quickly drops to 5 mV, and the transepithelial resistance plummets to 3.2 KΩcm (Fig. 2B). Low transepithelial voltages and resistances are the properties of so-called "leaky epithelia" with characteristically high rates of transporting salt and water. In contrast, high transepithelial voltages and resistance are the properties of so-called "tight epithelia" with characteristically low rates of transepithelial transport that reflect largely storage functions. Thus, the Malpighian tubules in the yellow fever mosquito can function as both tight and leaky epithelia depending on instructions from extracellular and intracellular signals.

In *Aedes* Malpighian tubules, leucokinin-VIII significantly increases the transepithelial secretion of NaCl and KCl, and consequently water (Fig. 2A). The stimulation of both NaCl and KCl secretion together with the low values of the transepithelial voltage and resistance suggest that leucokinin-VIII opens the paracellular permeability to Cl$^-$. Symmetrical transepithelial Cl$^-$ diffusion potentials in the presence of leucokinin-VIII confirm this hypothesis.[16,17] Moreover, a large increase in the paracellular Cl$^-$ permeability is expected to produce an electrical short circuit across the epithelium, which is confirmed by the drop of the transepithelial voltage and resistance to values close to zero in the presence of leucokinin (Fig. 2B).

Probing the nature of the paracellular Cl$^-$ conductance activated by insect kinins, we found that leucokinin not only increases the anion conductance but it also shifts the anion selectivity of the paracellular pathway.[35] Under control conditions, the paracellular pathway behaves like an aqueous pathway for anions with the selectivity sequence I$^-$ > Br$^-$ > Cl$^-$ > F$^-$. In the presence of leucokinin-VIII, the paracellular pathway behaves like an anion channel with the selectivity sequence Br$^-$ > Cl$^-$ > I$^-$ > F$^-$.[35]

The presence of two natural diuretic agents in the yellow fever mosquito, MNP targeting Na$^+$ and kinins targeting Cl$^-$ for excretion, suggests that both are released upon the enormous volume expansion from the blood meal. The potential for additive or synergistic effects is obvious. The release of both natriuretic and chloruretic agents during the blood meal is expected to lower the electrochemical potential against which Na$^+$ must be transported across the epithelium, and reduce the resistance of

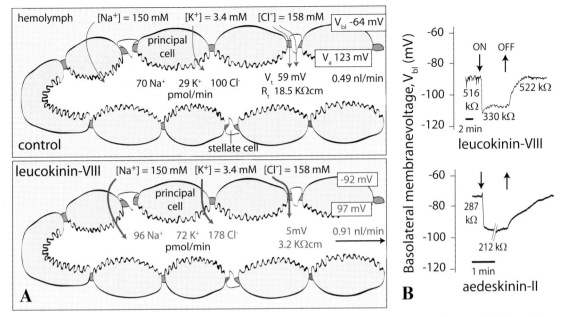

Figure 2. The effect of insect kinins on the isolated Malpighian tubules of the yellow fever mosquito *Aedes aegypti*. (A) Leucokinin-VIII, one of the kinins of *Leucophaea*, significantly increases the rate of fluid secretion by stimulating the transepithelial secretion of both NaCl and KCl. At the same time, leucokinin-VIII depolarizes the transepithelial voltage (V_t) and reduces the transepithelial resistance (R_t); V_{bl} and V_a are, respectively, the basolateral membrane voltage and apical membrane voltage of a principal cell. Red letters indicate minimum statistical significance at $P < 0.05$; data from Ref. 17. (B) Time course of the on/off effects of leucokinin-VIII and aedeskinin-II, one of the kinins of *Aedes aegypti*. Principal cells of the Malpighian tubules were studied by the two-electrode voltage clamp method.[39] Both kinins depolarize the V_{bl} and decrease the cell-input resistance, reflecting in part the opening of Ca^{2+} channels in the basolateral membrane.[19]

the transport route taken by Cl^-. At least additive effects are expected on thermodynamic grounds, but synergistic effects have been observed by others.[36–38]

The discovery of insect kinins that target the paracellular pathway was all the more remarkable in that the transport pathway between epithelial cells had been considered rather stable and not subject to the rapid regulation of permeability. Moreover, rapid transitions between "tight" and "leaky" states, reflecting posttranslational mechanisms, had not been observed before in epithelia. It raised our curiosity about receptor-mediated signaling to the paracellular pathway and the molecular structure of the paracellular pathway.

Signaling to the paracellular pathway

The insect kinins, including leucokinin and aedeskinin, are considered multifunctional peptides with effects on metabolism, smooth muscle, and salt and water balance.[18,40,41] Their activities are attributed to a critical C-terminal pentapeptide (Fig. 3A). As

shown in Figure 3B, the pentapeptide is thought to fold in order to reach binding sites in the receptor pocket.[42]

In the isolated Malpighian tubules of *Aedes aegypti*, leucokinin-VIII triggers diuretic activity when it is presented from the peritubular side and not the luminal side, indicating the presence of the kinin receptor on the hemolymph side of epithelial cells.[17] Physiological studies in our laboratory find the leucokinin receptor to localize to at least the principal cells.[43] More recently, the laboratory of Pietrantonio found immunological evidence for the location of the receptor in stellate cells as well.[44] Regardless of the cellular localization of the kinin receptor, there is good agreement that insect kinin receptors are coupled to G proteins.[42,45–47] In *Aedes* Malpighian tubules, high peritubular concentrations of fluoride duplicate the effects of leucokinin.[35] The effect is attributed to aluminum tetrafluoride AlF_4^-, which is thought to form in the presence of trace quantities of aluminum. AlF_4^- is known to activate G proteins by mimicking the γ-phosphate of GTP.[48–50]

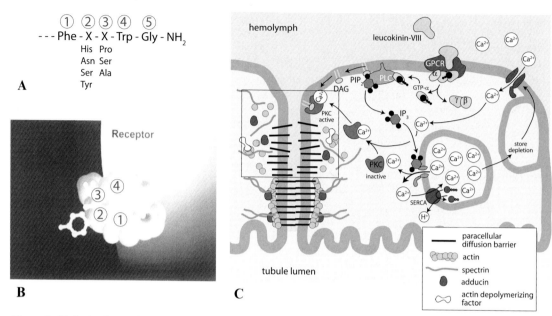

Figure 3. Kinin signaling to the paracellular pathway of *Aedes* Malpighian tubules. (A) The active C-terminal amide pentapeptide sequence of insect kinins.[68] (B) Binding of insect kinin to its receptor. The critical C-terminal pentapeptide sequence is thought to fold for reaching into the receptor pocket.[42,69] (C) G protein–coupled receptor signaling of leucokinin-VIII to SJs. Binding of leucokinin-VIII to its receptor increases intracellular Ca^{2+} concentrations. The Ca^{+2} activation of protein kinase C phosphorylates adducing, which destabilizes the cytoskeleton associated with the paracellular pathway. The destabilization opens paracellular diffusion barriers (box). GPCR, G–protein coupled receptor; α, β, γ, subunits of G protein; GTP guanosine-5′-triphosphate; PLC, phospholipase C; PIP_2, phosphatidylinositol (4,5)-triphosphate; IP_3, inositol triphosphate; DAG, diacylglycerol; SERCA, sacra/endoplasmic reticulum Ca^{2+}-ATPase.

The Malpighian tubules of *Drosophila*, *Aedes*, and *Anopheles* express only one kinin receptor gene, which encodes a G protein–coupled receptor.[46,47,51] Stimulation of this single receptor activates more than one signaling pathway, as indicated by the effects of synthetic derivatives of kinins.[19] For example, some kinin derivatives stimulate transepithelial fluid secretion in the isolated *Aedes* Malpighian tubules but have no effect on tubule electrophysiology (voltage and resistance). Other derivatives have the opposite effect, and all three aedeskinins, the natural kinins of *Aedes aegypti*, significantly stimulate both fluid secretion and tubule electrophysiology.[19] The partial activation of a receptor is well known for G protein-coupled receptors as agonist-directed signaling, ligand-directed trafficking, conformation-specific agonism, or functional selectivity.[52,53]

There is good agreement that Ca^{2+} is one of the second messengers of insect kinins.[35–38,54–59] Both intra- and extracellular Ca^{2+} are needed for leucokinin signaling in the Malpighian tubules of *Aedes aegypti* as illustrated in the signaling model of Fig. 3C.[56,60] In brief, the binding of leucokinin or aedeskinin[61] to the G protein–coupled receptor activates phospholipase C (Fig. 3B and C). One product of PLC activity is inositol triphosphate (IP_3), which increases in the presence of aedeskinins, the kinins of *Aedes aegypti*.[62] IP_3 is expected to release Ca^{2+} from intracellular stores. Store depletion in turn opens a nifedipine-sensitive Ca^{2+} channel in the basolateral membrane,[56] which admits extracellular Ca^{2+} to the cytoplasm. Another product of PLC activity is membrane-resident diacylglycerol (DAG). Together, DAG and Ca^{2+} recruit cytosolic protein kinase C (PKC) to the plasma membrane, which activates the kinase (Fig. 3C). Two inhibitors of PKC, staurosporine and bisindolylmaleimide, block the effect of aedeskinin-III on tubule electrophysiology and on fluid secretion.[63] The blockade of the electrophysiological effects are significant in that they point to a role of PKC in the activation of the paracellular Cl^- conductance by insect kinins.

To identify proteins that signal to the paracellular pathway, we undertook a proteomic analysis of the cytosol of *Aedes* Malpighian tubules. One set of tubules served as controls; the other set was treated with aedeskinin-III for only 1 min in order to observe the posttranslational effects of kinin stimulation on cytosolic proteins.[64] The amounts of actin-depolymerizing factor, actin, adducin, and regucalcin significantly increase in the cytoplasm upon aedeskinin-III stimulation. The appearance of adducin in the cytoplasm as phospho-adducin is significant in that unphosphorylated adducin stabilizes actin and spectrin at sites of cell–cell contact,[65] and adducin (phosphorylated by PKC) destabilizes actin and spectrin.[66,67] Accordingly, the appearance of phospho-adducin in the cytoplasm suggests the reorganization of the cytoskeleton associated with paracellular diffusion barriers. Figure 3C presents a hypothetical scheme where the destabilization of the cytoskeleton destabilizes unknown extracellular structures in the paracellular pathway with the effect of increasing the paracellular permeability.

Although hypothetical, the model in Figure 3 mirrors the well-known interaction of extracellular and intracellular proteins in epithelial junctions of vertebrates. Here, transmembrane proteins of the tight junction mediate cell adhesion, prevent indiscriminate paracellular permeation, and define the transport properties of the paracellular pathway. The transmembrane proteins bind to scaffolding (adapter) proteins on the cytoplasmic side of the junction. In turn, scaffolding proteins bind to actin and spectrin. The assembled protein complex stabilizes the junctions and consequently the paracellular structure and function. The junctional protein complex is assembled during embryogenesis when it plays a critical role in the development of epithelial polarity.

As links between transmembrane proteins and the cytoskeleton, scaffolding proteins of the tight junction, such as ZO-1, ZO-2, and ZO-3, are the targets of regulatory proteins such as protein kinases and phosphatases, small GTPases and G proteins, cytoplasmic proteins, transcription factors, heat shock proteins, and even microRNA in vertebrate epithelia. These regulatory agents may further stabilize or destabilize the junction with obvious effects on the structure and function of the paracellular pathway. For example, the junctional complex can be disassembled entirely as in epithelial-mesenchymal transitions, or reorganized as part of the epithelial response to inflammation and infection, or radically changed in the development of epithelial tumors. In view of these dynamics observed in the vertebrate tight junctions, it seems that junctional proteins could also serve to mediate rapid, reversible, posttranslational changes in paracellular permeability.[70,71]

Tight and septate junctions

The paracellular pathway is delineated by tight junctions in vertebrate epithelia and by septate junctions (SJs) in invertebrate epithelia (Fig. 4). In the past, both junctions were considered static and to serve barrier functions. Today, the dynamic nature of junctions is increasingly appreciated in: embryonic development, the maturation of epithelial cells from stem cells, the crypt-to-tip migration of epithelial cells in the mammalian intestine, epithelial-mesenchymal transitions, inflammatory responses, and the posttranslational regulation of paracellular permeability.

Tight junctions surround epithelial cells at their most apical region. They extend into the paracellular pathway for only a short distance, typically less than 1 μm (Fig. 4A). Tight junctions are formed by at least four integral membrane proteins: claudins, occludins, tricellulins, and junctional adhesion molecules (JAMs).[72,73] Interactions between the extracellular loops of, for example, claudins, pull the outer membrane leaflets of adjoining cells so close to each other as to nearly fuse them[74,75] and to form a barrier to the paracellular permeation of most solutes. Tight junctions can be quite leaky, as in epithelia mediating isosmotic fluid transport, permselective to ions, and permeable to organic solutes.[75–78]

In contrast to tight junctions, SJs of insects (and arthropods in general) occupy the more basal portion of the paracellular pathway for a considerable length (Fig. 4B). In the Malpighian tubules of *Aedes aegypti*, the SJ spans the whole epithelium from apical to basal poles.[18] Rather than annealing the plasma membranes of neighboring cells, the septa of SJs appear to maintain a cleft of 15–20 nm between adjacent cells.[79] The septa present a ladder-like appearance in apico-basal sections where the rungs of the ladder may be regular and irregular. As illustrated in Figure 4B, adherens junctions (AJ) and/or

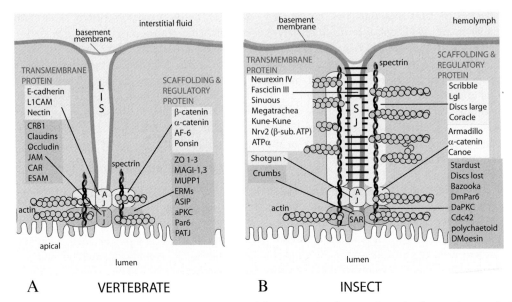

Figure 4. The vertebrate tight junction (A) and the insect SJ (B). A comparison between the molecular constituents of epithelial junctions in insects and vertebrates reveals a striking homology of proteins but remarkable organizational and anatomical differences.[80] The long paracellular cleft of the septa of the junction poses questions about functional roles. Gap junctions and desmosomes are omitted. Adapted from Refs. 75, 80, 88, and 89. AJ, adherens junction; LIS, lateral intercellular space; SAR, subapical region; SJ, septate junction; TJ, tight junction.

a subapical region (SAR) have been proposed at the apical border of the insect paracellular pathway.[75,80] The apico-basal SAR, AJ, and SJ organization of the paracellular pathway stems from the observations of epithelial polarization during development,[81–84] but it is not known to what extent this embryonic topology is maintained in the adult differentiated epithelium.

SJs are also found in vertebrates at paranodal junctions between myelinating Schwann cells and axons,[85] where they appear to function as a "fence" between excitable Na^+ channels at the node (for saltatory conduction) and K^+ channels in the internodes. The septal fence may preserve the axonal polarity of excitable and nonexcitable membrane stretches. In epithelia, the SJ could play a similar role in fencing basal from apical membrane domains. In the Malpighian tubules of *Aedes aegypti*, the apical membrane forms a thick brush border, with each microvillus housing a mitochondrion.[18] The V-type H^+ ATPase is exclusively and richly expressed at this membrane.[9,18] The physiological activity of Kir channels is richly and exclusively found at the basal membrane.[86] Given such a strong separation of apical and basal membrane domains by a long SJ, the question arises whether the SJ main-

tains a third, lateral plasma membrane domain with specialized functions?

Clefts generated by septa-like structures are also observed at neuromuscular junctions and neuronal synapses in invertebrates.[87] Here, the extracellular space serves the diffusion of neurotransmitters. To propose similar communicative interactions between epithelial cells baffles the imagination. A more conservative view has septa providing the structural stability for neuronal and epithelial cell contacts alike.

It is likely that the cleft in epithelial SJs is filled with an aqueous fluid, namely the extracellular fluid, similar to the fluid-filled lateral intercellular space of tight junctions (LIS, see Fig. 4A). However, in the case of SJs, a dense or regular spacing of septa would limit fluid in septal clefts and increase the electrical resistance of the paracellular pathway. Sparse and irregular septa would fill the paracellular space with conducting fluid and offer a low electrical paracellular resistance (Figs. 3C and 4). This model assigns a general barrier function to the SJ and the selectivity filter to other domains such as AJ and SAR (Fig. 4B). Moreover, the model of variable septa density is consistent with the large reversible changes in the paracellular electrical resistance we observe in *Aedes*

Malpighian tubules in the presence of leucokinin (Fig. 2). Measurements of specifically the paracellular electrical resistance reveal the remarkable drop from 18.6 KΩcm to 1.8 KΩcm in the presence of leucokinin.[17] Such a large resistance drop may reflect the opening of septa or a reduction in septa density. Moreover, an open SJ would provide a low paracellular resistance from the hemolymph to the most apical junctions where the channel-like permselectivity sequence $Br^- > Cl^- > I^- > F^-$ may be located.[35]

Figure 4 illustrates the paracellular pathway between two principal cells, which may be up to 40 μm long spanning the epithelium from the tubule lumen to the hemolymph. The length of the paracellular pathway between a stellate cell and a principal cell is much shorter because stellate cells are only 1–3 μm deep.[43] It would appear that septa at these short sites of cell-to-cell contact are particularly well suited for profound and rapid changes in paracellular permeability.

In a recent study that sought to identify the proteins of the SJ, the Furuse laboratory found a new a small tetraspan protein of 17 kDa in the midgut of *Drosophila*.[90] The protein called "snakeskin" is associated with the SJ and is essential to the formation of the SJ in the midgut of *Drosophila*. The lack of snakeskin is lethal during the larval stage. Importantly, in a snakeskin-deficient midgut epithelium, the septa are few, and the midgut is permeable to dextran. These studies establish a barrier function of septa in the *Drosophila* midgut, and the loss of this barrier when septa are lost. The expected electrical correlate of losing septa is the loss of paracellular electrical resistance. Whether snake skin also mediates rapid changes in septa density and paracellular resistance remains to be seen.

Of particular interest is that snakeskin is expressed in the Malpighian tubules as well as the midgut—namely two epithelia with a high dynamic range of epithelial transport. Here, epithelial transport activity depends on the loads presented. A normal extracellular fluid volume and composition places as little work load on the Malpighian tubules as an empty gut places on intestinal transport. The situation changes suddenly with the ingestion of a meal when transport mechanisms in the midgut (and the Malpighian tubules) must be quickly revved up. Significantly, snakeskin is not found in epithelial cells of the hindgut,[90] where transport rates are expected to be low. It would ap-

pear, therefore, that the expression of snakeskin in epithelia correlates well with the capacity for high rates of transport in general and with the presence of dynamic septa and paracellular functions in particular. The proteins that form the septa associated with snakeskin in *Drosophila* midgut are unknown as are those that mediate the dynamic resistance changes of the paracellular pathway in *Aedes* Malpighian tubules.

The reader who has come this far may wonder whether a dynamic paracellular pathway may be unique to gorging hematophagous insects and therefore irrelevant to other epithelia. On the contrary, the ability to commence a diuresis on short notice is also observed in phytophagous insects that gorge on the sap of plants.[91] Moreover, insects leaving their aquatic habitats with first flight undergo a so-called eclosion diuresis that lightens the load.[92] Thus, all insects appear to have a need of natural diuretic neuropeptides or hormones consistent with the wide distribution of insect kinins. Natural diuretic agents may be particularly well expressed in small animals, such as flies and bugs, where gorging meals can quickly increase body weight, up to 10 times in *Rhodnius prolixus*.[93] A dynamic paracellular pathway that can be activated quickly is therefore well suited for eliminating excess solute and water.

A dynamic paracellular pathway would also be well suited for extending the functional range of vertebrate epithelia. Secretory glands that produce tears, sweat, saliva, and digestive secretions may operate for hours or days at modest rates, but then be suddenly called upon to produce fluids profusely. Under these conditions, a paracellular pathway responding to posttranslational modification would serve well the transport challenges in a timely manner.

Acknowledgments

The author thanks the National Science Foundation and the National Institutes of Health for supporting our work through Grants IBN 0078058 and R21 AI072102, respectively. Thank you also Peter Piermarini and Mikio Furuse for constructive comments.

Conflicts of interest

The author declares no conflicts of interest.

References

1. Klowden, M.J. 1995. Blood, sex and the mosquito. *Bioscience* **45**: 326–331.

2. Williams, J.C., H.H. Hagedorn & K.W. Beyenbach. 1983. Dynamic changes in flow rate and composition of urine during the post-bloodmeal diuresis in *Aedes aegypti* (L.). *J. Comp. Physiol. [A]* **153**: 257–265.

3. Roitberg, B.D., E.B. Mondor & J.G.A. Tyerman. 2003. Pouncing spider, flying mosquito: blood acquisition increases predation risk in mosqitoes. *Behav. Ecol.* **14**: 736–740.

4. Beyenbach, K.W. & P.M. Piermarini. 2011. Transcellular and paracellular pathways of transepithelial fluid secretion in Malpighian (renal) tubules of the yellow fever mosquito *Aedes aegypti*. Acta Physiol. (Oxf.) **202**: 387–407.

5. Piermarini, P.M. *et al.* 2010. A SCL4-like anion exchanger from renal tubules of the mosquito (*Aedes aegypti*): evidence for a novel role of stellate cells in diuretic fluid secretion. *Am. J. Physiol. Regul. Integr. Comp. Physiol.* **298**: 642–660.

6. Piermarini, P.M. *et al.* 2011. Role of an apical K, Cl cotransporter in urine formation by renal tubules of the yellow fever mosquito (*Aedes aegypti*). *Am. J. Physiol. Regul. Integr. Comp. Physiol.* **301**: R1318–1337.

7. Kosse, C. *et al.* 2011. Inward rectifier K⁺ (Kir) channels in Malpighian tubules of the yellow fever mosquito *Aedes aegypti*. Presented at Annual Meeting of Socety for Experimental Biology (SEB) Glasgow, Scotland, July 1–4, 2011.

8. Patrick, M.L. *et al.* 2006. P-type Na^+/K^+-ATPase and V-type H^+-ATPase expression patterns in the osmoregulatory organs of larval and adult mosquito *Aedes aegypti*. *J. Exp. Biol.* **209**: 4638–4651.

9. Weng, X.H. *et al.* 2003. The V-type H^+-ATPase in Malpighian tubules of *Aedes aegypti*: localization and activity. *J. Exp. Biol.* **206**: 2211–2219.

10. Okech, B.A. *et al.* 2008. Cationic pathway of pH regulation in larvae of *Anopheles gambiae*. *J. Exp. Biol.* **211**: 957–968.

11. Rheault, M.R. *et al.* 2007. Molecular cloning, phylogeny and localization of AgNHA1: the first Na^+/H^+ antiporter (NHA) from a metazoan, *Anopheles gambiae*. *J. Exp. Biol.* **210**: 3848–3861.

12. Beyenbach, K.W. 2001. Energizing epithelial transport with the vacuolar H^+-ATPase. *News Physiol. Sci.* **16**: 145–151.

13. Pullikuth, A.K. *et al.* 2006. Molecular characterization of sodium/proton exchanger 3 (NHE3) from the yellow fever vector, *Aedes aegypti*. *J. Exp. Biol.* **209**: 3529–3544.

14. Hegarty, J.L. *et al.* 1991. Dibutyryl cAMP activates bumetanide-sensitive electrolyte transport in Malpighian tubules. *Am. J. Physiol. Cell Physiol.* **261**: C521–C529.

15. O'Connor, K.R. & K.W. Beyenbach. 2001. Chloride channels in apical membrane patches of stellate cells of Malpighian tubules of *Aedes aegypti*. *J. Exp. Biol.* **204**: 367–378.

16. Beyenbach, K.W. 2003. Regulation of tight junction permeability with switch-like speed. *Curr. Opin. Nephrol. Hypertens.* **12**: 543–550.

17. Pannabecker, T.L., T.K. Hayes & K.W. Beyenbach. 1993. Regulation of epithelial shunt conductance by the peptide leucokinin. *J. Membr. Biol.* **132**: 63–76.

18. Beyenbach, K.W., H. Skaer & J.A. Dow. 2010. The developmental, molecular, and transport biology of Malpighian tubules. *Annu. Rev. Entomol.* **55**: 351–374.

19. Schepel, S.A. *et al.* 2010. The single kinin receptor signals to separate and independent physiological pathways in Malpighian tubules of the yellow fever mosquito. *Am. J. Physiol. Regul. Integr. Comp. Physiol.* **299**: R612–622.

20. Petzel, D.H., H.H. Hagedorn & K.W. Beyenbach. 1985. Preliminary isolation of mosquito natriuretic factor. *Am. J. Physiol.* **249**: R379–386.

21. Wheelock, G.D. *et al.* 1988. Evidence for hormonal control of diuresis after a blood meal in the mosquito *Aedes aegypti*. *Arch. Insect. Biochem. Physiol.* **7**: 75–90.

22. Petzel, D.H., M.M. Berg & K.W. Beyenbach. 1987. Hormone-controlled cAMP-mediated fluid secretion in yellow-fever mosquito. *Am. J. Physiol.* **253**: R701–711.

23. Sawyer, D.B. & K.W. Beyenbach. 1985. Dibutyryl-cAMP increases basolateral sodium conductance of mosquito Malpighian tubules. *Am. J. Physiol.* **248**: R339–R345.

24. Coast, G.M. *et al.* 2005. Mosquito natriuretic peptide identified as a calcitonin-like diuretic hormone in *Anopheles gambiae* (Giles). *J. Exp. Biol.* **208**: 3281–3291.

25. Williams, J.C. & K.W. Beyenbach. 1983. Differential effects of secretagogues on Na and K secretion in the Malpighian tubules of *Aedes aegypti* (L.). *J. Comp. Physiol.* **149**: 511–517.

26. Williams, J.C. & K.W. Beyenbach. 1984. Differential effects of secretagogues on the electrophysiology of the Malpighian tubules of the yellow fever mosquito. *J. Comp. Physiol. [B]* **154**: 301–309.

27. Beyenbach, K.W. & D.H. Petzel. 1987. Diuresis in mosquitoes: role of a natriuretic factor. *News Physiol. Sci.* **2**: 171–175.

28. Holman, G.M., B.J. Cook & R.J. Nachman. 1987. Isolation, primary structure and synthesis of leucokinin-VII and VIII: the final members of the new family of cephalomyotropic peptides isolated from head extracts of *Leucophaea maderae*. Comp. *Biochem. Physiol. [C]* **88**: 31–34.

29. Hayes, T.K. *et al.* 1989. Leucokinins, a new family of ion transport stimulators and inhibitors in insect Malpighian tubules. *Life Sci.* **44**: 1259–1266.

30. Coast, G.M., G.M. Holman & R.J. Nachman. 1990. The diuretic activity of a series of cephalomyotropic neuropeptides, the achetakinins, on isolated Malpighian tubules of the house cricket *Acheta domesticus*. *J. Insect Physiol.* **36**: 481–488.

31. Thompson, K.S. *et al.* 1995. Cellular colocalization of diuretic peptides in locusts: a potent control mechanism. *Peptides* **16**: 95–104.

32. Blackburn, M.B. *et al.* 1995. The isolation and identification of three diuretic kinins from the abdominal ventral nerve cord of adult *Helicoverpa zea*. *J. Insect Physiol.* **41**: 723–730.

33. Terhzaz, S. *et al.* 1999. Isolation and characterization of a leucokinin-like peptide of *Drosophila melanogaster*. *J. Exp. Biol.* **202**: 3667–3676.

34. Iaboni, A. *et al.* 1998. Immunocytochemical localisation and biological activity of diuretic peptides in the housefly, *Musca domestica*. *Cell Tissue Res.* **294**: 549–560.

35. Yu, M. & K.W. Beyenbach. 2001. Leucokinin and the modulation of the shunt pathway in Malpighian tubules. *J. Insect Physiol.* **47**: 263–276.

36. Coast, G.M. 1995. Synergism between diuretic peptides controlling ion and fluid transport in insect Malpighian tubules. *Regul. Pept.* **57**: 283–296.

37. Maddrell, S.H.P. *et al.* 1993. Synergism of hormones controlling epithelial fluid transport in an insect. *J. Exp. Biol.* **174:** 65–80.

38. O'Donnell, M.J. & J.H. Spring. 2000. Modes of control of insect Malpighian tubules: synergism, antagonism, cooperation and autonomous regulation. *J. Insect Physiol.* **46:** 107–117.

39. Masia, R. *et al.* 2000. Voltage clamping single cells in intact Malpighian tubules of mosquitoes. *Am. J. Physiol.* **279:** F747–F754.

40. Goldsworthy, G.J. *et al.* 1992. The structural and functional activity of neuropeptides. *In* Royal Entomological Society Symposium on Insect Molecular Science. J.M. Crampton & P. Eggleton, Eds.: 205–225. Academic Press. London.

41. Seinsche, A. *et al.* 2000. Effect of helicokinins and ACE inhibitors on water balance and development of *Heliothis virescens* larvae. *J. Insect Physiol.* **46:** 1423–1431.

42. Taneja-Bageshwar, S. *et al.* 2008. Comparison of insect kinin analogs with cis-peptide bond, type VI-turn motifs identifies optimal stereochemistry for interaction with a recombinant arthropod kinin receptor from the southern cattle tick *Boophilus microplus. Peptides* **29:** 295–301.

43. Yu, M.J. & K.W. Beyenbach. 2004. Effects of leucokinin-VIII on *Aedes* Malpighian tubule segments lacking stellate cells. *J. Exp. Biol.* **207:** 519–526.

44. Lu, H.L., C. Kersch & P.V. Pietrantonio. 2011. The kinin receptor is expressed in the Malpighian tubule stellate cells in the mosquito *Aedes aegypti* (L.): a new model needed to explain ion transport? *Insect Biochem. Mol. Biol.* **41:** 135–140.

45. Holmes, S.P. *et al.* 2003. Functional analysis of a G protein-coupled receptor from the southern cattle tick *Boophilus microplus* (Acari: Ixodidae) identifies it as the first arthropod myokinin receptor. *Insect Mol. Biol.* **12:** 27–38.

46. Pietrantonio, P.V. *et al.* 2005. The mosquito *Aedes aegypti* (L.) leucokinin receptor is a multiligand receptor for the three *Aedes* kinins. *Insect Mol. Biol.* **14:** 55–67.

47. Radford, J.C. *et al.* 2004. Functional characterisation of the *Anopheles* leucokinins and their cognate G-protein coupled receptor. *J. Exp. Biol.* **207:** 4573–4586.

48. Sternweis, P.C. & A.G. Gilman. 1982. Aluminum: a requirement for activation of the regulatory component of adenylate cyclase by fluoride. *Proc. Natl. Acad. Sci. USA* **79:** 4888–4891.

49. Bigay, J. *et al.* 1985. Fluoroaluminates activate transducin-GDP by mimicking the gamma-phosphate of GTP in its binding site. *FEBS Lett.* **191:** 181–185.

50. Bigay, J. *et al.* 1987. Fluoride complexes of aluminium or beryllium act on G-proteins as reversibly bound analogues of the gamma phosphate of GTP. *Embo J.* **6:** 2907–2913.

51. Radford, J.C., S.A. Davies & J.A. Dow. 2002. Systematic G-protein-coupled receptor analysis in *Drosophila melanogaster* identifies a leucokinin receptor with novel roles. *J. Biol. Chem.* **277:** 38810–38817.

52. Berg, K.A. & W.P. Clarke. 2006. Development of functionally selective agonists as novel therapeutic agents. *Drug Discov. Today Ther. Strateg.* **3:** 421–428.

53. Simmons, M.A. 2005. Functional selectivity, ligand-directed trafficking, conformation-specific agonism: what's in a name? *Mol. Interv.* **5:** 154–157.

54. Beyenbach, K.W. 2003. Transport mechanisms of diuresis in Malpighian tubules of insects. *J. Exp. Biol.* **206:** 3845–3856.

55. Lu, H.L. *et al.* 2011. A calcium bioluminescence assay for functional analysis of mosquito (*Aedes aegypti*) and tick (*Rhipicephalus microplus*) G protein-coupled receptors. *J. Vis. Exp.* **50:** 2732.

56. Yu, M.J. & K.W. Beyenbach. 2002. Leucokinin activates Ca^{2+}-dependent signal pathway in principal cells of *Aedes aegypti* Malpighian tubules. *Am. J. Physiol. Renal Physiol.* **283:** F499–508.

57. Holman, G.M., R.J. Nachman & G.M. Coast. 1999. Isolation, characterization and biological activity of a diuretic myokinin neuropeptide from the housefly, *Musca domestica. Peptides.* **20:** 1–10.

58. Holtzhausen, W.D. & S.W. Nicolson. 2007. Beetle diuretic peptides: the response of mealworm (Tenebrio molitor) Malpighian tubules to synthetic peptides, and cross-reactivity studies with a dung beetle (Onthophagus gazella). *J. Insect Physiol.* **53:** 361–369.

59. Te Brugge, V.A., D.A. Schooley & I. Orchard. 2002. The biological activity of diuretic factors in *Rhodnius prolixus. Peptides* **23:** 671–681.

60. Clark, T.M. *et al.* 1998. The concentration-dependence of CRF-like diuretic peptide: mechanisms of action. *J. Exp. Biol.* **201:** 1753–1762.

61. Schepel, S.A. *et al.* 2010. The single kinin receptor signals to separate and independent physiological pathways in Malpighian tubules of the yellow fever mosquito. *Am. J. Physiol. Regul. Integr. Comp. Physiol.* **299:** R612–R622.

62. Cady, C. & H.H. Hagedorn. 1999. Effects of putative diuretic factors on intracellular second messenger levels in the Malpighian tubules of *Aedes aegypti. J. Insect Physiol.* **45:** 327–337.

63. Miyauchi, J.T. *et al.* 2011. The role of adducin in the diuresis triggered by aedeskinin III in Malpighian tubules of the yellow fever mosquito. Glasgow, Scotland, July 1–4, 2011.

64. Beyenbach, K.W. *et al.* 2009. Signaling to the apical membrane and to the paracellular pathway: changes in the cytosolic proteome of *Aedes* Malpighian tubules. *J. Exp. Biol.* **212:** 329–340.

65. Kaiser, H.W. *et al.* 1993. Localization of adducin in epidermis. *J. Invest. Dermatol.* **101:** 783–788.

66. Matsuoka, Y., X. Li & V. Bennett. 1998. Adducin is an in vivo substrate for protein kinase C: phosphorylation in the MARCKS-related domain inhibits activity in promoting spectrin–actin complexes and occurs in many cells, including dendritic spines of neurons. *J. Cell Biol.* **142:** 485–497.

67. Matsuoka, Y., X. Li & V. Bennett. 2000. Adducin: structure, function and regulation. *Cell Mol Life Sci.* **57:** 884–895.

68. Taneja-Bageshwar, S. *et al.* 2009. Biostable agonists that match or exceed activity of native insect kinins on recombinant arthropod GPCRs. *Gen. Comp. Endocrinol.* **162:** 122–128.

69. Nachman, R.J. & P.V. Pietrantonio. 2010. Interaction of mimetic analogs of insect kinin neuropeptides with arthropod receptors. *Adv. Exp. Med. Biol.* **692:** 27–48.

70. Duffey, M.E. *et al.* 1981. Regulation of epithelial tight junction permeability by cyclic AMP. *Nature* **294:** 451–453.

71. Bentzel, C.J. *et al.* 1980. Cytoplasmic regulation of tight-junction permeability: effect of plant cytokinins. *Am. J. Physiol.* **239:** C75–89.

72. Furuse, M. 2010. Molecular basis of the core structure of tight junctions. *Cold Spring Harb. Perspect. Biol.* **2:** a002907.

73. Anderson, J.M. & C.M. Van Itallie. 2009. Physiology and function of the tight junction. *Cold Spring Harb. Perspect. Biol.* **1:** a002584.

74. Diamond, J.M. 1974. Tight and leaky junctions of epithelia: a perspective on kisses in the dark. *Fed. Proc.* **33:** 2220–2224.

75. Furuse, M. & S. Tsukita. 2006. Claudins in occluding junctions of humans and flies. *Trends Cell Biol.* **16:** 181–188.

76. Diamond, J.M. 1977. Twenty-first Bowditch lecture. The epithelial junction: bridge, gate, and fence. *Physiologist* **20:** 10–18.

77. Schneeberger, E.E. & R.D. Lynch. 2004. The tight junction: a multifunctional complex. *Am. J. Physiol. Cell Physiol.* **286:** C1213–1228.

78. Van Itallie, C.M. & J.M. Anderson. 2004. The molecular physiology of tight junction pores. *Physiology* **19:** 331–338.

79. Lane, N.J. *et al.* 1994. Electron microscopic structure and evolution of epithelial junctions. *In* Molecular Mechanisms of Epithelial Cell Junctions: From Development to Disease. S. Citi, Ed.: 23–43. R.G. Landes. Austin, Texas.

80. Knust, E. & O. Bossinger. 2002. Composition and formation of intercellular junctions in epithelial cells. *Science* **298:** 1955–1959.

81. Tepass, U. *et al.* 2001. Epithelial cell polarity and cell junctions in *Drosophila*. *Annu. Rev. Genet.* **35:** 747–784.

82. Behr, M., D. Riedel & R. Schuh. 2003. The claudin-like megatrachea is essential in septate junctions for the epithelial barrier function in *Drosophila*. *Dev. Cell.* **5:** 611–620.

83. Wu, V.M. *et al.* 2004. Sinuous is a *Drosophila* claudin required for septate junction organization and epithelial tube size control. *J. Cell Biol.* **164:** 313–323.

84. Nelson, K.S. & G.J. Beitel. 2009. Cell junctions: lessons from a broken heart. *Curr. Biol.* **19:** R122–123.

85. Hortsch, M. & B. Margolis. 2003. Septate and paranodal junctions: kissing cousins. *Trends Cell Biol.* **13:** 557–561.

86. Beyenbach, K.W. & R. Masia. 2002. Membrane conductances of principal cells in Malpighian tubules of *Aedes aegypti*. *J. Insect Physiol.* **48:** 375–386.

87. Prokop, A. 1999. Integrating bits and pieces: synapse structure and formation in *Drosophila* embryos. *Cell Tissue Res.* **297:** 169–186.

88. Nelson, K.S., M. Furuse & G.J. Beitel. 2010. The *Drosophila* claudin kune–kune is required for septate junction organization and tracheal tube size control. *Genetics* **185:** 831–839.

89. Wedlich, D. 2002. The polarising role of cell adhesion molecules in early development. *Curr. Opin. Cell Biol.* **14:** 563–568.

90. Yanagihashi, Y. *et al.* 2012. A novel smooth septate junction-associated membrane protein, Snakeskin, is required for intestinal barrier function in *Drosophila*. *J. Cell Sci.* Feb 10. [Epub ahead of print].

91. Bushman, D.W., A.K. Raina & J.O. Nelson. 1989. Posteclosion diuresis in adult *Heliothis zea*. *Physiol. Entomology* **14:** 391–396.

92. Coast, G.M. 1996. Neuropeptides implicated in the control of diuresis in insects. *Peptides* **17:** 327–336.

93. Orchard, I. 2006. Serotonin: a coordinator of feeding-related physiological events in the blood-gorging bug, Rhodnius prolixus. *Comp. Biochem. Physiol. A Mol. Integr. Physiol.* **144:** 316–324.

Ann. N.Y. Acad. Sci. ISSN 0077-8923

ANNALS OF THE NEW YORK ACADEMY OF SCIENCES

Issue: *Barriers and Channels Formed by Tight Junction Proteins*

Impaired paracellular ion transport in the loop of Henle causes familial hypomagnesemia with hypercalciuria and nephrocalcinosis

Lea Haisch and Martin Konrad

Department of General Pediatrics, University Hospital Münster, Münster, Germany

Address for correspondence: Martin Konrad, University Hospital Münster, Department of General Pediatrics, Pediatric Nephrology, Waldeyerstr 22, D-49149 Münster, Germany. konradma@uni-muenster.de

Familial hypomagnesemia with hypercalciuria and nephrocalcinosis (FHHNC) is a rare tubular disorder caused by mutations in genes coding for tight junction (TJ) proteins. TJs define the paracellular path between adjacent cells and thereby play a pivotal role for the regulation of the paracellular ion permeability of epithelia. The family of TJ proteins comprise a variety of transmembrane proteins, including the claudins. Multiple distinct mutations in the genes for claudin-16 and -19 have been described to be responsible for FHHNC. Both encoded proteins are especially important for the paracellular reabsorption of Mg^{2+} and Ca^{2+} in the thick ascending limb of Henle's loop. Interestingly, in addition to ion disturbances, FHHNC leads to chronic renal failure and may be associated with extrarenal symptoms.

Keywords: magnesium; calcium; tight junctions; chronic renal failure; hyperparathyroidism; urinary tract infections

Introduction

The kidneys are of crucial importance for the regulation of ion homeostasis and water balance. After glomerular filtration, the renal tubules mediate ion reabsorption and secretion from primary urine into the tubular fluid, hence defining the urine composition and the final excretion of ions. The transport modalities vary widely between different nephron segments. Reabsorption of ions and water can occur either actively through epithelial cells mediated by ion channels, transporters or pumps, or passively along the paracellular pathway driven by an electrical or chemical gradient. This pathway depends on specific characteristics of the epithelial cell layer, in which adjacent cells are connected by specialized cell-to-cell contacts, the tight junctions (TJs).

The importance of this paracellular pathway for electrolyte handling in the kidney was emphasized by the discovery of mutations in two members of the claudin multigene family encoding TJ proteins claudin-16 and -19. Mutations in either *CLDN16* or *CLDN19* cause hereditary renal Mg^{2+} and Ca^{2+} wasting, known as familial hypomagnesemia with hypercalciuria and nephrocalcinosis (FHHNC).

Mg^{2+} handling in the kidney

Mg^{2+} is the second most abundant intracellular ion in the human body.[1] Less than 1% of total body Mg^{2+} is circulating in the blood. Serum levels are kept in a narrow range by balancing intestinal absorption and renal excretion. These processes are regulated by metabolic and hormonal mechanisms.[2] Approximately 80% of the circulating Mg^{2+} is filtered through the glomeruli, but only 3–5% is finally excreted in the urine. Mg^{2+} reabsorption in the kidney differs substantially between the different nephron segments: in the proximal tubule, around 15–20% of filtered Mg^{2+} is reabsorbed, whereas the main reabsorption of approximately 70% occurs in the loop of Henle, primarily in the thick ascending limb (TAL). Only 5–10% of filtered Mg^{2+} is reabsorbed in the distal convoluted tubule, the part of the nephron, where the fine-tuning of renal excretion is achieved.[3]

doi: 10.1111/j.1749-6632.2012.06544.x

Claudins: paracellular pores in the kidney

Tight junctions

TJs are located in the junctional complex between two adjacent epithelial cells. They surround each cell like a belt and interact with each other. Thereby they form a continous net-like structure that defines the apical and the basolateral side of the cell layer.[2] The TJs are composed of several groups of integral membrane proteins such as occludin[4] and its splice variants,[5,6] the junctional adhesion molecules (JAMs),[7,8] tricellulin,[9] and finally the family of claudins and their splice variants.[10,11]

Since the discovery of claudin-1 and -2 by Furuse et al., 24 additional members have been identified in humans.[12–15] The encoded proteins range in molecular weight from 20 kDa to 33 kDa. All claudins consist of four transmembrane helices, two extracellular loops, and intracellular amino and carboxy termini. The first extracellular loop is believed to influence paracellular charge selectivity, whereas the second loop is probably responsible for the interaction between opposing claudins of adjacent cells. The carboxy terminus of most claudins ends with a sequence known as the PDZ-binding motif.[16] Through this binding motif, claudins are capable of interacting with the most important member of the ZO (Zonula occludens) proteins, ZO-1, thus linking themselves to the actin cytoskeleton of the cell.[17] All claudins also share the highly conserved amino acid sequence W-GLW-C-C in the first extracellular loop.

The function of different claudins has been addressed by numerous overexpression studies in different epithelial cell lines[11,18–20] as well as in different mouse models.[21–23]

Claudins are key components of the paracellular pathway. In general, they may either seal the intercellular space or allow selective transport of charged solutes and water. The paracellular pores formed by claudins exhibit characteristics similar to conventional transmembrane ion channels. Thus, the properties of the paracellular pore are selective ion permeability, pH dependency, and specific mole fraction effects.[2]

Electrophysiological studies have shown that there are at least two types of paracellular pores. One high-capacity pathway with a small pore radius of about 4 Å and a low-capacity channel with a pore radius of at least 7–8 Å. Interestingly, it is believed

that the former allows remodeling of the whole-TJ structure. Because it depends on the rate at which the TJ strands break apart and reseal, the paracellular transport occurs slower than through the small high-capacity pore.[24,25] As an alternative hypothesis, Krug et al. investigated the TJ protein, tricellulin, and suggested that the low-capacity transport occurs along the tricellular TJ and is specifically blocked by tricellulin.[26]

Claudin-16 and claudin-19

Among other TJ proteins claudin-16 (paracellin-1) and claudin-19 are expressed in the thick ascending limb of Henle's loop (TAL). Mutations in their genes could be linked to a rare hereditary tubular disorder, known as familial hypomagnesemia with hypercalciuria and nephrocalcinosis (FHHNC).[27,28] (Fig. 1)

More than 40 different *CLDN16* mutations have been identified in patients with the clinical signs of FHHNC.[29–36] Genotype–phenotype correlations suggest that the genotype may help to predict the clinical course in affected individuals.[29] In a subset of FHHNC patients with severe visual impairment, no mutations in *CLDN16* were detected indicating genetic heterogeneity. Subsequent molecular analysis revealed that mutations in another claudin gene, namely *CLDN19*, may also cause FHHNC. To date, eight different *CLDN19* mutations have been described[28,37–40] *CLDN19* mutations are associated with ocular abnormalities, such as severe myopia, nystagmus, and/or macular coloboma[28,41–44] (Fig. 2).

Whereas in humans claudin-16 is expressed in the kidney and salivary glands, claudin-19 is expressed in various tissues, with the highest expression in kidney and eye.[45–49] These expression patterns explain the different extrarenal phenotype of FHHNC caused by either *CLDN16* or *CLDN19* mutations. As described for other claudins, claudin-16 and -19 both play a role in organ development and various cancers.[50–52]

The human *CLDN16* gene on chromosome 3 has two in-frame start codons: the first encodes a 305 amino acid protein (33 kDa), and the second is located at the methionine 71 amino acids downstream (27 kDa).[18] Different genetic analyses and cell culture studies determined the second translational site as the correct initiation start site in humans, which

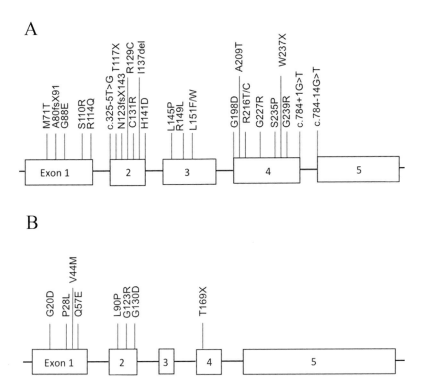

Figure 1. Schematic model of mutation locations in claudin-16 (A) and claudin-19 (B).

corresponds to the start codon previously described in other species.[18,34]

The human *CLDN19* gene on chromosome 1 encodes a 224 amino acid protein (23.8 kDa) with the typical characteristics of claudin family members.[47] Through alternative splicing two different isoforms can be formed in humans. The importance of these isoforms has to be investigated.

As previously mentioned, the TJ is believed to be a flexible, adjustable structure. The recruitment of several claudins to the TJ is performed throughout phosphorylation. The sites of phosphorylation and the enzymes involved differ between the claudins.[53,54] Claudin-16's subcellular localization seems to be determined by the phosphorylation at Ser217 by protein kinase A as a S217A mutant of claudin-16 was mislocalized to the lysosome.[55]

Interaction properties of claudins
As described above, all claudins have two extracellular loops, which configure the paracellular pore. These extracellular loops are essential contact points for the formation of TJ strands. Several claudins not only can provide homotypic interaction with identical claudins, but also can establish heterotypic inter-

actions with different claudins. However, not every theoretical combination is possible.[56] In addition to the intercellular interaction between claudins of adjacent cells (*trans*-interaction), claudins copolymerize side-by-side within the plasma membrane of one cell (*cis*-interaction).[57,58] Expression studies in murine L fibroblasts, which normally cannot form TJs, showed that exogenously expressed claudin-16 and claudin-19 do lead to the formation of well-developed TJ strands. Different missense mutations in either *CLDN19* (L90P and G123R) or *CLDN16* (F232C) that cause FHHNC lead to a loss of interaction between claudin-16 and claudin-19.[59] Several analyses of different mouse models support the importance of this interaction. In claudin-16 and -19 knockdown mice, the loss of claudin-16 or claudin-19 expression prevents the respective interacting claudin from reaching the TJ.[59] In an analysis of *Cldn16* knockout mice, significantly lower claudin-19 expression was observed.[23]

Paracellular transport in the TAL
The phenotype and the clinical implications resulting from mutations in these claudins can be derived from the physiological function of the TJ in the TAL.

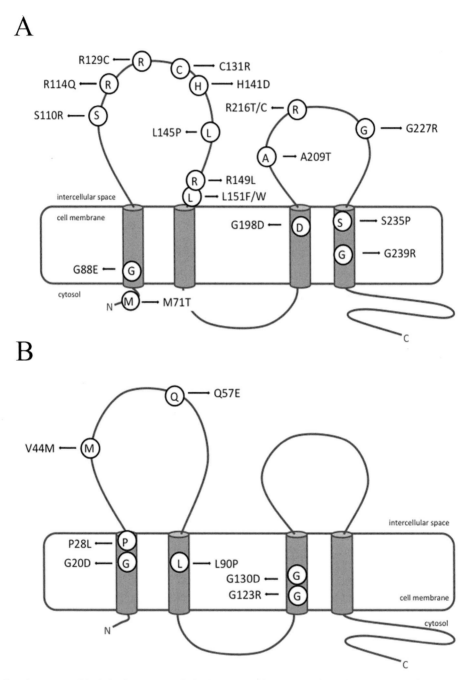

Figure 2. Schematic model of claudin-16 (A) and claudin-19 (B). The arrows indicate the amino acid changes caused by the different missense mutations.

In general, the TAL has two main functions: (1) absorption of salt, Mg^{2+}, and Ca^{2+}; and (2) generation of an osmotic gradient, which is required for water reabsorption in the following nephron segments. The sodium transport occurs through two different pathways: an active transcellular transport mediated by the sodium–potassium–chloride cotransporter 2 (NKCC2) in the apical membrane and the basolateral adenosine triphosphatase sodium–potassium pump, and passively along the paracellular route

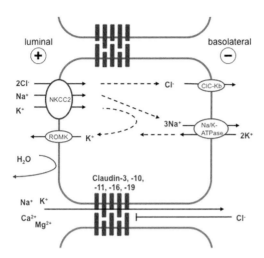

Figure 3. Schematic model of ion transport in the TAL. The lumen-positive potential is indicated by the circled plus. NKCC2 = sodium, potassium, two chloride cotransporter; ROMK = renal outer medullary potassium channel; ClC-Kb = chloride channel; Na/K-ATPase = sodium–potassium pump.

with the lumen-positive transepithelial potential as driving force. In addition to the sodium absorption, NKCC2 provides the potassium and chloride uptake into the cell. Whereas chloride ions leave the cell through chloride channels (ClC-Ka/b) at the basolateral side into the interstitium, only a small amount of the potassium is released at the basolateral membrane. The larger quantity is recycled back into the tubular fluid through specific potassium channels (ROMK) in the apical membrane of the TAL cells. This recycling mechanism is important for the driving force for the paracellular reabsorption of cations, because it enforces the lumen-positive potential (Fig. 3).

FHHNC

The typical triad of hypomagnesemia, hypercalciuria, and nephrocalcinosis characterizes the phenotype of FHHNC. This clinical presentation was first described in 1972 by Michelis *et al.* and named Michelis–Castrillo syndrome.[60] Clincal observations and ion excretion studies showed indirect evidence that the primary defect is related to an impaired Mg^{2+} handling in the TAL.[43] Genome-wide linkage analysis localized the gene responsible for the disease on chromosome 3. And finally, the gene was identified as *CLDN16*, formerly called *paracellin-1*.[27] The mode of inheritance is autosomal recessive.

Clinical presentation and course

In patients with *CLDN16* mutations, the first onset of symptoms occurs between infancy and early childhood, the mean age at first presentation ranged between 0.1 and 3 years in three different patient cohorts, and the mean age at diagnosis ranged between 0.5 and 12 years.[41,61,62] Recurrent urinary tract infections and polyuria and/or polydipsia are the most frequent clinical symptoms at presentation. In addition, the renal function is often already impaired and the patients develop chronic kidney disease that progresses to end-stage renal failure in the second to third decade of life.[29,41] Two different case studies reported that 30–75% of patients with a *CLDN16* mutation required renal replacement therapy within a decade from first diagnosis.[41,42] Interestingly, serum intact parathyroid hormone levels tend to be increased before the stage of advanced chronic kidney disease.

Additional clinical symptoms of FHHNC include kidney stone formation, failure to thrive, and vomiting, as well as abdominal pain. Less common are clinical signs of severe hypomagnesemia such as tetanic episodes or generalized seizures.[2] Other laboratory findings may include incomplete distal renal tubular acidosis, hypocitraturia, and hyperuricemia. As an extrarenal manifestation, a severe ocular phenotype has been reported repeatedly. Especially in the patients with *CLDN19* mutations visual impairment may be very disabling due to severe myopia, progressive retinopathy with the development of macular coloboma, and nystagmus.[28,34,42,63] A recently published case series suggested a neuromuscular involvement in *CLDN19* patients. These patients with genetically confirmed *CLDN19* mutations showed muscular exercise intolerance with limb stiffness and cramps.[39] This observation can be supported by findings in the *CLDN19* knockout mouse, which exhibited disorganized TJs in the Schwann cells of the peripheral nervous system, and knockout animals showed abnormal behavior and peripheral neuropathy.[21]

Therapeutic approaches

There is no specific therapy for FHHNC. Conservative management aims at the supplementation of Mg^{2+}, the prevention of recurrent urinary tract infections, and the treatment of sequelae resulting from progressive renal failure. To date, a satisfying explanation for the loss of GFR is still missing; thus, a specific treatment is currently not available. In the past, a close correlation between the progression of renal failure and the degree of nephrocalcinosis was reported. Therefore, therapeutic approaches aim at decreasing urinary calcium excretion. One short-term clinical study showed that treatment with thiazides decreases urinary calcium excretion in patients with FHHNC.[64] However, it is still unclear whether thiazides truly stop or slow down the decline of renal function observed in patients with FHHNC, as no correlation to the progression of GFR loss was made.

Interestingly, other hereditary tubular disorders associated with severe nephrocalcinosis, for example antenatal Bartter syndrome, do not uniformly lead to progressive loss of GFR. Conclusively, nephrocalcinosis alone cannot explain the clinical course in FHHNC.[65] An alternative explanation for the deterioration in renal function might be the progressive tubulointerstitial nephritis associated with FHHNC and/or the recurrent febrile urinary tract infections. Finally, the loss of renal function might be related to the gene defect itself, which is supported by the naturally occurring bovine claudin-16 knockout phenotype associated with an early onset of end-stage renal disease without histological signs for significant nephrocalcinosis.[66] Genotype/phenotype correlation studies performed in FHHNC patients support this idea since a more rapid decline in renal function could be observed in mutations, which lead to a complete loss of function compared to mutations with a residual function.[29] Recurrence of FHHNC after kidney transplantation has never been observed, because the primary defect resides in the kidney.

Pathophysiological principles

The key question is how the TJ complex, including claudin-16 and claudin-19, provides the specific ion handling of the TAL under physiological conditions and how mutations impair this function. In the last years, various electrophysiological studies investigated the function of claudin-16 and claudin-19 in different cell lines and mice models. However, these studies yielded contradictory results, at the first glance or at least to some extent. But it has to be kept in mind that one important factor in determining the electrophysiological function of claudins in cell culture is the endogenous expression of other claudins, as several different claudins configure the

TJ complex and therefore define the paracellular pathway. Thus, the conflicting data could just reflect the different TJ protein composition and the different interaction patterns in the investigated cell lines.

Two alternative hypotheses have been proposed to explain the pathophysiology of FHHNC. The first suggests that claudin-16 and/or claudin-19 function mainly as a cation-selective paracellular pore with an increased permeability especially for divalent ions as Ca^{2+} and Mg^{2+}.[18,30] The second model proposes that claudin-16 functions primarily as a sodium-specific paracellular pore, whereas claudin-19 acts as a chloride block preventing the paracellular backleak of chloride into the lumen along the lumen-positive potential, especially at the end of the TAL. This passage would lead to a decline of the lumen-positive potential and therefore of the driving force for the reabsorption of cations. Hence, claudin-16 and -19 work synergistically by establishing the electrophysiological environment, which is necessary for the reabsorption of cations in the TAL. In this model, a loss-of-function mutations in claudin-16 and -19 would be predicted to abrogate the electrical driving force for the cation reabsorption in the TAL.[37,67,68]

Interestingly, this variability can also be observed in different FHHNC mice models with a knockout or knockdown approach respectively. The phenotype in each model is compatible with the clinical presentation of FHHNC in humans, although they differ slightly from each other: while the claudin-16 knockdown shows a similar effect on Mg^{2+} and Ca^{2+} excretion, the knockout model reveals a more pronounced effect on the Ca^{2+} than on Mg^{2+} excretion.[22,23] In conclusion, the results of the different *in vitro* as well as *in vivo* approaches propose that a claudin cannot be seen isolated from other TJ proteins in its physiological function[69] and it probably do not hold one specific function. Furthermore, while the pathomechanism in FHHNC may result from one mutated protein, there are various consequences for other claudins and throughout that for the overall TJ function.

Summary

In conclusion, the molecular dissection of the underlying cause of a rare hereditary tubular disorder, FHHNC, has provided an opportunity to understand the role of claudins in the function of the TAL. While there is information about their physiological role in regulating tubular transport of fluid and electrolytes in various cell lines and different mouse models, there are still many open questions and unexplained symptoms.

Importantly, despite all of these investigations and new insights into the physiological function of claudins in the TAL, the clinical management of FHHNC has not changed over the last decade, and the clinical outcome of patients has not improved. This underscores the need for further research to fully understand the disease pathophysiology. This knowledge will hopefully serve as a starting point for the development of specific therapeutic approaches; the main goal of such interventions will be the prevention of chronic renal failure at young ages.

Conflicts of interest

The authors declare no conflicts of interest.

References

1. Swaminathan, R. 2003. Magnesium metabolism and its disorders. *Clin. Biochem. Rev.* **24:** 47–66.
2. Hou, J. 2010. Chapter 7: claudins and renal magnesium handling. *Curr. Topic Membr.* **65:** 151–176.
3. Quamme, G.A. & C. de Rouffignac. 2000. Epithelial magnesium transport and regulation by the kidney. *Front Biosci.* **5:** D694–D711.
4. Furuse, M. *et al.* 1993. Occludin: a novel integral membrane protein localizing at tight junctions. *J. Cell Biol.* **123**(6 Pt 2): 1777–1788.
5. Ghassemifar, M.R. *et al.* 2002. Occludin TM4(–): an isoform of the tight junction protein present in primates lacking the fourth transmembrane domain. *J. Cell Sci.* **115**(Pt 15): 3171–3180.
6. Muresan, Z., D.L. Paul & D.A. Goodenough. 2000. Occludin 1B, a variant of the tight junction protein occludin. *Mol. Biol. Cell* **11:** 627–634.
7. Ebnet, K. *et al.* 2000. Junctional adhesion molecule interacts with the PDZ domain-containing proteins AF-6 and ZO-1. *J. Biol. Chem.* **275:** 27979–27988.
8. Vestweber, D. 2000. Molecular mechanisms that control endothelial cell contacts. *J. Pathol.* **190:** 281–291.
9. Ikenouchi, J. *et al.* 2005. Tricellulin constitutes a novel barrier at tricellular contacts of epithelial cells. *J. Cell Biol.* **171:** 939–945.
10. Gunzel, D. *et al.* 2009. Claudin-10 exists in six alternatively spliced isoforms that exhibit distinct localization and function. *J. Cell Sci.* **122**(Pt 10): 1507–1517.
11. Van Itallie, C.M. *et al.* 2006. Two splice variants of claudin-10 in the kidney create paracellular pores with different ion selectivities. *Am. J. Physiol. Renal Physiol.* **291:** F1288–F1299.
12. Furuse, M. *et al.* 1998. Claudin-1 and -2: novel integral membrane proteins localizing at tight junctions with no sequence similarity to occludin. *J. Cell Biol.* **141:** 1539–1550.
13. Tsukita, S., M. Furuse & M. Itoh. 2001. Multifunctional strands in tight junctions. *Nat. Rev. Mol. Cell Biol.* **2:** 285–293.
14. Itoh, M. *et al.* 2001. Junctional adhesion molecule (JAM) binds to PAR-3: a possible mechanism for the recruitment of PAR-3 to tight junctions. *J. Cell Biol.* **154:** 491–497.
15. Mineta, K. *et al.* 2011. Predicted expansion of the claudin multigene family. *FEBS Lett.* **585:** 606–612.
16. Morita, K. *et al.* 1999. Claudin multigene family encoding four-transmembrane domain protein components of tight junction strands. *Proc. Natl. Acad. Sci. USA* **96:** 511–516.
17. Itoh, M. *et al.* 1999. Direct binding of three tight junction-associated MAGUKs, ZO-1, ZO-2, and ZO-3, with the COOH termini of claudins. *J. Cell Biol.* **147:** 1351–1363.
18. Hou, J., D.L. Paul & D.A. Goodenough. 2005. Paracellin-1 and the modulation of ion selectivity of tight junctions. *J. Cell Sci.* **118:** 5109–5118.
19. Angelow, S., E.E. Schneeberger & A.S. Yu. 2007. Claudin-8 expression in renal epithelial cells augments the paracellular barrier by replacing endogenous claudin-2. *J. Membr. Biol.* **215:** 147–159.
20. Gunzel, D. *et al.* 2009. Claudin-16 affects transcellular Cl-secretion in MDCK cells. *J. Physiol.* **587**(Pt 15): 3777–3793.
21. Miyamoto, T. *et al.* 2005. Tight junctions in Schwann cells of peripheral myelinated axons: a lesson from claudin-19-deficient mice. *J. Cell Biol.* **169:** 527–538.
22. Hou, J. *et al.* 2007. Transgenic RNAi depletion of claudin-16 and the renal handling of magnesium. *J. Biol. Chem.* **282:** 17114–17122.
23. Will, C. *et al.* 2010. Targeted deletion of murine Cldn16 identifies extra- and intrarenal compensatory mechanisms of Ca2+ and Mg2 +wasting. *Am. J. Physiol. Renal Physiol.* **298:** F1152–F1161.
24. Tang, V.W. and D.A. Goodenough. 2003. Paracellular ion channel at the tight junction. *Biophys. J.* **84:** 1660–1673.
25. Steed, E., M.S. Balda & K. Matter. 2010. Dynamics and functions of tight junctions. *Trends Cell Biol.* **20:** 142–149.
26. Krug, S.M. *et al.* 2009. Tricellulin forms a barrier to macromolecules in tricellular tight junctions without affecting ion permeability. *Mol. Biol. Cell* **20:** 3713–3724.
27. Simon, D.B. *et al.* 1999. Paracellin-1, a renal tight junction protein required for paracellular Mg2+ resorption. *Science* **285:** 103–106.
28. Konrad, M. *et al.* 2006. Mutations in the tight-junction gene claudin 19 (CLDN19) are associated with renal magnesium wasting, renal failure, and severe ocular involvement. *Am. J. Hum. Genet.* **79:** 949–957.
29. Konrad, M. *et al.* 2008. CLDN16 genotype predicts renal decline in familial hypomagnesemia with hypercalciuria and nephrocalcinosis. *J. Am. Soc. Nephrol.* **19:** 171–181.
30. Kausalya, P.J. *et al.* 2006. Disease-associated mutations affect intracellular traffic and paracellular Mg2+ transport function of Claudin-16. *J. Clin. Invest.* **116:** 878–891.
31. Blanchard, A. *et al.* 2001. Paracellin-1 is critical for magnesium and calcium reabsorption in the human thick ascending limb of Henle. *Kidney Int.* **59:** 2206–2215.
32. Muller, D. *et al.* 2006. Unusual clinical presentation and possible rescue of a novel claudin-16 mutation. *J. Clin. Endocrinol. Metab.* **91:** 3076–3079.
33. Weber, S. *et al.* 2000. Familial hypomagnesaemia with hypercalciuria and nephrocalcinosis maps to chromosome 3q27

and is associated with mutations in the PCLN-1 gene. *Eur. J. Hum. Genet* **8:** 414–422.

34. Weber, S. *et al.* 2001. Primary gene structure and expression studies of rodent paracellin-1. *J. Am. Soc. Nephrol.* **12:** 2664–2672.

35. Kasapkara, C.S. *et al.* 2011. A novel mutation of the claudin 16 gene in familial hypomagnesemia with hypercalciuria and nephrocalcinosis mimicking rickets. *Genet. Couns.* **22:** 187–192.

36. Gunzel, D. *et al.* 2009. Claudin function in the thick ascending limb of Henle's loop. *Ann. N. Y. Acad. Sci.* **1165:** 152–162.

37. Hou, J. *et al.* 2008. Claudin-16 and claudin-19 interact and form a cation-selective tight junction complex. *J. Clin. Invest.* **118:** 619–628.

38. Haisch, L. *et al.* 2011. The role of tight junctions in paracellular ion transport in the renal tubule: lessons learned from a rare inherited tubular disorder. *Am. J. Kidney Dis.* **57:** 320–330.

39. Naeem, M., S. Hussain, & N. Akhtar. 2011. Mutation in the tight-junction gene claudin 19 (CLDN19) and familial hypomagnesemia, hypercalciuria, nephrocalcinosis (FHHNC) and severe ocular disease. *Am. J. Nephrol.* **34:** 241–248.

40. Faguer, S., D. Chauveau, P. Cintas, I. *et al.* 2011. Renal, ocular, and neuromuscular involvements in patients with CLDN19 mutations. *Clin. J. Am. Soc. Nephrol.* **6:** 355–360. doi:10.2215/CJN.02870310.

41. Benigno, V. *et al.* 2000. Hypomagnesaemia-hypercalciuria-nephrocalcinosis: a report of nine cases and a review. *Nephrol. Dial. Transplant* **15:** 605–610.

42. Praga, M. *et al.* 1995. Familial hypomagnesemia with hypercalciuria and nephrocalcinosis. *Kidney Int.* **47:** 1419–1425.

43. Rodriguez-Soriano, J., A. Vallo & M. Garcia-Fuentes. 1987. Hypomagnesaemia of hereditary renal origin. *Pediatr. Nephrol.* **1:** 465–472.

44. Rodriguez-Soriano, J. & A. Vallo. 1994. Pathophysiology of the renal acidification defect present in the syndrome of familial hypomagnesaemia-hypercalciuria. *Pediatr. Nephrol.* **8:** 431–435.

45. Hewitt, K.J., R. Agarwal & P.J. Morin. 2006. The claudin gene family: expression in normal and neoplastic tissues. *BMC Cancer* **6:** 186.

46. Van Itallie, C.M. & J.M. Anderson. 2006. Claudins and epithelial paracellular transport. *Annu. Rev. Physiol.* **68:** 403–429.

47. Lee, N.P. *et al.* 2006. Kidney claudin-19: localization in distal tubules and collecting ducts and dysregulation in polycystic renal disease. *FEBS Lett.* **580:** 923–931.

48. Angelow, S. *et al.* 2007. Renal localization and function of the tight junction protein, claudin-19. *Am. J. Physiol. Renal Physiol.* **293:** F166–F177.

49. Kriegs, J.O. *et al.* 2007. Identification and subcellular localization of paracellin-1 (claudin-16) in human salivary glands. *Histochem. Cell Biol.* **128:** 45–53.

50. Martin, T.A. *et al.* 2008. Claudin-16 reduces the aggressive behavior of human breast cancer cells. *J. Cell Biochem.* **105:** 41–52.

51. Ozden, O. *et al.* 2010. Developmental profile of claudin-3, -5, and -16 proteins in the epithelium of chick intestine. *Anat. Rec.* **293:** 1175–1183.

52. Holmes, J.L. *et al.* 2006. Claudin profiling in the mouse during postnatal intestinal development and along the gastrointestinal tract reveals complex expression patterns. *Gene. Expr. Patterns* **6:** 581–588.

53. D'Souza, T., F.E. Indig & P.J. Morin. 2007. Phosphorylation of claudin-4 by PKCepsilon regulates tight junction barrier function in ovarian cancer cells. *Exp. Cell Res.* **313:** 3364–3375.

54. Andreeva, A.Y. *et al.* 2006. Assembly of tight junction is regulated by the antagonism of conventional and novel protein kinase C isoforms. *Int. J. Biochem. Cell Biol.* **38:** 222–233.

55. Ikari, A. *et al.* 2006. Phosphorylation of paracellin-1 at Ser217 by protein kinase A is essential for localization in tight junctions. *J. Cell Sci.* **119:** 1781–1789.

56. Tsukita, S. & M. Furuse. 2002. Claudin-based barrier in simple and stratified cellular sheets. *Curr. Opin. Cell Biol.* **14:** 531–536.

57. Piontek, J. *et al.* 2008. Formation of tight junction: determinants of homophilic interaction between classic claudins. *FASEB J.* **22:** 146–158.

58. Hou, J. & D.A. Goodenough. 2010. Claudin-16 and claudin-19 function in the thick ascending limb. *Curr. Opin. Nephrol. Hypertens* **19:** 483–488.

59. Hou, J. *et al.* 2009. Claudin-16 and claudin-19 interaction is required for their assembly into tight junctions and for renal reabsorption of magnesium. *Proc. Natl. Acad. Sci. USA* **106:** 15350–15355.

60. Michelis, M.F. *et al.* 1972. Decreased bicarbonate threshold and renal magnesium wasting in a sibship with distal renal tubular acidosis. (Evaluation of the pathophysiological role of parathyroid hormone). *Metabolism* **21:** 905–920.

61. Weber, S. *et al.* 2001. Novel paracellin-1 mutations in 25 families with familial hypomagnesemia with hypercalciuria and nephrocalcinosis. *J. Am. Soc. Nephrol.* **12:** 1872–1881.

62. Kari, J.A., M. Farouq & H.O. Alshaya. 2003. Familial hypomagnesemia with hypercalciuria and nephrocalcinosis. *Pediatr. Nephrol.* **18:** 506–510.

63. Nicholson, J.C. *et al.* 1995. Familial hypomagnesaemia–hypercalciuria leading to end-stage renal failure. *Pediatr. Nephrol.* **9:** 74–76.

64. Zimmermann, B. *et al.* 2006. Hydrochlorothiazide in CLDN16 mutation. *Nephrol. Dial. Transplant* **21:** 2127–2132.

65. Brochard, K. *et al.* 2009. Phenotype-genotype correlation in antenatal and neonatal variants of Bartter syndrome. *Nephrol. Dial. Transplant* **24:** 1455–1464.

66. Ohba, Y. *et al.* 2000. A deletion of the paracellin-1 gene is responsible for renal tubular dysplasia in cattle. *Genomics* **68:** 229–236.

67. Angelow, S. & A.S. Yu. 2007. Claudins and paracellular transport: an update. *Curr. Opin. Nephrol. Hypertens.* **16:** 459–464.

68. Li, J., W. Ananthapanyasut & A.S. Yu. 2011. Claudins in renal physiology and disease. *Pediatr. Nephrol.* **26:** 2133–2142.

69. Shan, Q. *et al.* 2009. Insights into driving forces and paracellular permeability from claudin-16 knockdown mouse. *Ann. N. Y. Acad. Sci.* **1165:** 148–151.

Ann. N.Y. Acad. Sci. ISSN 0077-8923

ANNALS OF THE NEW YORK ACADEMY OF SCIENCES
Issue: *Barriers and Channels Formed by Tight Junction Proteins*

The yin and yang of claudin-14 function in human diseases

Jianghui Hou

Renal Division, Washington University Medical School, St. Louis, Missouri

Address for correspondence: Jianghui Hou, Department of Internal Medicine—Renal Division, Washington University School of Medicine, 660 South Euclid Avenue, Campus Box 8126, St. Louis, MO 63110. jhou@dom.wustl.edu

Claudins are tight junction integral membrane proteins that are key regulators of the paracellular pathway. The paracellular pathways in the inner ear and in the kidney are predominant routes for transepithelial cation transport. Mutations in claudin-14 cause nonsyndromic recessive deafness DFNB29. A recent genome-wide association study has identified claudin-14 as a major risk gene of hypercalciuric nephrolithiasis. *In vitro* analyses show that claudin-14 functions as a cation barrier in epithelial cells. The barrier function of claudin-14 is crucial for generating the K^+ gradient between perilymph and endolymph in the inner ear. However, neither homozygous individuals with DFNB29 mutations nor claudin-14 knockout mice show any renal dysfunction. In this short review, I discuss several possible mechanisms to integrate the physiological function of claudin-14 in the inner ear and the kidney.

Keywords: tight junction; ion channel; kidney; inner ear

Introduction

The tight junction (zonula occludens; TJ) is the most apical member of the junctional complex[1] found in vertebrate epithelia and is responsible for the barrier to movement of ions and molecules between apical and basal compartments, the paracellular pathway.[2] The known integral membrane proteins of the TJ include occludin (a 65 kDa membrane protein bearing four transmembrane domains and two uncharged extracellular loops),[3] the junctional adhesion molecules (JAMs),[4] a four-member group of glycosylated proteins, and the claudins. Claudins (CLDNs) are tetraspan proteins consisting of a family of at least 22 members.[5–6] They range in molecular mass from 20–28 kD with charged extracellular loops. The cytoplasmic C-terminus of most claudins ends with a PDZ (*p*ostsynaptic density 95/*d*iscs large/*z*onula occludens-1) binding domain that is critical for interaction with the submembrane scaffold protein ZO-1 and correct localization in the TJ.[7–8]

Claudin mutations have serious consequences, suggesting defects in epithelial ion flux. CLDN1-deficient mice die within one day of birth and show a loss of the water barrier function of skin.[9] CLDN2 knockout mice show salt-wasting defects,

presumably through leaky junctions in the proximal tubules of the kidney.[10] Targeted deletion of CLDN5, which is known to be expressed in vascular endothelia as well as other locations,[11] results in a selective increase in brain vascular permeability to molecules <800 daltons.[12] Targeted disruption of the CLDN11 gene results in severe demyelination and male sterility, consistent with the presence of this protein at the nodes of Ranvier and in Sertoli TJs, leading to disrupted ionic balances.[13] CLDN16, also known as paracellin-1, has been genetically linked to the inherited disorder FHHNC (familial hypomagnesemia with hypercalciuria and nephrocalcinosis OMIM 248250).[14] Many different FHHNC mutations have been identified in the CLDN16 gene.[15–16] The expression of CLDN16 is restricted to the thick ascending limb (TAL) of the nephron in the kidney.[14] CLDN19 mutations have also been associated with human FHHNC and renal Mg^{2+} loss.[17] While targeted deletion of CLDN19 in mice initially focused on its role in peripheral myelin,[18] promoter analysis[19] and subsequent studies[20] have emphasized the presence of CLDN19 in the TAL of the nephron (colocalizing with CLDN16 in the kidney).

In renal epithelia, claudins have been shown to confer ion selectivity to the paracellular pathway,

doi: 10.1111/j.1749-6632.2012.06529.x

resulting in differences in TER and paracellular permeabilities. Studies have shown that CLDN4, -5, -8, -11, and -14 selectively decrease the permeability of cations through TJs,[21–25] specifically to Na$^+$, K$^+$, H$^+$, and ammonium. CLDN2 and -15 increase cation permeability.[26–28] These properties have been attributed to charged amino acids in the first extracellular domain.[29] These and other studies[30] have led to models of the claudins forming paracellular channels, a novel class of channels oriented perpendicular to the membrane plane and serving to join two extracellular compartments.[31] Measurement of paracellular permeability using cell membrane–impermeable tracers indicate that there are 4–7 Å diameter channels in the TJ.[30,32–33] The paracellular channels in the TJ have properties of ion selectivity, pH dependence, and anomalous mole fraction effects, similar to conventional transmembrane channels.[30]

CLDN14 function in the inner ear

The cochlea maintains two distinct fluid compartments, the scala vestibuli/tympani and the scala media, which are filled with the perilymph and endolymph fluids. The perilymph fluid resembles the extracellular fluid, while the endolymph fluid has the characteristics of an intracellular fluid, with high K$^+$ and low Na$^+$ concentrations. The TJ barrier separates perilymph from endolymph fluid and maintains the ionic composition of each fluid. Mutations in CLDN14 cause nonsyndromic recessive deafness DFNB29 in humans.[34] CLDN14 knockout mice are also deaf, owing to rapid degeneration of the cochlear outer hair cells (OHC).[23] The OHC loss in CLDN14 KO mice starts at P8–9 and coincides with a critical developmental stage when the endocochlear potential is generated. *In vitro* experiments have demonstrated that CLDN14, when overexpressed in epithelial cells, selectively blocks the paracellular permeation of cations, including K$^+$.[23] The paracellular barrier provided by CLDN14 would be required for maintaining the K$^+$ gradient between perilymph and endolymph. A loss of CLDN14 would result in elevated K$^+$ concentration in the space of Nuel, which now becomes toxic to the OHCs.

CLDN14 function in the kidney

There are segment-specific claudin expression profiles along the length of the nephron. Northern analysis of mouse kidneys using probes specific for CLDN1–19 reveal that only CLDN6, -9, and -13 are not detectable, and CLDN5 and -15 only in endothelial cells; the rest are specifically expressed in different segments of the nephron.[35] Using antisera available at the time to perform immunostaining on mouse kidneys,[20,35] CLDN3, -10, -11, -16, and -19 were observed in the thick ascending limb (TAL); CLDN3 and -8 in the distal convoluted tubule; and CLDN3, -4, and -8 in the collecting duct (CLDN4 was also observed in the thick ascending limb,[35] although absent in the bovine TAL[36]). CLDN2 is highly expressed in the "leaky" proximal nephron,[37] consistent with its high cation permselectivity when expressed in MDCK cells.[26–27] CLDN4 and -8 are expressed primarily along the aldosterone-sensitive distal nephron, and in inner medullary segments of the thin descending limbs of juxtamedullary nephrons.[38–39] Immunofluorescence analysis has shown that CLDN7 is expressed in the thick ascending limbs of Henle's loop and collecting ducts of porcine and rat kidneys,[40] although another study described CLDN7 in the distal nephron as located primarily on the basolateral membrane.[39] In summary, while there are still some conflicting published data, CLDN2, -10, -11, -17, and -18 are expressed in proximal tubules, while CLDN3, -4, -7, -8, -10, -16, and -19 have been reported in the thick ascending limbs and the distal nephron. The renal localization of CLDN14 has been controversial. Immunofluorescence analysis showed CLDN14 gene expression in the TAL and the proximal tubules of mouse kidneys,[41] while another study reported no CLDN14 expression in the kidney.[35] The thick ascending limb reabsorbs a major percentage of filtered divalent cations (30–35% Ca^{2+} and 50–60% Mg^{2+}).[42] At this segment, Ca^{2+} and Mg^{2+} are passively reabsorbed from the lumen to the interstitial space through the paracellular channel, driven by lumen-positive transepithelial voltage (V_{te}). V_{te} is generated by two mechanisms: (1) the active transport V_{te} owing to apical K$^+$ recycling through ROMK and basolateral Cl$^-$ exit through ClC-Kb, coupled with NaCl reabsorption via the apical Na$^+$2Cl$^-$K$^+$ cotransporter (NKCC2); and (2) the diffusion V_{te} generated by a transepithelial NaCl concentration gradient through the cation selective paracellular channel of the TAL. A run of *in vitro*[43–44] and *in vivo*[45–46] studies have shown that CLDN16 and -19 form a

A

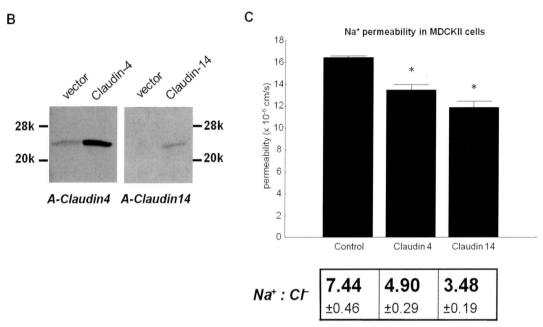

Figure 1. Influences of CLDN4 and -14 on Na⁺ permeation through the TJ. (A) Alignment of amino acid sequences of the first extracellular loop of CLDN4, -14, and -16 (paracellin-1). Note the relative abundance of positively charged amino acids in the first extracellular loop of CLDN4 and -14 (labeled in blue) and the abundance of negatively charged amino acids in CLDN16 (labeled in red). (B) Expression of CLDN4 and -14 in MDCK cell membranes. CLDN4 and -14 migrate as a 21–22 kDa band. Note that MDCK cells express endogenous CLDN4 but not CLDN14. (C) Both CLDN4 and -14 suppress the permeability of Na⁺ and show discrimination against cation (Na⁺) over anion (Cl⁻). *$P < 0.01$; $n = 4$.

heteromeric cation channel, which confers cation selectivity including Ca^{2+} and Mg^{2+} and generates the diffusion V_{te}. Although mutations in CLDN16 or -19 have been linked to a severe renal phenotype–familial hypomagnesemia with hypercalciuria and nephrocalcinosis (FHHNC), CLDN14 mutations found in recessive deafness DFNB29 cause no renal defect in affected homozygous individuals.[34]

Analyses of CLDN14 function in kidney epithelial cells

Ben-Yosef et al.[23] first determined the electrophysiological properties of CLDN14 channel in transfected kidney MDCK cells. Overexpression of CLDN14 induced a sixfold increase in transepithelial resistance (TER), accompanied by a significant decrease in cation selectivity (P_{Na}/P_{Cl}).[23] The calculated Na⁺ permeability (P_{Na}) through the CLDN14 channel was reduced by 72.5% compared to mock trans-

fection, while the Cl⁻ permeability (P_{Cl}) was not affected. Bi-ionic potentials found the permeability sequence to be $K^+ > Na^+ > Rb^+ > Li^+ > Cs^+$ that resembled the Eisenman selectivity sequence V–VIII,[23] suggesting high field strength within the CLDN14 channel pore. In a preliminary study, I have expressed CLDN14 in MDCK cells with a retroviral infection approach and compared its electrophysiological properties with the CLDN4 channel (Fig. 1). Similar to CLDN4, the first extracellular loop (ECL1) of CLDN14 is enriched with positively charged amino acids while the ECL1 of CLDN16 is abundant with negatively charged amino acids (Fig. 1A). The MDCK cells express a low level of endogenous CLDN4 proteins but no CLDN14 (Fig. 1B). Ectopic expression induces significant increases in expression of both claudins. The CLDN14 protein has a similar molecular weight of 21–22 kDa compared to CLDN4 (Fig. 1B). Both claudins show discrimination against cation (Na$^+$) over an-ion (Cl$^-$), reflected by a significant decrease in dilution potential (Fig. 1C). CLDN14 is a more potent blocker to Na$^+$ permeation than CLDN4 (Fig. 1C), although the protein level of CLDN14 seems lower than CLDN4 (Fig. 1B).

The CLDN14 paradox

A recent genome-wide association study has identified CLDN14 as a major risk gene of hypercalciuric nephrolithiasis.[47] Metabolic abnormalities such as hypercalciuria, metabolic acidosis, and bone mineral loss, in addition to kidney stones, have been associated with common synonymous sequence variants (i.e., single-nucleotide polymorphisms, SNPs) in the CLDN14 gene.[47] The previously discovered rare mutations (398delT and T254A) that associate with deafness were not identified in any kidney stone patient.[34] The 398delT mutation causes a frameshift within the codon of Met133, which substitutes 23 incorrect amino acids and prematurely terminates translation 69 nucleotides later.[34] The T254A mutation substitutes aspartic acid for valine (V85D).[34] Valine 85 is a conserved site within the second transmembrane domain (TM2) of CLDN14 protein. Aspartic acid at position 85 is predicted to affect hydrophobicity and disrupt the secondary structure of TM2. Therefore, both 398delT and T254A are loss-of-function mutations. Nevertheless, none of the deaf individuals homozygous for these mutations shows signs of renal dysfunction. CLDN14 KO mice show normal renal functions, ruling out any direct renal requirement of CLDN14. If the physiological function of CLDN14 is to provide a cation barrier against K$^+$ permeation between perilymph and endolymph in the inner ear, this similar function will not be required for maintaining normal kidney physiology. On the other hand, the mechanism provided by CLDN14 in kidney stone diseases must not affect the physiological function of CLDN14 in the inner ear, that is, the cation barrier function. Because the formation of kidney stones is a common condition and deafness a rare disease, the functional role played by CLDN14 in the two diseases should be different. There are two hypotheses. First, the function of CLDN14 depends on its binding partner. It is well known that different claudins interact to form an oligomer. Because the binding partners are different in the inner ear and the kidney, the CLDN14

functions vary accordingly. Second, the function of CLDN14 is the same in the two organs but its physiological roles are different. The CLDN14 channel is required for the barrier function in the inner ear. The same barrier function plays a negative regulatory role in the kidney. CLDN14 may actually block Ca^{2+} permeation in the kidney, unlike CLDN16, and CLDN19 that permeate Ca^{2+}. Knocking out CLDN14 will not damage normal Ca^{2+} reabsorption but instead protect against hypercalciuria. The mechanism underlying kidney stone pathogenesis is provided by a gain-of-function effect of CLDN14 rather than a loss-of-function effect. These two hypotheses are now being tested in my lab. We will ask: (1) What are the binding partners for CLDN14? (2) What is the function of CLDN14 in the kidney? (3) Is CLDN14 required for renal Ca^{2+} homeostasis?

In conclusion, CLDN14 is an important TJ molecule for a broad range of epithelial physiologies. Mutations in CLDN14 have been found in both rare (monogenic) and common (polygenic) forms of human genetic diseases. Although preliminary studies have suggested a role for CLDN14 as cation blocker in the inner ear, its precise role in the kidney is still elusive. I have hypothesized two mechanisms to unite the seemingly different functions of CLDN14 in the physiology of the inner ear and the kidney.

Conflicts of interest

The author declares no conflicts of interest.

References

1. Farquhar, M.G. & G.E. Palade. 1963. Junctional complexes in various epithelia. *J. Cell Biol.* **17:** 375–412.
2. Miller, F. 1960. Hemoglobin absorption by the cell of the proximal convoluted tubule in mouse kidney. *J. Biophys. Biochem. Cytol.* **8:** 689–718.
3. Furuse, M., T. Hirase, M. Itoh, *et al.* 1993. Occludin—a novel integral membrane protein localizing at tight junctions. *J. Cell Biol.* **123:** 1777–1788.
4. Ebnet, K., A. Suzuki, S. Ohno & D. Vestweber. 2004. Junctional adhesion molecules (JAMs): more molecules with dual functions? *J. Cell Sci.* **117:** 19–29.
5. Tsukita, S., M. Furuse & M. Itoh. 2001. Multifuctional strands in tight junctions. *Nat. Rev. Molecul. Cell Biol.* **2:** 285–293.
6. Morita, K., M. Furuse, K. Fujimoto & S. Tsukita. 1999. Claudin multigene family encoding four-transmembrane domain protein components of tight junction strands. *Proc. Natl. Acad. Sci. USA* **96:** 511–516.
7. Hamazaki, Y., M. Itoh, H. Sasaki, *et al.* 2001. Multi-PDZ-containing protein 1 (MUPP1) is concentrated at tight

junctions through its possible interaction with claudin-1 and junctional adhesion molecule (JAM). *J. Biol. Chem.* **277:** 455–461.

8. Itoh, M., M. Furuse, K. Morita, *et al.* 1999. Direct binding of three tight junction-associated MAGUKs, ZO-1, ZO-2, and ZO-3, with the COOH termini of claudins. *J. Cell Biol.* **147:** 1351–1363.

9. Furuse, M., M. Hata, K. Furuse, *et al.* 2002. Claudin-based tight junctions are crucial for the mammalian epidermal barrier: a lesson from claudin-1-deficient mice. *J. Cell Biol.* **156:** 1099–1111.

10. Muto, S., M. Hata, J. Taniguchi, *et al.* 2010. Claudin-2-deficient mice are defective in the leaky and cation-selective paracellular permeability properties of renal proximal tubules. *Proc. Natl. Acad. Sci. USA* **107:** 8011–8016.

11. Morita, K., H. Sasaki, K. Furuse, *et al.* 2003. Expression of claudin-5 in dermal vascular endothelia. *Exp. Dermatol.* **12:** 289–295.

12. Nitta, T., M. Hata, S. Gotoh, *et al.* 2003. Size-selective loosening of the blood-brain barrier in claudin-5-deficient mice. *J. Cell Biol.* **161:** 653–660.

13. Gow, A., C.M. Southwood, J.S. Li, *et al.* 1999. CNS myelin and sertoli cell tight junction strands are absent in OSP/claudin-11 null mice. *Cell* **99:** 649–659.

14. Simon, D.B., Y. Lu, K.A. Choate, *et al.* 1999. Paracellin-1, a renal tight junction protein required for paracellular Mg^{2+} resorption. *Science* **285:** 103–106.

15. Weber, S., L. Schneider, M. Peters, *et al.* 2001. Novel paracellin-1 mutations in 25 families with familial hypomagnesemia with hypercalciuria and nephrocalcinosis. *J. Am. Soc. Nephrol.* **12:** 1872–1881.

16. Konrad, M., J. Hou, S. Weber, *et al.* 2008. The *CLDN16* genotype predicts the progression of renal failure in familial hypomagnesemia with hypercalciuria and nephrocalcinosis. *J. Am. Soc. Nephrol.* **19:** 171–181.

17. Konrad, M., A. Schaller, D. Seelow, *et al.* 2006. Mutations in the tight-junction gene claudin 19 (CLDN19) are associated with renal magnesium wasting, renal failure, and severe ocular involvement. *Am. J. Hum. Genet.* **79:** 949–957.

18. Miyamoto, T., K. Morita, D. Takemoto, *et al.* 2005. Tight junctions in Schwann cells of peripheral myelinated axons: a lesson from claudin-19-deficient mice. *J. Cell Biol.* **169:** 527–538.

19. Luk, J.M., M.K. Tong, B.W. Mok, *et al.* 2004. Sp1 site is crucial for the mouse claudin-19 gene expression in the kidney cells. *FEBS Lett.* **578:** 251–256.

20. Angelow, S., R. El-Husseini, S.A. Kanzawa & A.S. Yu. 2007. Renal localization and function of the tight junction protein, claudin-19. *Am. J. Physiol. Renal. Physiol.* **293:** F166–F177.

21. Van Itallie, C., C. Rahner & J.M. Anderson. 2001. Regulated expression of claudin-4 decreases paracellular conductance through a selective decrease in sodium permeability. *J. Clin. Invest.* **107:** 1319–1327.

22. Colegio, O.R., C.M. Van Itallie, H.J. McCrea, *et al.* 2002. Claudins create charge-selective channels in the paracellular pathway between epithelial cells. *Am. J. Physiol. Cell Physiol.* **283:** C142–C147.

23. Ben-Yosef, T., I.A. Belyantseva, T.L. Saunders, *et al.* 2003. Claudin 14 knockout mice, a model for autosomal recessive deafness DFNB29, are deaf due to cochlear hair cell degeneration. *Hum. Mol. Genet.* **12:** 2049–2061.

24. Yu, A.S., A.H. Enck, W.I. Lencer & E.E. Schneeberger. 2003. Claudin-8 expression in MDCK cells augments the paracellular barrier to cation permeation. *J. Biol. Chem.* **278:** 17350–17359.

25. Wen, H., D.D. Watry, M.C. Marcondes & H.S. Fox. 2004. Selective decrease in paracellular conductance of tight junctions: role of the first extracellular domain of claudin-5. *Mol. Cell Biol.* **24:** 8408–8417.

26. Furuse, M., K. Furuse, H. Sasaki & S. Tsukita. 2001. Conversion of zonulae occludentes from tight to leaky strand type by introducing claudin-2 into Madin-Darby canine kidney I cells. *J. Cell Biol.* **153:** 263–272.

27. Amasheh, S., N. Meiri, A.H. Gitter, *et al.* 2002. Claudin-2 expression induces cation-selective channels in tight junctions of epithelial cells. *J.Cell Sci.* **115:** 4969–4976.

28. Van Itallie, C.M., A.S. Fanning & J.M. Anderson. 2003. Reversal of charge selectivity in cation or anion selective epithelial lines by expression of different claudins. *Am. J. Physiol. Renal. Physiol.* **285:** F1078–F1084.

29. Colegio, O.R., C. Van Itallie, C. Rahner & J.M. Anderson. 2003. Claudin extracellular domains determine paracellular charge selectivity and resistance but not tight junction fibril architecture. *Am. J. Physiol. Cell Physiol.* **284:** C1246–C1254.

30. Tang, V.W. & D.A. Goodenough. 2003. Paracellular ion channel at the tight junction. *Biophys. J.* **84:** 1660–1673.

31. Tsukita, S. & M. Furuse. 2000. Pores in the Wall. Claudins constitute tight junction strands containing aqueous pores. *J. Cell Biol.* **149:** 13–16.

32. Watson, C.J., M. Rowland & G. Warhurst. 2001. Functional modeling of tight junctions in intestinal cell monolayers using polyethylene glycol oligomers. *Am. J. Physiol. Cell Physiol.* **281:** C388–C397.

33. Van Itallie, C.M., J. Holmes, A. Bridges, *et al.* 2008. The density of small tight junction pores varies among cell types and is increased by expression of claudin-2. *J. Cell Sci.* **121:** 298–305.

34. Wilcox, E.R., Q.L. Burton, B. Naz, *et al.* 2001. Mutations in the gene encoding tight junction claudin-14 cause autosomal recessive deafness DFNB29. *Cell* **104:** 165–172.

35. Kiuchi-Saishin, Y., S. Gotoh, M. Furuse, *et al.* 2002. Differential expression patterns of claudins, tight junction membrane proteins, in mouse nephron segments. *J. Am. Soc. Nephrol.* **13:** 875–886.

36. Ohta, H., H. Adachi, M. Takiguchi & M. Inaba. 2006. Restricted localization of claudin-16 at the tight junction in the thick ascending limb of henle's loop together with claudins 3, 4, and 10 in bovine nephrons. *J. Vet. Med. Sci.* **68:** 453–463.

37. Enck, A.H., U.V. Berger & A.S. Yu. 2001. Claudin-2 is selectively expressed in proximal nephron in mouse kidney. *Am. J. Physiol. Renal. Physiol.* **281:** F966–F974.

38. Le Moellic, C., S. Boulkroun, D. Gonzalez-Nunez, *et al.* 2005. Aldosterone and tight junctions: modulation of claudin-4 phosphorylation in renal collecting duct cells. *Am. J. Physiol. Cell Physiol.* **289:** C1513–C1521.

39. Li, W.Y., C.L. Huey & A.S. Yu. 2004. Expression of claudins 7 and 8 along the mouse nephron. *Am. J. Physiol. Renal. Physiol.* **286:** F1063–F1071.

40. Alexandre, M.D., Q. Lu & Y.H. Chen. 2005. Overexpression of claudin-7 decreases the paracellular Cl⁻ conductance and increases the paracellular Na⁺ conductance in LLC-PK1 cells. *J. Cell Sci.* **118:** 2683–2693.

41. Elkouby-Naor, L., Z. Abassi, A. Lagziel, *et al.* 2008. Double gene deletion reveals lack of cooperation between claudin 11 and claudin 14 tight junction proteins. *Cell Tissue Res.* **333:** 427–438.

42. Greger, R. 1985. Ion transport mechanisms in thick ascending limb of Henle's loop of mammalian nephron. *Physiol. Rev.* **65:** 760–797.

43. Hou, J., D.L. Paul & D.A. Goodenough. 2005. Paracellin-1 and the modulation of ion selectivity of tight junctions. *J. Cell Sci.* **118:** 5109–5118.

44. Hou, J., A. Renigunta, M. Konrad, *et al.* 2008. Claudin-16 and claudin-19 interact and form a cation-selective tight junction complex. *J. Clin. Invest.* **118:** 619–628.

45. Hou, J., Q. Shan, T. Wang, *et al.* 2007. Transgenic RNAi depletion of claudin-16 and the renal handling of magnesium. *J. Biol. Chem.* **282:** 17114–17122.

46. Hou, J., A. Renigunta, A.S. Gomes, *et al.* 2009. Claudin-16 and claudin-19 interaction is required for their assembly into tight junctions and for renal reabsorption of magnesium. *Proc. Natl. Acad. Sci. USA* **106:** 15350–15355.

47. Thorleifsson, G., H. Holm, V. Edvardsson, *et al.* 2009. Sequence variants in the CLDN14 gene associate with kidney stones and bone mineral density. *Nat. Genet.* **41:** 926–930.

Appendix

Materials and Methods

The following antibodies were used in this study: rabbit polyclonal anti-CLDN14 antibody (Zymed Laboratories); mouse monoclonal anti-CLDN4 (Zymed Laboratories). MDCK-II cells were cultured in Dulbecco's modified Eagle's medium supplemented with 10% FBS, penicillin/streptomycin, and 1 mM sodium pyruvate. The claudin expressing MDCK-II cells were lysed in 50 mM Tris (pH 8.0) by 25–30 repeated passages through a 25-gauge needle, followed by centrifugation at 5,000 g. The membranes of lysed cells were extracted using a CSK buffer (150 mM NaCl; 1% Triton X-100; 50 mM Tris, pH 8.0; and protease inhibitors). Lysates (containing 20 μg total membrane proteins) were subjected to SDS-PAGE and immunoblotting. Electrophysiological studies were performed on cell monolayers grown on porous filters (Transwell). Voltage and current clamps were performed using the EVC4000 Precision V/I Clamp (World Precision Instruments) with Ag/AgCl electrodes and an Agarose bridge containing 3 M KCl. Transepithelial resistance (TER) was measured using the Millicell-ERS and chopstick electrodes (Millipore). All experiments were conducted at 37°C. TER of the confluent monolayer of cells was determined in buffer A (145 mM NaCl, 1 mM CaCl₂, 1 mM MgCl₂, and 10 mM HEPES, pH 7.4) and the TER of blank filters was subtracted. Dilution potentials were measured when buffer B (80 mM NaCl, 130 mM mannitol, 1 mM CaCl₂, 1 mM MgCl₂, and 10 mM HEPES, pH 7.4) replaced buffer A on the apical side or basal side of filters. Electrical potentials obtained from blank inserts were subtracted from those obtained from inserts with confluent growth of cells. The ion permeability ratio (P_{Na}/P_{Cl}) for the monolayer was calculated from the dilution potential using the Goldman-Hodgkin-Katz equation. The absolute permeabilities of Na⁺ (P_{Na}) and Cl⁻ (P_{Cl}) were calculated by using the Kimizuka–Koketsu equation.

Ann. N.Y. Acad. Sci. ISSN 0077-8923

ANNALS OF THE NEW YORK ACADEMY OF SCIENCES
Issue: *Barriers and Channels Formed by Tight Junction Proteins*

Erratum for Ann. N. Y. Acad. Sci. 1253: 206–221

Ledeen, R.W., G. Wu, S. André, D. Bleich, G. Huet, H. Kaltner, J. Kopitz & H-J. Gabius. 2012. Beyond glycoproteins as galectin counterreceptors: effector T cell growth control of tumors via ganglioside GM1. *Ann. N.Y. Acad. Sci.* 1253: 206–221.

The correct title of the article cited above is listed below:

Beyond glycoproteins as galectin counterreceptors: tumor-effector T cell growth control via ganglioside GM1

doi: 10.1111/j.1749-6632.2012.06676.x